ADVANCED TECHNOLOGIES FOR WATER QUALITY TREATMENT AND MANAGEMENT

ADVANCED TECHNOLOGIES FOR WATER QUALITY TREATMENT AND MANAGEMENT

Edited by

Mehraj U. Din Dar, PhD

Aamir Ishaq Shah, PhD

Shakeel Ahmad Bhat, PhD

Syed Rouhullah Ali, PhD

AAP | APPLE ACADEMIC PRESS

First edition published 2023

Apple Academic Press Inc.
1265 Goldenrod Circle, NE,
Palm Bay, FL 32905 USA

760 Laurentian Drive, Unit 19,
Burlington, ON L7N 0A4, CANADA

CRC Press
6000 Broken Sound Parkway NW,
Suite 300, Boca Raton, FL 33487-2742 USA

4 Park Square, Milton Park,
Abingdon, Oxon, OX14 4RN UK

Library and Archives Canada Cataloguing in Publication

Title: Advanced technologies for water quality treatment and management / edited by Mehraj U. Din Dar, PhD, Aamir Ishaq Shah, MTech, Shakeel Ahmad Bhat, PhD, Syed Rouhullah Ali, PhD.
Names: Dar, Mehraj U. Din, editor. | Shah, Aamir Ishaq, editor. | Bhat, Shakeel Ahmad, editor. | Ali, Syed Rouhullah, editor.
Description: First edition. | Includes bibliographical references and index.
Identifiers: Canadiana (print) 20220444056 | Canadiana (ebook) 20220444145 | ISBN 9781774911778 (hardcover) | ISBN 9781774911785 (softcover) | ISBN 9781003314578 (ebook)
Subjects: LCSH: Water quality management. | LCSH: Sewage—Purification.
Classification: LCC TD365 .A38 2023 | DDC 628.1—dc23

Library of Congress Cataloging-in-Publication Data

CIP data on file with US Library of Congress

ISBN: 978-1-77491-177-8 (hbk)
ISBN: 978-1-77491-178-5 (pbk)
ISBN: 978-1-00331-457-8 (ebk)

About the Editors

Mehraj U. Din Dar, PhD
Punjab Agricultural University, Punjab, India

Mehraj U. Din Dar, PhD, holds a PhD in Soil and Water Engineering from Punjab Agricultural University, Punjab, India, with specialization in climate change studies, crop modeling, groundwater management studies, watershed conservation, wastewater treatment, hydrological modeling, and agricultural drainage. He worked on nitrate movement dynamics through subsurface drainage systems during his doctorate. Dr. Dar has been actively involved in research activities for the last several years. He has authored and published research articles, book chapters, and review articles in national and international journals, and he has presented and participated at various national and international seminars, conferences, and symposia. He was awarded a CSIR-SRF fellowship for his doctorate and has qualified ICAR JRF and SRF with 32 and 16 all India ranks respectively. Dr. Dar is a member of various scientific associations dealing with soil conservation, climate assessment, and wastewater treatment. He continues to explore new and innovative ways of wastewater treatment to meet the increasing demand for water in near future.

Aamir Ishaq Shah, PhD
Punjab Agricultural University, Punjab, India

Aamir Ishaq Shah, PhD, holds a PhD in Soil and Water Engineering from Punjab Agricultural University, Ludhiana, India. He worked on evaluating the performance of bioretention cells in relation to hydrology and water quality. He has published several research articles and book chapters in the field of water quality. He earned his MTech in Hydrology from IIT Roorkee, India with an MHRD fellowship, and was an ICAR senior research fellowship holder for his PhD. He has qualified several national competitive exams including GATE, ASRB-NET, and ICAR-SRF.

Shakeel Ahmad Bhat, PhD
Sher-e-Kashmir University of Agricultural Sciences and Technology, Kashmir, India

Shakeel Ahmad Bhat, PhD, holds a PhD in Soil and Water Engineering from the College of Agricultural Engineering at Sher-e-Kashmir University of Agricultural Sciences and Technology (SKUAST-K), India. He has worked on hydroponics technology during his PhD program. He is an author of more than 15 scientific articles and five book chapters. He has presented and participated at numerous state, national, and international conferences, seminars, and workshops. He is also a reviewer of various international journals and holds lifetime memberships in various international organizations.

Syed Rouhullah Ali, PhD
Sher-e-Kashmir University of Agricultural Sciences and Technology, Kashmir, India

Syed Rouhullah Ali, PhD, holds a PhD in Soil and Water Engineering from Sher-e-Kashmir University of Agricultural Sciences and Technology of Kashmir (SKUAST-K), Srinagar, J&K, India. He has qualified several national competitive examinations including ASRB-NET and ICAR-JRF and SRF. Dr. Ali has published several research/review articles and book chapters in national and international peer-reviewed journals in the field of surface/groundwater hydrology, climate change, and water quality. He has participated, presented, and organized several national and international conferences, seminars, consortiums, workshops, and training sessions. The author has expertise in the field of hydrological modeling (surface/groundwater), hydrogeology, remote sensing and GIS, climate change, etc.

Contents

Contributors

Syed Rouhullah Ali
College of Agricultural Engineering and Technology, SKUAST–K, Jammu and Kashmir – 190025, India, E-mail: agasyedrouhullah@gmail.com

Afnan Ashraf
Department of Soil and Water Engineering, PAU, Ludhiana, Punjab – 141004, India

Vishnu Ji Awasthi
Govind Ballabh Pant University of Agriculture and Technology, Pantnagar, Uttarakhand, India

Shakeel Ahmad Bhat
Department of Soil and Water Engineering, College of Agricultural Engineering and Technology, Sher-e-Kashmir University of Agricultural Sciences and Technology of Kashmir, Shalimar Campus, Srinagar – 190025, Jammu and Kashmir, India, E-mail: wakeelbhat@gmail.com

Mehraj U. Din Dar
Department of Soil and Water Engineering, Punjab Agricultural University, Ludhiana – 141004, Punjab, India, E-mails: mehrajudindar24@gmail.com; mehrajdar417@gmail.com

Susanta Das
Punjab Agricultural University, Ludhiana, Punjab, India

Sunil Garg
Department of Soil and Water Engineering, Punjab Agricultural University, Ludhiana – 141004, Punjab, India

Pooja Goyal
Department of Soil and Water Engineering, Punjab Agricultural University, Ludhiana – 141004, Punjab, India

Sheeza Haroon
College of Agricultural Engineering and Technology, SKUAST–Kashmir, Jammu and Kashmir, India, E-mail: sheeza.haroon11@gmail.com

Aamir Shah Ishaq
Punjab Agricultural University, Ludhiana, Punjab, India, E-mail: shahaamir2613@gmail.com

Md. Towfiqul Islam
Department of Disaster Management, Begum Rokeya University, Rangpur – 5400, Bangladesh

Maryam Sadat Jaafarzadeh
Department of Watershed Management Engineering, Faculty of Agriculture and Natural Resources, Lorestan University, Iran

Bodasingi Krishna Kanth
Department of Farm Machinery and Power Engineering, Dr. NTR College of Agricultural Engineering, ANGRAU, Bapatla, Andhra Pradesh, India

Samanpreet Kaur
Department of Soil and Water Engineering, Punjab Agricultural University, Ludhiana – 141004, Punjab, India

N. L. Kushwaha
Scientist, Division of Agricultural Engineering ICAR-IARI, New Delhi, India,
E-mail: nand.kushwaha@icar.gov.in

Mahrukh
College of Agricultural Engineering and Technology, SKUAST–K, Jammu and Kashmir – 190025, India

Anurag Malik
Regional Research Station, Bathinda – 151001, Punjab, India

Rajat Mishra
Govind Ballabh Pant University of Agriculture and Technology, Pantnagar, Uttarakhand, India

Rohit Pratap Ojha
Govind Ballabh Pant University of Agriculture and Technology, Pantnagar, Uttarakhand, India

Jaripiti Trivikrama Raju
Department of Soil and Water Engineering, College of Technology and Engineering, MPUAT, Udaipur, Rajasthan, India, E-mail: trivikramraju@gmail.com

Ratnala Sudha Rani
Department of Soil and Water Engineering, Dr. NTR College of Agricultural Engineering, ANGRAU, Bapatla, Andhra Pradesh, India

Nethi Naga Hari Sairam
Department of Soil and Water Engineering, Dr. NTR College of Agricultural Engineering, ANGRAU, Bapatla, Andhra Pradesh, India

Leila Shafea
Department of Soil Science, Kassel University, Germany

Mudasir Shafi
College of Agricultural Engineering and Technology, SKUAST–K, Jammu and Kashmir – 190025, India

Aamir Ishaq Shah
Department of Soil and Water Engineering, Punjab Agricultural University, Ludhiana – 141004, Punjab, India

Puneet Sharma
Assistant Professor, Soil, and Water Engineering, Krishi Vigyan Kendra, Pathankot (Gho) – 145023, Punjab Agricultural University, Ludhiana – 141004, Punjab, India,
E-mail: sharmapuneet840@gmail.com

J. P. Singh
Department of Soil and Water Engineering, Punjab Agricultural University, Ludhiana – 141004, Punjab, India

Kuldip Singh
Department of Soil Science, Punjab Agricultural University, Ludhiana – 141004, Punjab, India

Neha Singhal
Department of Soil and Water Engineering, Punjab Agricultural University, Ludhiana – 141004, Punjab, India

Pakkiranna Sivamma
Department of Agricultural Processing and Food Engineering, Dr. NTR College of Agricultural Engineering, ANGRAU, Bapatla, Andhra Pradesh, India

Vallu Tejaswini
Department of Soil and Water Engineering, Dr. NTR College of Agricultural Engineering, ANGRAU, Bapatla, Andhra Pradesh, India, E-mail: vallutejaswini@gmail.com

Abbreviations

ACP	anaerobic contact process
ACT	automated chemostat treatment
AFX	ammonia fiber-explosion
Ag	silver
Al	aluminum
ANNs	artificial neural network
AOPs	advanced oxidation process
APEX	agricultural policy/environmental extender
ARP	ammonia-recycle percolation
Au	gold
AuNPs	gold nanoparticles
BOD	biological oxygen demand
BPNN	back propagation neural network
BTEX	benzene toluene ethylbenzene and xylene
CA	cluster analysis
CAWST	center for affordable water and sanitation technology
CDOM	colored dissolved organic matter
chl-a	chlorophyll-a
CNR	continuous nitrification rate
CNTs	carbon nanotubes
COD	chemical oxygen demand
Cr	chromium
CRP	conservation reserve program
CSTR	continuous stirred tank reactor
CT	carbon tetrachloride
CTAB	hexadecylcetyl trimethyl ammonium bromide
Cu	copper
DO	dissolved oxygen
DOC	dissolved organic carbon
DOE	department of energy
DOM	dissolved organic matter
E. coli	*Escherichia coli*
ECM	export coefficient model

EDCs	endocrine disrupting compounds
EFB	ecological floating bed
EFDC	environmental fluid dynamics code
EPA	Environmental Protection Agency
ESGB	expanded granular sludge bed reactor
ETM	enhanced thematic mapper
FC	fecal coliform
Fe	iron
FS	fecal streptococci
GCPC	Gujarat Cleaner Production Center
GOCI	geostationary ocean color imager
H2S	hydrogen sulfide
HER	hydrological effective rainfall
HFS	horizontal flow systems
Hg	mercury
HRT	hydraulic retention time
HRU	hydrologic response unit
IMBR	immersed membrane biomass rejection
IOPs	inherent optical properties
ISO	International Standard Organization
LSWT	lake surface water temperature
MC	methylene chloride
MIM	matrix inversion method
MIRAS	microwave imaging radiometer using aperture synthesis
Mn	manganese
Mo	molybdenum
MOS	modular optical scanner
MSS	multispectral scanner
MUSLE	modified universal soil loss equation
MWR	microwave radiometers
N	nitrogen
NAC	nitroaromatic compounds
NaCl	sodium chloride
NAPs	non-algal particles
NIR	near infrared
NPs	nanoparticles
OCTS	ocean color, temperature scanner
OLI	operational land imager

OLR	organic loading rates
OPH	organophosphorus hydrolase
P	phosphorus
PAH	polycyclic aromatic hydrocarbons
Pb	lead
PCA	principal component analysis
PCBs	polychlorinated biphenyls
PCE	tetrachloroethylene
PCP	pentachlorophenol
Pd	palladium
PET	polyethylene terephthalate
PETR	Pentaerythritol tetranitrate reductase
PFC	proper functioning condition
PHAs	polyhydroxy alkenoates
PSS	poly(styrene sulfonate)
Pt	platinum
QAA	quasi-analytical algorithm
RBC	rotating biological contractor
RDX	detonating explosive
REMM	riparian ecosystem management model
RTE	radiative transfer equation
SAR	synthetic aperture radar
SDD	Secchi disk depth
Se	selenium
SMOS	soil moisture and ocean salinity
SS	suspended sediments
SS	suspended solids
SSS	sea surface salinity
SST	sea surface temperature
STPs	sewage treatment plants
SVE	Saint Venant equations
SWAT	soil and water assessment tool
TC	total coliform
TCE	trichloroethylene
TIR	thermal infrared
TM	thematic mapper
TMDL	total maximum daily load
TN	total nitrogen

TNT	trinitrotoluene
TOMCAT	temporal/overall model for catchments
TP	total phosphorus
TSS	total suspended solids
UASB	up-flow anaerobic sludge blank reactor
UAV	unmanned aerial vehicle
USDA-ARS	USDA Agricultural Research Service
USEPA	U.S. Environmental Protection Agency
VFAs	volatile fatty acids
VFS	vertical flow systems
WASP	water quality analysis simulation program
WES	waterways experiment station
Zn	zinc

Preface

Water forms the basis of life. More than 75% of the earth's surface is occupied with water. The freshwater availability is, however, very limited, and only a very small percentage (2.5%) of total water present on the earth's surface may be classified as freshwater. Various anthropogenic activities result in the release of harmful pollutants directly into the natural environment, thereby degrading the already constrained water resources. Water quality management has emerged as a serious issue in the environment in recent decades. It is estimated that nearly 80% of the water resources in the world are contaminated. Water quality management would provide additional water for reuse in irrigation, recreational activities, and other uses, thereby making wastewater a valuable resource.

In recent decades, researchers and environmentalists have adopted several techniques for water quality improvement. With the rapid advancement in technologies, several new and advanced techniques for water quality management have been suggested and recommended. Conventional water treatment technologies have also evolved over the years and are in a continual stage of development. Nanomaterials, ecological floating beds (EFB), eco-roofs, bioenergy landscapes, and the use of water quality models are some of the interesting advances in the field of water quality management.

Despite the advancement in techniques for controlling water pollution, the problem still seems not sufficiently controlled as the world continues to suffer from varying degrees of pollution. There may be several reasons for the degrading water quality; one of them is the lack of dissemination of information and, hence, the lack of adoption of new technologies for water quality improvement and management. The scale of the water pollution problem and economic constraints remains. This book has been written with the aim of providing a deep insight into the recent approaches and the various issues pertaining to water quality management. This book summarizes the recent advanced technologies for water quality improvement and management. The book would prove beneficial to students and researchers and provide technological support and a scientific foundation for achieving optimum water quality management.

CHAPTER 1

Ecological Floating Bed (EFB) for Decontamination of Polluted Water Bodies

AFNAN ASHRAF,[1] SHAKEEL AHMAD BHAT,[2] ANURAG MALIK,[3] and LEILA SHAFEA[4]

[1]*Department of Soil and Water Engineering, PAU, Ludhiana, Punjab – 141004, India*

[2]*College of Agricultural Engineering, SKUAST–K, Jammu and Kashmir, India, E-mail: wakeelbhat@gmail.com*

[3]*Regional Research Station, Bathinda – 151001, Punjab, India*

[4]*Department of Soil Science, Kassel University, Germany*

ABSTRACT

The pollution of water bodies is an alarming situation worldwide. Although various techniques have been employed for the decontamination of water bodies, but there has to be some technique that not only efficiently decontaminates the water body but also maintains an ecological balance between various flora and fauna present in it. The development of an ecological floating bed (EFB) is one such alternative. EFB very efficiently treats the wastewater and is a sustainable environment friendly technique. The main idea is based on the self-cleaning ability of nature. This technology does not yield any toxic by-product. This chapter focuses on the various pre requites for the development of an EFB and its working.

Advanced Technologies for Water Quality Treatment and Management. Mehraj U. Din Dar, PhD, Aamir Ishaq Shah, MTech, Shakeel Ahmad Bhat, PhD & Syed Rouhullah Ali, PhD (Eds.)

1.1 INTRODUCTION

In the present scenario, the quality of water is under threat and is degrading worldwide. Most of the water reservoirs are getting polluted, the main pollutant being the heavy load of contaminants from the industrial sector. In addition to this, the increasing growth of population and rapid urbanization are continuously putting the freshwater reservoirs under high stress. Furthermore, the appropriate handling and treatment of the huge wastewater that is generated is a matter of big concern. As per the studies done by Slutsky and Yen (1997), the annual rate of precipitation recorded worldwide is 119,000 km^3. Out of this, 61% goes to evaporation and 39% goes as stormwater runoff. The quantity of runoff is showing an increasing trend due to the increase in the development of the impermeable surfaces as well as heavy urbanization. When the infiltration capacity of the land decreases due to its impervious nature, the rainwater has no other option than to flow as runoff. This problem is aggravated when heavy rainfall occurs since it erodes away with itself a large part of the land and carries with itself all the pollutants on the land towards the aquatic and terrestrial ecosystems, thereby polluting them.

The rapid urbanization has resulted in the increase in the toxicity of the water bodies as well as the air since a large amount of vehicular emissions and all the harmful substances present on the land surface are washed away into the rivers, lakes, oceans that increase the toxic nature of sediments and water. The industrial emissions add to the toxicity of the water bodies since they are mostly discharged into them without any appropriate treatment (Hwang et al., 2016; Schwammberger et al., 2017; Shah et al., 2020). These pollutants that are discharged into the water bodies in one way or the other not only destroy it, but the pollutants are transferred from one organism to the other and result in biomagnification of the toxic pollutants inside the organisms that is the reason of the death of many of them.

There are many techniques that have been used to mitigate the pollution of the water bodies. The main focus in doing so is not only to prevent the pollution of water bodies but also to protect the habitat of flora and fauna residing in them, thereby protecting the overall biodiversity. For this reason, an ecological floating bed (EFB) is a good alternative to decontaminate the contaminated water body, since it does both the jobs simultaneously.

1.2 ECOLOGICAL FLOATING BEDS (EFBS)

Ecological floating bed (EFB) is an environment friendly and economical technique having the capacity of very effectively restoring the polluted lake water. It takes its idea from the nature's self-cleaning ability and the most impactful feature of this technology is that it does not yield any toxic product. EFB uses emergent plants that grow in the form of a floating mat that rests over a frame that remains afloat in the water. The roots of the plants descend down into the water in order to suck the dissolved pollutants and the stem of the plants remain above the water level. There is an intertwined system of roots, rhizomes, and attached biofilms that emerge out from the floating mat serving as an active surface area for physio and biochemical like filtering and entrapment (Bi et al., 2019). Figure 1.1 displays a schematic diagram of EFB.

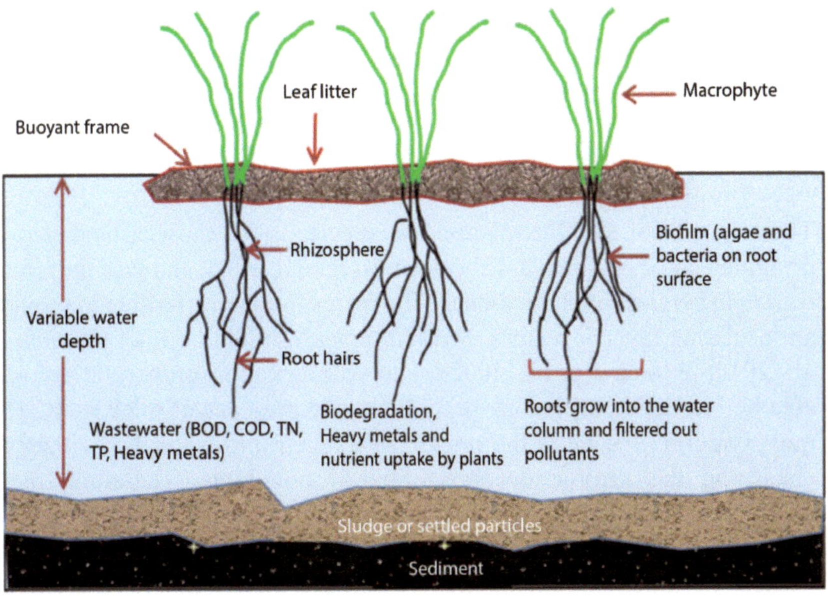

FIGURE 1.1 Schematic diagram of an EFB.
Source: Reprinted with permission from Ref. Samal et al., 2019. © Elsevier.

1.2.1 WORKING PRINCIPLE OF AN ECOLOGICAL FLOATING BED (EFB)

The water bodies get permanently contaminated once the slow and continuous addition of pollution causing agents occurs from various sources. When the rate with which the pollutants are added exceeds its self-cleaning, the water body is said to be polluted. After the inspection of the polluted water body and surveying all the pros and cons of all the material, a particular plant species is selected for effective decontamination of the polluted water body and an appropriate plant bed is built in the contaminated lake in EFB network. This bed comprises of buoyant materials capable of providing a continuous upward force to the plants and the surface area so that it can support the growth of microbes (Lynch et al., 2015; Xian et al., 2010). The plant species used in an EFB, to ensure efficiency in removal of pollutants, have dense root system. Table 1.1 illustrates the use of EFB for handling different forms of wastewater.

1.2.2 FACTORS IMPACTING THE EFFICACY OF AN ENVIRONMENTALLY SAFE FLOATING BED TO DECONTAMINATE THE CONTAMINATED WATER BODIES

1.2.2.1 CHOOSING A SUITABLE MACROPHYTE SPECIES

The selection of suitable macrophyte species is the most critical step in designing an EFB. The plant chosen must remain afloat and thus they must be rich in aerenchymatous tissues. The plants used must be able to provide an insulating layer in winter. Nutrient uptake should be high (Waajen et al., 2016; Williams et al., 2002). The sedimentation process is accelerated by increasing the water speed due to the presence of thick roots. The most important usage is the release of oxygen into water body, thereby enhancing the aerobic decomposition of pollutants. Therefore, these plants absorb the nutrients found in wastewater and are retained in their tissues. There are numerous varieties of macrophyte species that differ in wastewater treatment capabilities owing to the difference in the growth rate, rooting type, and accumulation capacity (Samal et al., 2017; Zhu et al., 2011). The most commonly used plants globally in EFB are *Cyperus*, *Canna*, *Lollium*, *Typha*, and *Chrysopogon* species.

TABLE 1.1 Usage of Floating Aquatic Beds for Handling Different Forms of Wastewater

Sl. No.	Wastewater	Plant Species	Organic Removal (%)	Nutrient Removal (%)	Dimension (cm)	Operation Period (Days)	References
1.	River water spiked with KNO_3 and NH_4Cl	*Iris pseudacorus*	–	TN 53.3–90.9%	85×40× 90	3	Gao et al. (2018)
2.	Secondary effluent	*Cyperus papyrus* *Miscanthidium violaceum*	BOD 83.1, BOD 47.8	TN 66.8, NH_4 60.2, TP 61.8, TN 56, NH_4 47.1, TP 40.8	300×250×20 300×250×35	2.7	Kyambadde et al. (2004)
3.	Lake water	*Ipomoea aquatic*	–	TN 66.4–76.5, NH_4 58.7–68.9, TP 45.7–61.7	100×100×110	–	Li et al. (2010)
4.	Aquaculture wastewater	*Ipomoea aquatic*	–	TN 30.6, TP 18.2	200×300	–	Li and Li (2009)
5.	Polluted river water	*Phragmites australis, Canna indica*	BOD 51.8, COD 40.8	NH_4^+-N 44.2, NO_3^-–N 30.2, TP 23.4	300×58	–	Saeed et al. (2016)
6.	Synthetic graywater	*Phragmites australis*	BOD 33.4, COD 27.2	PO_4^{3-} P 18.9	14-L plastic buckets	7	Abed et al. (2017)
7.	Domestic sewage	*Typha domingensis*	BOD 56, COD 55, TSS 78	TKN 41, TP 37	1,700×1,700×250	11.5	Benvenuti et al. (2018)

1.2.2.2 PRESENCE OF A BIOFILM

Artificial biofilm carriers increase the biofilm biomass resulting in the increase in the filter efficacy. Biofilms are an amalgam of micro-organisms (bacteria, fungi, algae) which both bind to each other and to stable surfaces. Biofilms' slimy and sticky nature helps contain the suspended solids (SS) found in wastewater (Färm, 2002; Meuleman et al., 2002).

1.2.2.3 SELECTION OF AN APPROPRIATE MEDIA

In order to allow air exchanges and maintain aerobic condition, enough pore space should be present in the growth media. The elements that should be taken into account while choosing an appropriate growth media are water retention, porosity, capillarity, and fertility (Brisson and Chazarenc, 2009). Products such as soil, coconut fiber, compost, coarse peat, charcoal, bamboo, etc., were used in floating beds as growth media.

1.2.2.4 EFFECTIVE WATER DEPTH

We need to maintain a depth of 0.8–1.0 m of water. When the required depth is not preserved, the macrophyte roots reach down to the soil and become attached. There is a possibility that roots from the benthic zone cannot be removed, resulting in the submergence of the macrophyte in case of the rise (Ayaz and Saygin, 1996). This condition is lethal to the plants and result in damaging the floating bed. It is very important that the roots need to be afloat in the wastewater and thus the depth needs to be maintained accordingly (Chan et al., 2008).

1.2.2.5 BUOYANT NATURE

Buoyancy is maintained naturally with the help of the spongy and aerenchymatous tissues that retain air inside of them. CH_4 gas produced as a result of the anaerobic decomposition in the bottom that is clogged in the root mat ensures the floating state (Lynch et al., 2015; Sun et al., 2005). The age, growth mechanism, and rate of gas output all influence the plant's buoyancy. During the summer season, the metabolic activities of the plants and microbes occur at a higher rate due to the higher temperatures, which adds to the process of gas generation and the entrapment of air by the plant tissues, thereby increasing the buoyancy. Those materials should be used that are simple, cheap, and easily available so that they can be easily removed once the floating bed is fully grown and established. Bamboo, sealed PVC or PP pipes, inflatable vinyl pillows, etc., are some of the examples of the materials used for floatation. Table 1.2 shows different materials used to obtain buoyancy during EFB construction.

TABLE 1.2 Diverse Materials Used in EFB Design to Achieve Buoyancy

Sl. No.	Floating Frame	Raft Area (m^2)	References
1.	Bio heaven patent material	0.36	Tanner and Headley (2011)
2.	Foamy sheets	2.00	Yang et al. (2008)
3.	PVC pipe (Dia 38 mm)	0.29	Wang and Sample (2014)
4.	Polyethylene	0.025	Wen and Recknagel (2002)
5.	HDPE	0.20	Xian et al. (2010)
6.	Bamboos (Dia 1.0–1.5 cm)	–	Zhao et al. (2012)
7.	Tech-IA patent material	0.45	Stefani et al. (2011)
8.	Polyethylene terephthalate (PET) material (recycled)	50.0	Borne et al. (2013)
9.	Iron and timber	–	Kerr-Upal et al. (2000)

1.2.2.6 VEGETATION COVERAGE

Vegetation coverage ratio in EFB regulates the diffusion of atmospheric O_2 in the water body. Low coverage (9% to 18%) will add marginal treatment effects, while excessive vegetation coverage (>50%) due to wind movement restricts the entrance of O_2 into the water from the air, resulting in anoxic water conditions (Zhou and Wang, 2010). So, a lower DO amount in water is observed in EFB covered with more vegetation as compared to the floating bed without coverage.

1.2.2.7 OXYGEN TRANSPORT WITHIN THE WETLAND MACROPHYTES

Two processes are responsible for the oxygen transfer mechanism inside the wetland macrophyte and they are convection and passive molecular diffusion of air. The difference in the presence between the roots and the surface of the leaf induces convection of gases to below-ground biomass across the entire stem, while the spontaneous molecular movement results in the diffusion of gasses through the stem to the root (From greater partial pressure to reduced partial pressure). Such processes depend on the weight, temperature, and mean of the molecular gas.

1.2.2.8 HARVESTING OF BIOMASS

Harvesting of the floating plants at regular intervals is strongly recommended (Hubbard et al., 2004; Yeh et al., 2015). The modus operandi of EFB is shown in Figure 1.2. According to Zhou and Wang (2010), re-entering of nutrients stored in leaves and other aerial parts takes place as the decay initiates. Therefore, it is desirable that the harvesting is done before decay. Wang et al. (2015) researched that mining just the upper portion of water would not extract all nutrients from water, as large quantities of nutrients are stored in the biomass below water. Harvesting the surface water section only stops the plants from dying as new shooting from the cutting area occurs. They also concluded that, during the summer, above water biomass is rich in nutrients, but nutrients start translocating to below water biomass at around September. After harvesting, it can be used directly as animal fodder, as well as composting and vermicomposting processes (Brisson and Chazarenc, 2009; Islam et al., 2013). Temperature changes affect the activity of the microbials. High microbial activity is observed during the summer season while low activity occurs in the winter season (Bu and Xu, 2013). It can be assumed therefore that season and atmosphere have a major effect on the ecological performance of floating beds. Please keep an eye on floating bed maintenance to efficiently eliminate contaminants. The right time to harvest is thus a point to be taken into consideration.

1.3 COUPLING OF EFB'S WITH ELECTROLYSIS TO REDUCE WATER BODY EUTROPHICATION

EFBs have been commonly used as an *in-situ* ecological remediation technology for the treatment of surface water, as economical, simple in design and maintenance. However, due to problems with seasonal and anaerobic environmental constraints on the growth of macrophytes and microbes, restricted standing biomass, reduced adsorption ability of substrates, and a wider coverage area generally restrict the use of EFBs. Not only has this, Eutrophication, caused by enrichment of waters with excessive plant nutrients, become a matter of serious concern in aquatic ecological research because it leads to highly undesirable changes in ecosystem structure and function. Thus, the need of the hour is to devise those EFBs that are more stable and can nullify these constraints for better efficiency in decontamination. One such technique is integrating electrolysis with EFB's since it can yield excellent water treatment capacity.

Electrolysis in eutrophic bodies of water can simultaneously encourage the growth and reproduction of heterotrophic denitrifiers to enhance NO_3^-–N elimination. Besides this, the removal of PO_4^{3-}–P by sacrificial anode can also be improved and can be used as coagulant ion source. Some researchers have already used electrolysis reaction in developed wetlands (CWs) and biofilter after the anode was Fe or Al (Lacasa et al., 2012; Ju et al., 2014; Gao et al., 2017), in order to increase removal of N and P. Fe anode was used in the electrolysis process producing Fe^{2+} and Fe^{3+} ions that yielded Fe (III) oxide. Fe (III) oxide has a reddish-brown appearance and is often referred to as rust, which impacted water transparency when Bernal et al. (1959) got so much rust. It offered a greater P removal efficiency when Al was used as anode, as Al anode had a lower metal-to-P ratio in contrast to Fe anode, but more Al^{3+} ions have different bio toxicity effects on aquatic species such as fish, algae, and even humans (Parent et al., 2010; Kandimalla et al., 2016; Sharma et al., 2016). Besides Fe and Al, magnesium cations were often used to extract P due to corrosion and/or electrolytic dissolution, imparting the least impact on the aquatic surroundings.

Yan et al. (2020) in order to research the ability of nutrient removal in enhanced EEFBs integrated Mg-Al alloy anode, a graphite cathode, and biochar as substratum. Results showed that significantly higher N and P net removal rates than those rates representing conventional EFBs($p<0.05$), particularly for TN, NO_3^-–N, and TP removals (Table 1.3). The EEFB electrolysis method which also produces H_2 by enriching anodes in the biochar substratum biofilm for autotrophic hydrogen denitrification bacteria, e.g., the genera *Hydrogenophaga*, *Flavobacterium*, and *Dechloromonas* and improving the removal efficiency of NO_3^-–N (Figure 1.3). During the electrolysis step, Mg^{2+} and Al^{3+} ions were released to increase the flocculation potential of the eutrophic water, resulting in a stable high rate of P elimination. The absolute rate of growth of *I. Sibirica* was 10.82 ± 12.54 mg d^{-1}, which was substantially lower than EFB growth as the absolute growth rate was 101.71 ± 25.65 mg d^{-1} ($p < 0.05$), which suggested that electrolysis affected plant growth and plant uptake levels were lower than EFB development.

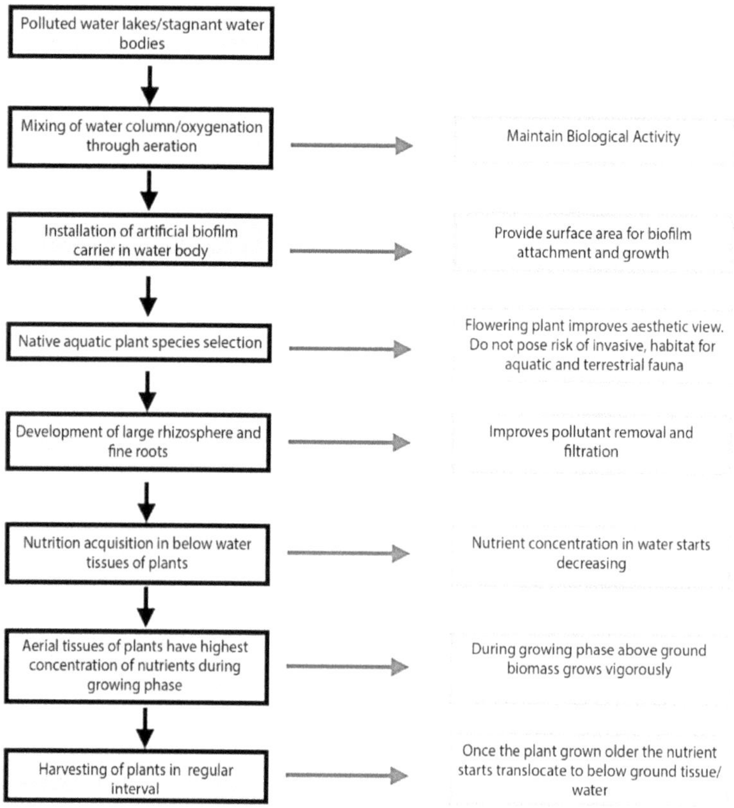

FIGURE 1.2 Modus operandi of ecological floating bed (EFB).
Source: Reprinted with permission from Ref. Samal et al., 2019. © Elsevier.

TABLE 1.3 Specific Treatments Include Rate of Removal and Net Rate of Removal of TN, NH_3–N, TP, and PO_4^{3-}–P*

The treatment	TN		NH_3^-–N		TP		PO_4^{3-}–P	
	Removal Rate (%)	**Net removal Rate (%)**	**Removal Rate (%)**	**Net removal Rate (%)**	**Removal Rate (%)**	**Net removal Rate (%)**	**Removal Rate (%)**	**Net removal Rate (%)**
Control	12.2 ± 1.28a	–	95.6 ± 0.04a	–	15.8 ± 1.26a	–	13.6 ± 1.93a	–
EFB's	26.7 ± 1.19b	14.5 ± 1.19a	94.5 ± 3.00a	–	10.54 ± 1.21b	–	8.9 ± 0.01b	–
EEFB's	53.1 ± 1.13c	40.9 ± 1.13b	96.5 ± 0.03a	0.9 ± 0.03a	76.5 ± 1.59c	60.7 ± 1.59a	74.5 ± 2.09c	60.9 ± 2.09a

*Data are means accompanied by regular (± SE) errors (n = 3). Similar letters suggest slightly different removals at $p<0.05$ between the three treatments.

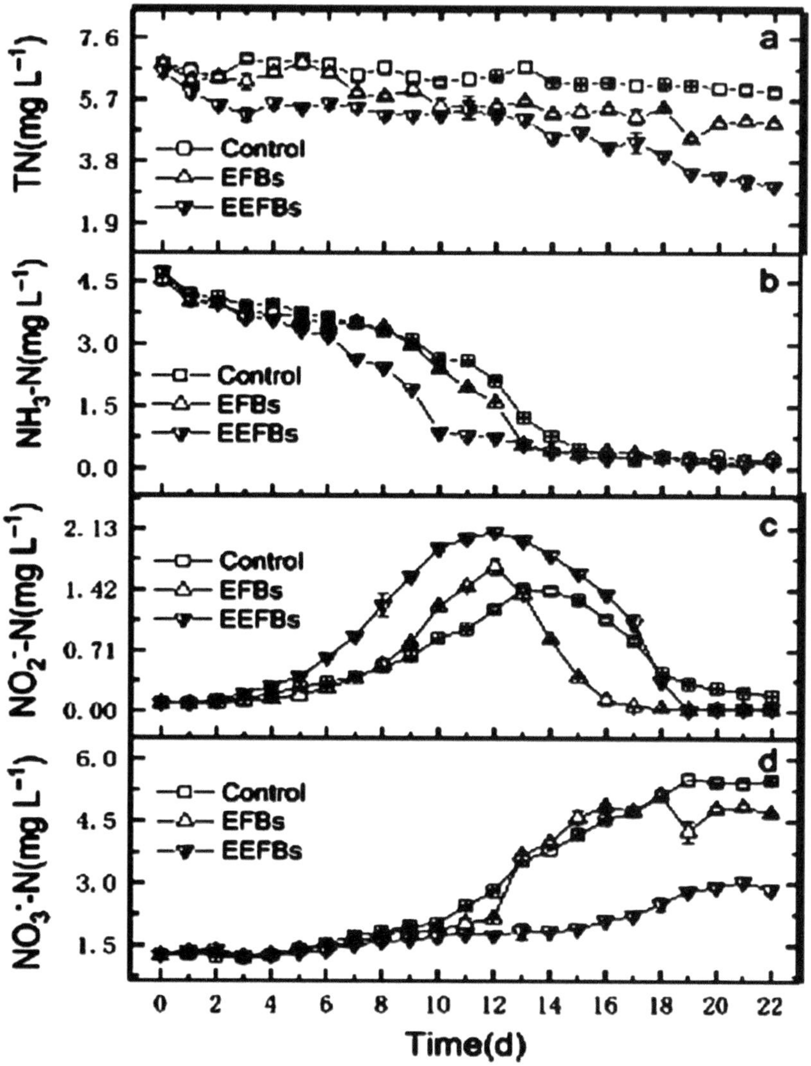

FIGURE 1.3 Data points indicate sample mean ± standard mean error in the concentrations of TN (a); NH_3–N (b); NO_2^-–N (c); and NO_3^-–N (d) in the EEFBs, the EFBs and control during the treatment of eutrophic water.

Source: Yan et al. (2020).

1.4 CONCLUSIONS

The literature available shows that EFB is one of the safe and environmentally friendly green technologies to handle different forms of wastewater. For the design of the EFB, the most important consideration is to choose the correct plant species that are rich in aerenchymatous tissues so that oxygen from the upper part of biomass can be transferred to lower part of biomass while at the same instant ensuring the buoyant condition of the plant species for which various bubbling materials using such as PVC pipes, sheets of polystyrene, bamboo, and various floating patent pieces. The water depth and plant coverage ratio influence the atmospheric diffusion rate to the water body, and thus is important in deciding the nature of a floating wetland. In addition to the above factors, the other pre-requirements are forms of macrophyte organisms, biofilm, and growth media. Filtration and aerobic degradation separate the organic and SS. Through nitrification-denitrification and plant absorption nitrogen is largely eliminated. The majority of phosphorus and heavy metals are absorbed by sedimentation, with a small amount removed by plant absorption. To avoid nutrient translocation from the stem and leaves to the water, plants must be harvested on a regular basis.

KEYWORDS

- **biofilm**
- **contaminants**
- **ecological floating bed (EFB)**
- **electrolysis**
- **macrophytes**
- **polyethylene terephthalate**
- **purification**

REFERENCES

Ayaz, S. C., & Saygin, Ö., (1996). Hydroponic tertiary treatment. *Water Res.*, *5*, 1295–1298.

Bernal, J. D., (1959). The oxides and hydroxides of iron and their structural interrelationships. *Clay. Miner.*, *4*(21), 15–30.

Bi, R., Zhou, C., Jia, Y., Wang, S., Li, P., Reichwaldt, E. S., & Liu, W., (2019). Giving water bodies the treatment, they need: A critical review of the application of constructed floating wetlands. *J. Environ. Manag., 238*, 484–498.

Borne, K. E., Fassman, E. A., & Tanner, C. C., (2013). Floating treatment wetland retrofit to improve stormwater pond performance for suspended solids, copper and zinc. *Ecol. Eng., 54*, 173–182.

Brisson, J., & Chazarenc, F., (2009). Maximizing pollutant removal in constructed wetlands: Should we pay more attention to macrophyte species selection? *Sci. Total Environ., 407*, 3923–3930.

Bu, F., & Xu, X., (2013). Planted floating bed performance in treatment of eutrophic river water. *Environ. Monit. Assess., 185*, 9651–9662.

Chan, S. Y., Tsang, Y. F., Cui, L. H., & Chua, H., (2008). Domestic wastewater treatment using batch-fed constructed wetland and predictive model development for NH_3-N removal. *Process Biochem., 43*, 297–305.

Chen, Y. H., Zhang, Y., Huang, M. S., Zhang, Y. F., Zhao, F., & Luo, J. H., (2012). Ecological restoration by a multistage floating-bed system in a eutrophic urban river. *China. Appl. Mech. Mater., 157, 158,* 945–949.

Färm, C., (2002). Metal sorption to natural filter substrates for stormwater treatment–column studies. *Sci. Total Environ., 298*, 17–24.

Gao, L., Zhou, W., Huang, J., He, S., Yan, Y., Zhu, W., Wu, S., & Zhang, X., (2017). Nitrogen removal by the enhanced floating treatment wetlands from the secondary effluent. *Bioresour. Technol., 234,* 243–252.

Garbutt, P., (2004). An investigation into the application of floating reed bed and barley straw techniques for the remediation of eutrophic waters. *Water Environ. J., 19*, 174–180.

Hubbard, R. K., Gascho, G. J., & Newton, G. L., (2004). Use of floating vegetation to remove nutrients from swine lagoon wastewater. *Trans. ASAE, 47*, 1963–1972.

Hwang, H. M., Fiala, M. J., Park, D., & Wade, T. L., (2016). Review of pollutants in urban road dust and stormwater runoff: Part 1. Heavy metals released from vehicles. *Int. J. Urban Sci., 20*, 334–360.

Islam, M. S., Ueno, Y., Sikder, M. T., & Kurasaki, M., (2013). Phytofiltration of arsenic and cadmium from the water environment using *Micranthemum umbrosum* (J.F. Gmel) S.F Blake as a hyperaccumulator. *Int. J. Phytoremediation, 15*, 1010–1021.

Ju, X., Wu, S., Zhang, Y., & Dong, R., (2014). Intensified nitrogen and phosphorus removal in a novel electrolysis-integrated tidal flow constructed wetland system. *Water. Res., 59*, 37–45.

Kandimalla, R., Vallamkondu, J., Corgiat, E. B., & Gill, K. D., (2016). Understanding aspects of aluminum exposure in Alzheimer's disease development. *Brain. Pathol., 26*, 139–154.

Kerr-Upal, M., Seasons, M., & Mulamoottil, G., (2000). Retrofitting a stormwater management facility with a wetland component. *J. Environ. Sci. Health A., 35*, 1289–1307.

Kyambadde, J., Kansiime, F., Gumaelius, L., & Dalhammar, G., (2004). A comparative study of *Cyperus papyrus* and *Miscanthidium violaceum* based constructed wetlands for wastewater treatment in a tropical climate. *Water Res., 38*, 475–485.

Lacasa, E., Cañizares, P., Llanos, J., & Rodrigo, M. A., (2012). Effect of the cathode material on the removal of nitrates by electrolysis in non chloride media. *J. Hazard. Mater., 213*, 478–484.

Li, X. N., Song, H. L., Li, W., Lu, X. W., & Nishimura, O., (2010). An integrated ecological floating-bed employing plant, freshwater clam and biofilm carrier for purification of eutrophic water. *Ecol. Eng., 36*, 382–390.

Lynch, J., Fox, L. J., Owen, J. S., & Sample, D. J., (2015). Evaluation of commercial floating treatment wetland technologies for nutrient remediation of stormwater. *Ecol. Eng., 75*, 61–69.

Meuleman, A. F. M., Beekman, J. P., & Verhoeven, J. T. A., (2002). Nutrient retention and nutrient- use efficiency in *Phragmites australis* stands after wastewater application. *Wetlands, 22*, 712–721.

Parent, L., & Campbell, P. G. C., (2010). Aluminum bioavailability to the green alga *Chlorella pyrenoidosa* in acidified synthetic soft water. *Environ. Toxicol. Chem., 13*(4), 587–598.

Samal, K., Dash, R. R., & Bhunia, P., (2017). Treatment of wastewater by vermifiltration integrated with macrophyte filter: A review. *J. Environ. Chem. Eng., 5*, 2274–2289.

Samal, K., Kar, S., & Trivedi, S., (2019). Ecological floating bed (EFB) for decontamination of polluted water bodies: Design, mechanism and performance. *Journal of Environmental Management,* 251.

Schwammberger, P., Walker, C., & Lucke, T., (2017). Using floating wetland treatment systems to reduce stormwater pollution from urban developments. *Int. J. Geom., 12*, 45–50.

Shah, A. I., Din, D. M. U., Bhat, R. A., Singh, J. P., Singh, K., & Bhat, S. A., (2020). Prospectives and challenges of wastewater treatment technologies to combat contaminants of emerging concerns. *Ecol. Eng., 152*, 105882. https://doi.org/10.1016/j.ecoleng.2020.105882.

Sharma, D. R., et al., (2016). Quercetin attenuates neuronal death against aluminum-induced neurodegeneration in the rat hippocampus. *Neuroscience, 324*, 163–176.

Slutsky, A. H., & Yen, B. C., (1997). A macro-scale natural hydrologic cycle water availability model. *J. Hydrol., 201*, 329–347.

Stefani, G. D., Tocchetto, D., Salvato, M., & Borin, M., (2011). Performance of a floating treatment wetland for in-stream water amelioration in NE Italy. *Hydrobiol., 674*, 157–167.

Sun, A., Zhao, Y., & Allen, S., (2005). Enhanced removal of organic matter and ammoniacal nitrogen in a column experiment of tidal flow constructed wetland systems. *J. Biotechnol., 115*, 189–197.

Tanner, C. C., & Headley, T. R., (2011). Components of floating emergent macrophyte treatment wetlands influencing removal of stormwater pollutants. *Ecol. Eng., 37*, 474–486.

Waajen, G. W. A. M., Van, B. N. C. B., Pires, L. M. D., Lengkeek, W., & Lürling, M., (2016). Biomanipulation with quagga mussels (*Dreissena rostriformis* bugensis) to control harmful algal blooms in eutrophic urban ponds. *Ecol. Eng., 90*, 141–150.

Wang, C. Y., Sample, D. J., Day, S. D., & Grizzard, T. J., (2015). Floating treatment wetland nutrient removal through vegetation harvest and observations from a field study. *Ecol. Eng., 78*, 15–26.

Wen, L., & Recknagel, F., (2002). In situ removal of dissolved phosphorus in irrigation drainage water by planted floats: Preliminary results from growth chamber experiment. *Agric. Ecosyst. Environ. Times, 90,* 9–15.

Williams, P., Whitfield, M., Biggs, J., Fox, G., Nicolet, P., Shillabeer, N., Sherratt, T., et al., (2002). How realistic are outdoor microcosms? A comparison of the biota of microcosms and natural ponds. *Environ. Toxicol. Chem., 21*, 143–150.

Xian, Q., Hu, L., Chen, H., Chang, Z., & Zou, H., (2010). Removal of nutrients and veterinary antibiotics from swine wastewater by a constructed macrophyte floating bed system. *J. Environ. Manag., 91,* 2657–2661.

Yan, C., Wang, M., Ma, T., Yang, S., Kong, M., Shen, J., Yang, L., & Gao, Y., (2020). Study on the experimental performance by electrolysis integrated ecological floating bed for nitrogen and phosphorus removal in eutrophic water. *Scientific Reports, 10*, 7619. www.nature.com/scientific reports/

Yang, Z., Zheng, S., Chen, J., & Sun, M., (2008). Purification of nitrate-rich agricultural runoff by a hydroponic system. *Bioresour. Technol., 99*, 8049–8053.

Yeh, N., Yeh, P., & Chang, Y. H., (2015). Artificial floating islands for environmental improvement. *Renew. Sustain. Energy Rev., 47*, 616–622.

Zhao, F., Xi, S., Yang, X., Yang, W., Li, J., Gu, B., & He, Z., (2012). Purifying eutrophic river waters with integrated floating island systems. *Ecol. Eng., 40*, 53–60.

Zhou, X., & Wang, G., (2010). Nutrient concentration variations during *Oenanthe javanica* growth and decay in the ecological floating bed system. *J. Environ. Sci., 22*, 1710–1717.

Zhu, L. D., Li, Z. H., & Ketola, T., (2011). Biomass accumulations and nutrient uptake of plants cultivated on artificial floating beds in China's rural area. *Ecol. Eng., 37*, 1460–1466.

CHAPTER 2

Advanced Functional Nanomaterials for Sustainable Water Treatment Technologies

SHEEZA HAROON,[1] SHAKEEL AHMAD BHAT,[1]
MD. TOWFIQUL ISLAM,[2] and MARYAM SADAT JAAFARZADEH[3]

[1]*College of Agricultural Engineering and Technology, SKUAST–K, Jammu and Kashmir, India, E-mail: sheeza.haroon11@gmail.com (S. Haroon)*

[2]*Department of Disaster Management, Begum Rokeya University, Rangpur – 5400, Bangladesh*

[3]*Department of Watershed Management Engineering, Faculty of Agriculture and Natural Resources, Lorestan University, Iran*

ABSTRACT

Water is the most important and necessary factor for the survival in all living things. In modern times, however, the world is struggling with water shortages. However, other microorganisms, gasses, and various toxins (heavy metals and chemicals) are introduced into our bodies during rainfall, rapid water flow, etc., thus causing water pollution. Applied nanoscale science and engineering science recommends that most of the day-to-day problems involving different liquids should be determined by the use of nano-sorbents, nano-catalysts, bioactive nanoparticles (NPs), composite membranes and nanoparticle to accelerate the evaporation and diversity of materials and materials arising from the growth of nanotechnology. New

Advanced Technologies for Water Quality Treatment and Management. Mehraj U. Din Dar, PhD, Aamir Ishaq Shah, MTech, Shakeel Ahmad Bhat, PhD & Syed Rouhullah Ali, PhD (Eds.)

techniques in the development of a science novel used to make water diversity are among the best and most exciting. In addition, the material taken from nanotechnology limits the focus of lethal computing to sub levels, which helps in obtaining high water quality requirements and durability specifications. There are different types of nanomatadium, which carried a large latent to deal with contaminated water very effectively due to their presence only as a high surface area, capable of operating at low latency, etc. With one step of a combination of gold (Au), silver (Ag), platinum (Pt), and palladium (Pd) NPs in water, Sodium Rhodizonate was once used as an effective mitigation agent as a stabilizing agent. The average diameter ranges from 2, 8, 26, and 39 mm to Pt, Au, Ag, and PdNPs used for exposure by transmission electron microscopy. With a strong discount of 4-NP ahead of Ag, Au, Pt, and Pd NPs, the public payers (kapp), accounted for 2.1482, 1.1167, 1.088×10^{-1}, and 1.65×10^{-2} $minute^{-1}$, respectively.

2.1 INTRODUCTION

Water is the most valuable commodity in human evolution, and a reliable water source is a basic human need. However, since our climate is less of a barrier to economic growth and urbanization, we are also far from the global issue of integration. As a result, this problem will worsen over time (Hillie and White, 2007). The rate is rising due to overpopulation, local climate change, and rising water levels (Ali and Aboul-Enein, 2004). Worldwide,> 700 million people no longer have access to water (WHO, 2014). This misery is rampant in the creation of sub-Saharan Africa. Existing sewage treatment infrastructure, such as in the case of stagnant water and in a complex way it is difficult to maintain a strong law and growing demand for excellent water, in each developed and old world. Therefore, in these affected areas water treatment needs to be done. The available science used in water treatment reaches its limit in providing the first sufficient level to meet human needs (Qu et al., 2013a; Shah et al., 2020). Therefore, reuse, reuse, and replace yourself with the needs of the "day."

Recent advances in nanomatadium understanding have helped the purpose of nanotechnology in the treatment of water and water. Decades ago, aqueous nanotechnology had an additional internal vision as a flower starter in traditional medicine methods. Construction materials with a minimum thickness of 100 nm are widely described as Nanomatadium

(Tesh and Scott, 2014). The resources of this scale consist of size-dependent structures, which differ in comparison to their larger counterparts. Nanomatadium is made up of elements, such as feature height, reactivity, hydrophilic, and hydrophobic interactions, which are useful in adsorption, catalysis, sensor, and optoelectronics (Das et al., 2014). Procedures powered by Nanotechnology for water treatment add significant challenges to existing approaches. Nanotechnology is also broad in the refinement and use of unconventional water resources in a financially viable manner. Treatment of industrial wastewater with nanomatadium is also important and comprehensive. Nanotechnologies are essential in wastewater treatment, as they remove contamination and help to obtain clean water during the recycling process. This results in decreased manpower, time, and costs when it comes to addressing industries with different environmental problems (Kanchi, 2014). Nanomatadium must be environmentally friendly and non-hazardous, for drinking water. Significant damage to vital organs in contact with the human body can be caused by unsafe particles (Figures 2.1 and 2.2; Table 2.1).

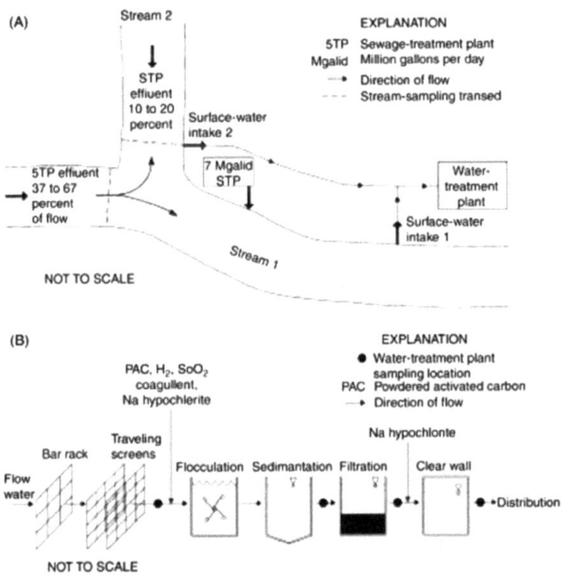

FIGURE 2.1 Schematic sketch displaying (A) area of flow sampling sites, drinking-water-treatment plant; and surface-water intakes (B) bodily and chemical approaches used in drinking-water-treatment plant.

Source: Reprinted with permission from Stackelberg et al., 2004. © Elsevier.

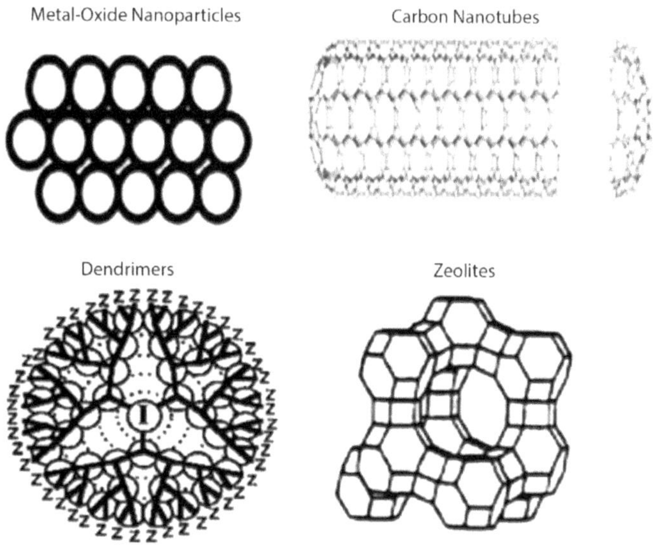

FIGURE 2.2 Selected nanomaterials now being evaluated as practical substances for water purification.

TABLE 2.1 Water Treatments Using Nanomaterial with the Aid of Exceptional Mechanisms

Nanoparticles (NPs)	Target Analyte	Treatment Mechanism	Positive Aspect	Limitation
Fe	Heavy metals, anions	Reduction, adsorption	*In situ* water remediation, cheap	Tough to recovery, disposal of sludge difficult, health risk
TiO_2	Organic pollutants	Photocatalysis	Less toxic, non-water soluble, photostability	The cost of operation is higher, it is impossible to heal., sludge generation
Metal-sorbing vesicles	Heavy metals	Adsorption	Maintaining continuity is difficult.	Recycling option, higher profile of selective absorption, improved metal affinity
Bimetallic NPs	Dichlorination	Reduction, adsorption	Reactivity is higher	Recovery is difficult, producing sludge

TABLE 2.1 *(Continued)*

Nanoparticles (NPs)	Target Analyte	Treatment Mechanism	Positive Aspect	Limitation
Magnetite NPs	Heavy metals	Adsorption	Separation is easy, and no sludge is generated.	Separation requires an external magnetic field.
Nanofiltration	Organic and inorganic materials	Nanofiltration	Pressure is low	Membrane fouling, high cost
Micelles	Organic pollutants	Adsorption	Expensive	Treatment is carried out on-site, and it has a high tolerance for hydrophobic materials.
Nanoclay	Heavy metals, organic pollutant, anions	Adsorption	Production of sludge	Cheaper, exclusive mechanisms, long-term reliability, and recycling better surface and pore depth, higher sorption capability, informal recovery
Nanotube	Heavy metals	Adsorption	The cost is high, and the adsorption process is inefficient. Recovery is difficult, and sludge processing and health risks are dangerous.	Deals with water waste, has strong, good chemical stability.
Dendrimers	Heavy metals	Encapsulation	Costly	Separation is easy, reusable, binding is high, and no sludge is generated.

Source: Mansoori et al. (2008).

Natural organic materials, man-made chemicals, and chemical processes such as polyaromatic hydrocarbons (PAH) and nitroaromatic compounds (NAC) pollute the marine environment. Various methods of stealing pollution from aquatic environments are being explored. The removal of these websites has led to the completeness of applied science novels that strengthen the deception of pollution. Therefore, the reduction in pollution reduction has been promoted as a means of prosperity for wastewater treatment technology. The catalytic discount can actually pollute the environment with faster prices with higher vigor and recycling.

Gold nanoparticles (AuNPs) are commonly available nanomatadiums due to their peculiar optical, physical, and chemical properties. AuNPs are designed for server performance but are not confined to catalysis, photocatalysis, photothermal photo voltaic distillation, sensor sensitivity, plasmonic photovoltaics, organic, and clinical shipping drug, biomedicine, bioimaging, and other therapy. In addition to the softening properties, AuNPs have an over-the-top molar makeup (108 to 1,010 M^{-1} cm^{-1}) on a transparent mirror due to a substance called floor plasmon resonance. By changing the size, texture, texture, performance of the floor, and diploma in the integration of nanoparticles (NPs) the color of the AuNPs response can be measured.

2.2 METHODS FOR THE PREPARATION OF NANOPARTICLES (NPS)

For the teaching of one-a-type NPs there are various methods. However, all power can be divided into categories under two categories, top-down processes, and bottom-up processes. In the top-down method, the fabric with macroscopic density is slightly split or reduced to nanoscopic sizes. A mixture of salts with iron such as oxidizer (e.g., nitrate of iron) and oils (e.g., urea, glycine, hydrazides) overheating at high temperatures, in the composite mixture, when heated to a temperature, the compound becomes independent of the uniform production response of iron oxide NPs while burning gas in the form of CO_2, H_2O, and different gasses. These mechanisms form nanostructured materials from atomic or molecular molecules.

2.2.1 SYNTHESIS OF NOBLE METAL NANOPARTICLES (NPS)

The two most commonly used uses of AuNPs are:

- The Turkevitch method, added in 1951, i.e., the apparent reduction in the stability of AuNPs by the use of sodium citrate in water; and
- Bi-phasic synthesis, $NaBH_4$ reduction and Tiols stabilization, produced using Brust-Schiffrin in 1994. Both of these techniques require heat. There is a need for a method for the formation of fine non-ferrous NPs that have been found to be fertile and have the potential to reduce the potent supporting structures.

2.2.2 SYNTHESIS OF METAL OXIDE NANOPARTICLES (NPS)

Other commonly used techniques for the production of iron oxide nanoparticles include sol-gel, moisture, hydro/solvothermal, laser dispersion, chemical application, mic emulsion techniques, and microwave assisting synthesis. Recently, the composite method has become the most modern and efficient method of forming iron oxide nanoparticles due to the flexibility, simplicity, and durability of the process. In combination, a mixture of iron-containing salts such as oxidizer (e.g., iron nitrate) and fuels (e.g., urea, glycine, hydrazides) is heated at high temperatures. When the temperature is warm, the compound produces effective responses to the production of iron oxide nanoparticles while burning gasoline in the form of CO_2, H_2O, and various gasses.

2.2.3 ROLE OF NANOMATERIALS IN TREATING AND PURIFYING WATER

Nanomatadium is growing rapidly as a solid collector of water treatment in the field of applied science, which can no longer tolerate its effectiveness, is very expensive and shows for a long time. This will be, above all, a major contribution to the growth of international destinations such as India and Bangladesh. Custom nanomatadiums can be replaced by traditional ones that require large raw materials, have great potential for production or are considered harmful to the environment. Using inexperienced chemical ideas for making nanoparticles can lead to a good discount on waste formation, less harmful chemicals in chemicals, and generally safer chemistry. However, to address these claims, many additional facts are also required that altering or altering common elements with nanoparticles does not without a doubt lead to a reduction in the strength and utilization of the fabric and the elimination of undesirable or sudden open competition outcomes. Additionally, there is wide-ranging debate about the safety of nanoparticles and their consideration can have an environmental impact. There is great hope that nanotechnology can play an important role in efficiently, cost-effectively, and sustainably delivering smooth water to developing nations. In contrast, the harmful effects of nanoparticles cannot be ruled out. In the event of a nanoparticle reinforcement effort, it can be helpful when applied to contaminants, however it can trigger a painful reaction when the cell is taken. So, this Janus personality of nanotechnology could prove to be a snag in its great acceptance. Nanotechnology can go a long way in reducing costs and as a result, grows more impressive than modern methods of removing water pollution when it comes in the long run. Nanoparticles can be used as heady sorbents, sources of media fragmentation, and catalysts for the sonochemical of contaminants in this perspective. A nanosized zerovalent iron that is used in nanofiltration membranes to isolate metals and natural compounds from water (Bhattacharya et al., 2013).

2.3 METHODS OF WATER TREATMENT

2.3.1 NANO-PHOTOCATALYSTS

The term "photocatalysis" is the silence of two Greek words for "image" and "catalysis" meaning the decay of computers in a light frame. Usually, in the technological world there is no study of photocatalysis (Saravanan et al., 2017). They are often used to purify wastewater, as they help to increase the recycling effect due to their rate of increase in structure and formwork (Chen et al., 2019). They can increase oxidation capacity due to the natural production of oxidizing species under the fabric which helps to reduce pollution from well-contaminated water (Gomez-Pastora et al., 2017). SiO_2, ZnO, TiO_2, Al_2O_3, etc., are the most common and extensive metal oxide nanophotocatalyst. Among them, TiO_2 is one of the largest images from all available fabric due to its innumerable reasons such as chief, nontoxic material, chemical durability, and its simple accessibility in the world. In addition, TiO_2 exists in three regions of herbs, anatase, rutile, and brookite. To date, anatase is considered a suitable nanophotocatalyst (Yamakata et al., 2019). With a UV lamp, the production of TiO_2 is done nonetheless given daylight or seen with gentle lights.

The liquid selection of nanoscale zero-valent iron (FeO) and bimetallic FeO particles has emerged as redox media to expose the pollution and impurities. These nanomatadium (10–100 nm) have significantly lower surface area and regenerate than most FeO particles (Schrick et al., 2002; Nurmi et al., 2005). Zhang (2003) provided an overview of the design, appearance, and application of nanoscale particles FeO and FeO/PdO, FeO/PtO, FeO/AgO, FeO/NiO and FeO/CoO in environmental guidance. List of environmental pollutants (e.g., alkanes in chlorine and alkenes, benzenes chlorine, pesticide, natural dye, nitro aromatics, PCBs) and various organics (e.g., nitrates) with water options which do not significantly reduce toxicity and repetitive products. These nanoparticles. Fe^0 and bimetallic Fe^0 nanoparticles have been further used to reduce redox in iron ions such as Cr (VI) is highly toxic and cellular types Cr (III). The suspension of metalloporphyrinogens in matric-gel matrices is detailed for use in combining redox and nanoparticles catalytically lively the defaultgenation of composed of natural complex (PCE, TCE, and carbon tetrachloride (CT)) with water options (Dror et al., 2005).

2.3.2 NANO AND MICROMOTORS

Skipping environmental problems such as water pollution treatment and environmental sensitivity/nano/micromotor recognition are used. Since functional nanomaterials have inbound habitats that make it more environmentally friendly to convert toxic pollution into a toxic environment, they are considered another alternative (Figure 2.3).

2.3.3 NANOMEMBRANES

Nanomembranes are an unparalleled type of membrane infused with certain nanofibers that are hired to eliminate the unsuitable nanoparticles present in the liquid phase. The system takes up close-knit areas that are critical properties and already acts as a pre-treatment system used for relapsing osmosis (Jhaveri et al., 2016). For water treatment, the porous membrane is subjected to nanofiltration, ultrafiltration, recurrent osmosis, etc. The membrane carries porous assist with composite layer. Typically, a large layer of compound is a carbon-based coating (graphene oxide/CNT) that is distributed in a polymer matrix that delivers promising progress and full size in amazing resistance to water transport. Separation of toxic particles from aquatic materials is an important objective of fluid application when entering water. To measure the waterproofing phase nanomembrane filters are used in addition (Bhat et al., 2018). These methods are viewed as state-of-the-art water purification technology and desalination technology.

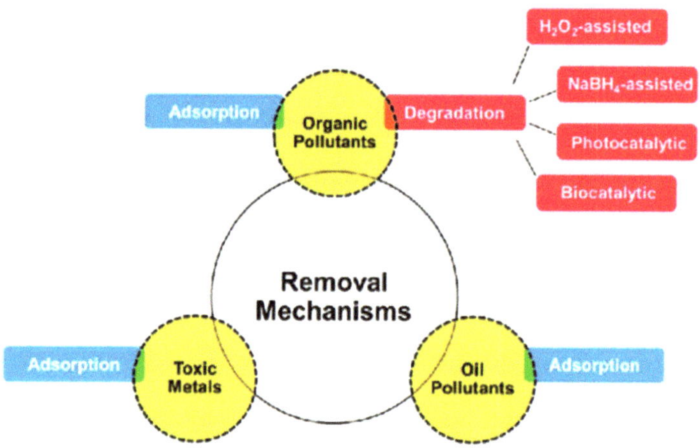

FIGURE 2.3 Different pollutant elimination mechanisms used through nano/micromotors.
Source: Reprinted from Eskandarloo et al., 2017. © Royal Society of Chemistry.

Nanomembranes such as NF and RO can filter out very small organisms (viruses and bacteria) and water-containing molecules that float faster than traditional filters because of their ability to differentiate by molecular level. They can be classified into two main categories: (a) non-synthetic filters for carbon nanotubes (CNTs) or nano-capillary arrays provide nanofiltration; and (b) as functional, functional nanoparticles useful filtration processes. The desalination membranes built into the CNTs offer exciting new opportunities. Nanomatadium such as CNTs, nanoparticles, and dendrimers contribute to improving the more environmentally friendly environment and within your water filtration processes. Improving advanced filtration methods for water purification and regeneration of precious metals such as uranium dendritic polymers provided opportunities.

2.3.4 NANOSORBENTS

Nanosorbents take up enough living space as an overactive sorption that makes nanosorbents more demanding and more capable of finding solutions. Sensitive homes make nanoparticles more productive as sorbents. On a larger scale, they have a much lower surface area as opposed to larger particles. In addition, they can be promoted by selecting reactor companies to increase their chemical relationships in achieving objective goals. Strengthening selected industries that prefer peace and environmental degradation in order to remove pollution and polluted water from these polluted areas these residences are a growing number of workers' exploitation. Many items have homes in terms of size. Hematite particles with a diameter of 7 nm, for example, Cu ion has a lower pH value than 25 particles or 88 nm in diameter, indicating a renewed grounding of iron oxide particles with decomposing size (Table 2.2) (Nowack et al., 2007).

TABLE 2.2 The Functions of Commonly Used Nanosorbents (Ns)

Different Nanosorbents	Treatment Function	References
Carbon-based Ns	In water, nickel ions present	Rodovalho et al. (2016)
Graphite oxide	Removal of dyes	Sun et al. (2015)
Regenerable polymeric Ns	Organic and inorganic contaminants	Lee et al. (2012)
Nanoclay	Hydrocarbons, dyes	Kyzas et al. (2015)
Nano-metal oxides	Different heavy metals	Wang et al. (2016)
Nano-aerogels	To take out the uranium from drinking water	Krstic et al. (2018)
Nano-iron oxides	To remove the hormones and toxic pharmaceuticals material from water	Unuabonah et al. (2014)
Polymer fibers	To eradicate the arsenic	Yadav et al. (2019)

In the extinction of various pollutants such as heavy metals from carbon dioxide are a flexible adsorbent used closely. Graphene is a modern member of the carbon family in appearance and is believed to be one of the most useful water treatment agents (Babel et al., 2003). It is a complete, sp^2-hybridized assembly, on the side of two carbon bees with the same carbon size (Park et al., 2009). Graphene and its compounds offer usability for a number of purposes due to its unique nature of size and related group cohesion. Factors such as large regional bases and the presence of low-cost companies that make them the best adsorbent water purification option (Sreeprasad et al., 2011). Mixtures of RGO-magnetite and GO-ferric hydroxide have been used to dissolve the aqueous arsenic RGO-magnetite and GO-ferric hydroxide composites (Chandra et al., 2010). Metals based entirely on oxides and hydroxides have already been proven to be highly efficient in removing arsenic from ingested water (Zhang et al., 2010). Unlike those free nanoparticles, RgO, and GO's supporting materials have a great ability to bind. Interestingly, the reduced content of graphene oxide has antibacterial properties and this property can help prevent the development of biofilm-based filters due to bacterial growth, which may remove unpleasant odors and excess odors or filter clogging.

2.4 APPLICATIONS OF NANOMATERIALS

- To dispose of arsenic from disposal and disposal of water, titanium nanoparticles and other fabrics used for resin are basically enriched with iron hydroxide material, nanomatadium highly efficient.
- To feel different from dirt, a sensor-based sensor is used because this one has ocular housings that make the sensor look sensible and tend to detect dirt.
- In natural reinforcement made of polymer-nanomaterial it helps to a greater extent.
- Nanomaterials can act as magnets such as CNTs, zeolites, and synthetic layers in a mesoporous assembly, which uses mesoporous clay with a sorbent that shows a system of non-alignment of metals and anions in depleted water.
- Some nanomaterials have very strong anti-damage properties. This type of fabric contains AgNPs, fullerene, TiO_2, CNT derived from nanocomposite completely, etc. They also include many habitats such as moderate oxidant, producing safer products and water repellent.

- Nanoparticles play a suitable function as catalyst in special modification of oxidation metallic oxides. They show strong resistance to pollutant molecules and sell these pollutants as an environmentally friendly material (Soppe et al., 2014; Bushra et al., 2014). At these nanomatadiums, other special houses exist such as nanosize, overuse, and expanded floor space. Specifically, TiO_2 photocatalysis performs an important function of disposing of separate impurities from groundwater as published in Figure 2.4.

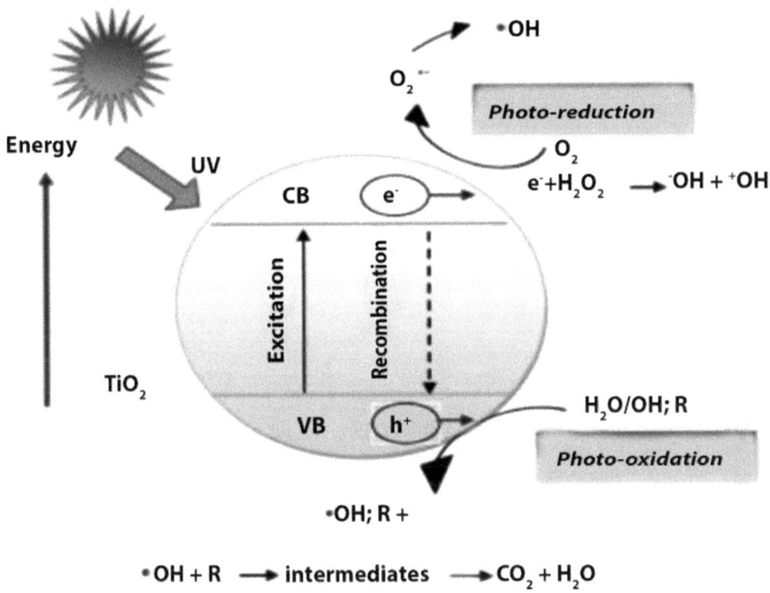

FIGURE 2.4 Schematic illustrating of TiO_2 photocatalytic method.
Source: Reprinted from Nor et al., 2013. © 2012 Penerbit UTM Press.

2.5 SAFETY, TOXICITY, AND ENVIRONMENTAL IMPACT OF NANOMATERIALS

Toxicity has become one of the burning problems of nanotechnology, due to our misalignment of the behavior and nanoparticular behavior in humans and the environment. The dangers of nanoparticles and fortune-telling are major problems associated with nanotechnology. The first important problem is the natural effects of chemicals on humans or the environment. Leakage, spillage, circulation, and attention to nanoparticles that may cause harm to humans or the environment is problem 2 (Gardner and Dhai, 2014). Nanoparticles are toxic due to their properties and can lead to excessive chemical reactions and ROS formation. ROS production is all possible with iron oxides and CNTs. The ROS produced will remove oxidative stress, inflammation, subsequent to the breakdown of proteins, membranes, and DNA. It is the agglomeration nanoparticles that show their toxicity in the environment (Table 2.3).

TABLE 2.3 Self-Toxic Effect of Some Commonly Used Nanomaterials

Nanomaterials	Observations	References
Gold nanorods doped poly(styrene sulfonate) (PSS) composite	Toxic-free	Wang et al. (2008)
TiO_2, Al_2O_3 composite	At micron particle-sized more toxic	Wang et al. (2008)
CdSe quantum dots doped with polyvinyl carbazole composite	Toxicity was discovered in an acute state.	Ray et al. (2009)
Fe_2O_3 and carbon nanotubes composite	Even at the lowest concentrations, toxic effects and DNA damage were observed.	Karlsson et al. (2008)
Metal, carbon, Al_2O_3 composite	Based on concentration and time.	Ajdary et al. (2018)
Au/carbon composite	For a cheaper price point toxic-free	Jia et al. (2005)
Carbon nanotubes with single and multiple walls	Toxicity increases when concentrations exceed 15 g/cm.	Taju et al. (2014)
Hexadecylcetyl trimethyl ammonium bromide (CTAB) doped Au nanorods composite	Highly poisonous at some concentrations	Takahashi et al. (2006)
Ag/carbon composite	It depends on the time and dosage, but it is poisonous.	Karlsson et al. (2008)

2.6 CONCLUSIONS

When compared to Pt and PdNPs, quicker kinetics are used for regular showing with the aid of AgNPs followed by AuNPs at a 4-NP discount with $NaBH_4$. In comparison to AuNPs and AgNPs, a MO disc with a formic acid made with Pt and PdNPs was obtained at a faster rate. Sodium Rhodizonate is a form of sodium rhodizonate. The use of a computer to mediate the synthesis of Au, Ag, Pt, and Pd nanoparticles demonstrates an increasing path of efficient computer acquisition for a wide range of useful materials such as environmental waste depletion and water treatment technologies.

Metal ions, anions, natural synterior, and organic organisms can be eliminated by the use of nanoparticles. The volumes of nanoparticles that are essential for water treatment are low, making their software extremely expensive. The prospect of nanoparticles in water treatment is slowing, however, it requires the combined efforts of educational and industrial resources to advance future technologies, economics, and water treatment technologies. We will be able to work together to find a global water supply.

KEYWORDS

- **carbon nanotubes**
- **gold nanoparticles**
- **nanoparticles**
- **nanotechnology**
- **nitroaromatic compounds**
- **wastewater**
- **water quality**

REFERENCES

Ajdary, M., Moosavi, M. A., Rahmati, M., Falahati, M., Mahboubi, M., Mandegary, A., & Varma, R. S., (2018). Health concerns of various nanoparticles: A review of their *in vitro* and *in vivo* toxicity. *Nanomaterials, 8*(9), 634.

Ali, I., & Aboul-Enein, H. Y., (2004). *Chiral Pollutants: Distribution, Toxicity and Analysis by Chromatography and Capillary Electrophoresis.* John Wiley & Sons.

Ali, M. G., Bastami, T. R., Ahmadpour, A., & Eshaghi, Z., (2008). Environmental application of nanotechnology. In: *Annual Review of Nano Research* (pp. 439–493).

Babel, S., & Kurniawan, T. A., (2003). Low-cost adsorbents for heavy metals uptake from contaminated water: A review. *Journal of Hazardous Materials, 97*(1–3), 219–243.

Bhat, A. H., Rehman, W. U., Khan, I. U., Khan, I., Ahmad, S., Ayoub, M., & Usmani, M. A., (2018). Nanocomposite membrane for environmental remediation. In *Polymer-based*

Nanocomposites for Energy and Environmental Applications (pp. 407–440). Woodhead Publishing.

Bhattacharya, S., Saha, I., Mukhopadhyay, A., Chattopadhyay, D., & Chand, U., (2013). Role of nanotechnology in water treatment and purification: Potential applications and implications. *Int. J. Chem. Sci. Technol., 3*(3), 59–64.

Bushra, R., Shahadat, M., Ahmad, A., Nabi, S. A., Umar, K., Oves, M., & Muneer, M. (2014). Synthesis, characterization, antimicrobial activity and applications of composite adsorbent for the analysis of organic and inorganic pollutants. *Journal of Hazardous Materials, 264,* 481–489.

Chandra, V., Park, J., Chun, Y., Lee, J. W., Hwang, I. C., & Kim, K. S., (2010). Water-dispersible magnetite-reduced graphene oxide composites for arsenic removal. *ACS Nano, 4*(7), 3979–3986.

Chen, W., Liu, Q., Tian, S., & Zhao, X., (2019). Exposed facet dependent stability of ZnO micro/nanocrystals as a photocatalyst. *Applied Surface Science, 470*, 807–816.

Das, R., Ali, M. E., Abd, H. S. B. A., Ramakrishna, S., & Chowhury, Z. Z., (2014). Carbon nanotube membranes for water purification: A bright future in water desalination. *Desalination, 336*, 97–109.

Dror, I., Baram, D., & Berkowitz, B., (2005). Use of nanosized catalysts for transformation of chloro-organic pollutants. *Environmental Science & Technology, 39*(5), 1283–1290.

Eskandarloo, H., Kierulf, A., and Abbaspourrad, A. 2017. Nano- and micromotors for cleaning polluted waters: focused review on pollutant removal mechanisms. Nanoscale, 37. https://pubs.rsc.org/en/content/articlelanding/2017/nr/c7nr05494g

Gardner, J., & Dhai, A., (2014). Nanotechnology and water: Ethical and regulatory considerations. *Application of Nanotechnology in Water Research,* 1.

Gómez-Pastora, J., Dominguez, S., Bringas, E., Rivero, M. J., Ortiz, I., & Dionysiou, D. D., (2017). Review and perspectives on the use of magnetic nanophotocatalysts (MNPCs) in water treatment. *Chemical Engineering Journal, 310*, 407–427.

Hillie, T., & Hlophe, M., (2007). Nanotechnology and the challenge of clean water. *Nature Nanotechnology, 2*(11), 663, 664.

Islam, M. T., Saenz-Arana, R., Wang, H., Berna, & Noveron, J. C., (2018). Green synthesis of gold, silver, platinum, and palladium nanoparticles reduced and stabilized by sodium rhodizonate and their catalytic reduction of 4-nitrophenol and methyl orange. *New Journal of Chemistry, 42*(8), 6472–6478.

Jhaveri, J. H., & Murthy, Z. V. P., (2016). A comprehensive review on anti-fouling nanocomposite membranes for pressure driven membrane separation processes. *Desalination, 379*, 137–154.

Jia, G., Wang, H., Yan, L., Wang, X., Pei, R., Yan, T., & Guo, X., (2005). Cytotoxicity of carbon nanomaterials: Single-wall nanotube, multi-wall nanotube, and fullerene. *Environmental Science & Technology, 39*(5), 1378–1383.

Kanchi, S., (2014). Nanotechnology for water treatment. *J. Environ. Anal. Chem, 1*(2), 1–3.

Karlsson, H. L., Cronholm, P., Gustafsson, J., & Moller, L., (2008). Copper oxide nanoparticles are highly toxic: A comparison between metal oxide nanoparticles and carbon nanotubes. *Chemical Research in Toxicology, 21*(9), 1726–1732.

Krstic, V., Urosevic, T., & Pesovski, B., (2018). A review on adsorbents for treatment of water and wastewaters containing copper ions. *Chemical Engineering Science, 192,* 273–287.

Kyzas, G. Z., & Matis, K. A., (2015). Nano adsorbents for pollutants removal: A review. *Journal of Molecular Liquids, 203*, 159–168.
Lee, X. J., Lee, L. Y., Foo, L. P. Y., Tan, K. W., & Hassell, D. G., (2012). Evaluation of carbon-Based nanosorbents synthesized by ethylene decomposition on stainless steel substrates as potential sequestrating materials for nickel ions in aqueous solution. *Journal of Environmental Science, 24*(9), 1559–1568.
Mansoori, A. G., Bastami, T. R., Ahmadpour, A., & Eshaghi, Z. (2008). Environmental application of nanotechnology. *Annual review of nano research*, 439-493.
Nor, N. A. M., Jaafar, J., Othman, M. H. D., & Rahman, M. A., (2013). A review study of nanofibers in photocatalytic process for wastewater treatment. *Jurnal Teknologi, 65*(4).
Nowack, B., & Bucheli, T. D., (2007). Occurrence, behavior and effects of nanoparticles in the environment. *Environmental Pollution, 150*(1), 5–22.
Nurmi, J. T., Tratnyek, P. G., Sarathy, V., Baer, D. R., Amonette, J. E., Pecher, K., & Driessen, M. D., (2005). Characterization and properties of metallic iron nanoparticles: Spectroscopy, electrochemistry, and kinetics. *Environmental Science & Technology, 39*(5), 1221–1230.
Park, S., & Ruoff, R. S., (2009). Chemical methods for the production of graphenes. *Nature Nanotechnology, 4*(4), 217–224.
Qu, X., Alvarez, P. J., & Li, Q., (2013). Applications of nanotechnology in water and wastewater treatment. *Water Research, 47*(12), 3931–3946.
Raval, H. D., & Gohil, J. M., (2010). Nanotechnology in water treatment: An emerging trend. *International Journal of Nuclear Desalination, 4*(2), 184–188.
Ray, P. C., Yu, H., & Fu, P. P., (2009). Toxicity and environmental risks of nanomaterials: Challenges and future needs. *Journal of Environmental Science and Health, Part C, 27*(1), 1–35.
Rodovalho, F. L., Capistrano, G., Gomes, J. A., Sodre, F. F., Chaker, J. A., Campos, A. F., & Sousa, M. H., (2016). Elaboration of magneto-Thermally recyclable nanosorbents for remote removal of toluene in contaminated water using magnetic hyperthermia. *Chemical Engineering Journal, 302*, 725–732.
Saravanan, R., Gracia, F., & Stephen, A., (2017). Basic principles, mechanism, and challenges of photocatalysis. In: *Nanocomposites for Visible Light-Induced Photocatalysis* (pp. 19–40). Springer, Cham.
Schrick, B., Blough, J. L., Jones, A. D., & Mallouk, T. E., (2002). Hydrochlorination of trichloroethylene to hydrocarbons using bimetallic nickel–iron nanoparticles. *Chemistry of Materials, 14*(12), 5140–5147.
Shah, A. I., Din, D. M. U., Bhat, R. A., Singh, J. P., Singh, K., & Bhat, S. A., (2020). Prospectives and challenges of wastewater treatment technologies to combat contaminants of emerging concerns. *Ecol. Eng., 152*, 105882. https://doi.org/10.1016/j.ecoleng.2020.105882.
Soppe, A. I. A., Heijman, S. G. J., Gensburger, I., Shantz, A., Van, H. D., Kroesbergen, J., & Smeets, P. W. M. H., (2015). Critical parameters in the production of ceramic pot filters for household water treatment in developing countries. *Journal of Water and Health, 13*(2), 587–599.
Sreeprasad, T. S., Maliyekkal, S. M., Lisha, K. P., & Pradeep, T., (2011). Reduced graphene oxide–metal/metal oxide composites: Facile synthesis and application in water purification. *Journal of Hazardous Materials, 186*(1), 921–931.

Stackelberg, P. E., Furlong, E. T., Meyer, M. T., Zuagg, S. D., Henderson, A. K., & Reissman, D. B., (2004). Persistence of pharmaceutical compounds and other organic wastewater contaminants in a conventional drinking-water-treatment plant. *Science of Total Environment, 329*(1–3), 99–113.

Sun, X., Liu, Z., Zhang, G., Qiu, G., Zhong, N., Wu, L., & Wu, Z., (2015). Reducing the pollution risk of pesticide using nano networks induced by irradiation and hydrothermal treatment. *Journal of Environmental Science and Health, Part B, 50*(12), 901–907.

Taju, G., Majeed, S. A., Nambi, K. S. N., & Hameed, A. S., (2014). *In vitro* assay for the toxicity of silver nanoparticles using heart and gill cell lines of *Catla* and gill cell line of *Labeo rohita. Comerative Biochemistry and Physiology Part C: Toxicology & Pharmacology, 161,* 41–52.

Takahashi, H., Niidome, Y., Niidome, T., Kaneko, K., Kawasaki, H., & Yamada, S., (2006). Modification of gold nanorods using phosphatidylcholine to reduce cytotoxicity. *Langmuir, 22*(1), 2–5.

Tchobanoglous, G., Burton, F. L., & Stensel, H. D., (1991). Wastewater engineering. *Management 7*(1991), 1–4.

Tesh, S. J., & Scott, T. B., (2014). Nano-composites for water remediation: A review. *Advanced Materials, 26*(35), 6056–6068.

Unuabonah, E. I., & Taubert, A., (2014). Clay-polymer nanocomposites (CPNs): Adsorbents of the future for water treatment. *Applied Clay Science, 99*, 83–92.

Wang, S., Lu, W., Tovmachenko, O., Rai, U. S., Yu, H., & Ray, P. C., (2008). Challenge in understanding size and shape dependent toxicity of gold nanomaterials in human skin keratinocytes. *Chemical Physics Letters, 463*(1–3), 145–149.

Wang, Y., Zhang, Y., Hou, C., & Liu, M., (2016). Mussel-inspired synthesis of magnetic polydopamine–chitosan nanoparticles as biosorbents for dyes and metals removal. *Journal of the Taiwan Institute of Chemical Engineers, 61*, 292–298.

WHO, U., (2014). *Progress on Drinking Water and Sanitation-2014 Update.* World Health Organization.

Yadav, V. B., Gadi, R., & Kalra, S., (2019). Clay based nanocomposites for removal of heavy metals from water: A review. *Journal of Environmental Management, 232*, 803–817.

Yaqoob, A. A., Parveen, T., Umar, K., & Mohamad, I. M. N., (2020). Role of nanomaterials in the treatment of wastewater: A review. *Water, 12*(2), 495.

Zhang, K., Dwivedi, V., Chi, C., & Wu, J., (2010). Graphene oxide/ferric hydroxide composites for efficient arsenate removal from drinking water. *Journal of Hazardous Materials, 182*(1–3), 162–168.

Zhang, S., Li, X. Y., & Chen, J. P., (2010). Preparation and evaluation of a magnetite-doped activated carbon fiber for enhanced arsenic removal. *Carbon, 48*(1), 60–67.

Zhang, W. X., (2003). Nanoscale iron particles for environmental remediation: An overview. *Journal of Nanoparticle Research, 5*(3, 4), 323–332.

CHAPTER 3

Bioenergy Landscapes for Sustainable Water Quality Management

MEHRAJ U. DIN DAR,[1] J. P. SINGH,[1] KULDIP SINGH,[2] SUNIL GARG,[1] and SAMANPREET KAUR[1]

[1]*Department of Soil and Water Engineering, Punjab Agricultural University, Ludhiana – 141004, Punjab, India,*
E-mail: mehrajudindar24@gmail.com (M. U. D. Dar)

[2]*Department of Soil Science, Punjab Agricultural University, Ludhiana – 141004, Punjab, India*

ABSTRACT

The aim of multi-functional bioenergy landscape management is to optimize potential ecosystem services viz., water, electricity, fuel, and feed besides, water quality control. In order to manage ecosystems in a way that contributes to increased ecosystem services, spatial decisions can be crucial. When compared to non-agricultural lands, agriculture has a negative impact on downstream water quality. Perennial watersheds can be developed to increase the performance of water quality measures when compared to agricultural watersheds lacking perennial biofuel feedstocks. When growing biofuel feedstocks, the concern arises which conservation practices should be adopted in the landscape design for their orientation so that better water quality goals are achieved. Different types of decisions pertaining to spatial variations, necessary in a bioenergy landscape design were focused in this chapter, which includes decisions about artificial drainage, wetlands, vegetated strip location, and residue removal.

Advanced Technologies for Water Quality Treatment and Management. Mehraj U. Din Dar, PhD, Aamir Ishaq Shah, MTech, Shakeel Ahmad Bhat, PhD & Syed Rouhullah Ali, PhD (Eds.)

Optimizations that aim to mitigate sediment and nutrient losses, reduce water temperature, and optimize farmer income are appropriate resources for answering spatial design questions. Since many biophysical processes are scale based, cost and benefits should be considered for both watershed-scale and field-scale while implementing conservation practices to achieve these goals. Based on our review, bioenergy ecosystems can be built to boost water quality and enhance other environmental resources by adopting spatial decisions on various conservation practices.

3.1 INTRODUCTION

Bioenergy Landscape design is a method of spatially explicit planning that involves different stakeholders who are concerned about the same things (Figure 3.1). Its goal is to find ways to better leverage capital for more sustainable service delivery while taking context, patterns, and current conditions into account (Jones et al., 2012). "Landscape architecture therefore incorporates supply chain optimization, horizontal, and vertical performance assessments, stakeholder processes, spatial planning, biomass production systems, education, monitoring, and adaptive management." The concept is derived from the study of spatial pattern and process interrelationships, which is the basis for the field having landscape-based ecology (Turner et al., 2001), which has origins in both "landscape architecture and environmental sciences" (Nassauer and Opdam, 2008). The parks, country estates, suburban neighborhoods, and campuses planned by Frederick Olmsted adopting his guiding principles for efficient space organization and management are early examples of these concepts (Beveridge et al., 1995). Olmsted's principles encourage the usage of a space's naturally occurring characteristics while concealing the overall design to create a harmonized unit that is more concerned with function than ornamentation. Landscape designs like Olmsted's, on the other hand, do not always produce conditions that maximize all values (Crow et al., 2006). Trade-offs are unavoidable, and how people communicate with the natural, controlled, and built environment is context-dependent and subject to change (Koh and Ghazoul, 2010). Landscape designs are a great option to indulge stakeholders and figure out how to best integrate a set of goals (Hazelton et al., 2013; Darshini et al., 2013; Sayer et al., 2013). Private organizations (e.g., "a family or corporate farmer, forester, or rancher"), public bodies (e.g., a federal or state agency), "or a non-governmental" entity might be involved as managers (e.g., an environmental assessment group or any professional member associations). Landscape design is based on the idea that management methods should explore solutions that span different economies, disciplines, and geographic areas. That is, landscape-based strategies are "better than the amount of their sector-specific parts" when it comes to achieving a shared objective (Holmgren, 2013). According to Holmgren, a landscape design strategy seeks to reduce or eliminate externalities between land-based sectors in terms of economics; in terms of planning, it considers a broader range of options by avoiding limited and fragmented solutions and promoting up-front involvement of a broader set of stakeholders who reflect a diverse set of objectives but also shared goals (Holmgren, 2013). In order to establish solutions that are ideal at larger scales, this process necessitates comprehensive cooperation between actors who want to compromise. When it comes to public lands, where land-use decisions are often the product of comprehensive stakeholder involvement, "elements of landscape design approaches are typical. The Pacific Remote Islands Marine National Monument, for example, was created after the US considered over 170,000 comments from researchers, environmentalists, fishermen, and other stakeholders. Although there is still space for development, the practices that have emerged in the public sector for park sand reserves provide insight into the costs and challenges that landscape designers face when designing for bioenergy and other facilities on private property. The landscape design approach to bioenergy necessitates consideration of the context under which the bioenergy system operates (Efroymson et al., 2013). At appropriate scales, developing landscape designs to incorporate bioenergy is dependent on ecological, social, biophysical, and economic conditions (Duvenage et al., 2013). Microclimate, geography, alignment, prior, and current land ownership and use, objectives of land owners, producers, and other stakeholders, site drainage and groundwater revival, municipal, and resource building codes, human, and vehicular access and circulation, property safety and security can all be factors to consider."

Landscape design could be used in various situations and bioenergy systems, but it is not a "one-size-fits-all" solution. Rather, it is a method for considering context-specific concepts and data from a collection of environmental, social, and economic measures that may be relevant to bioenergy systems (McBride et al., 2011; Dale et al., 2013). The significance and selection of each indicator can differ depending on the values and expectations of stakeholders (Sinclair et al., 2015; Dale et al., 2015). To ensure that the most suitable data is collected and analyzed, it is important to understand how stakeholders weigh these various metrics. Feedstock grown for both bioenergy and remediation of former industrial or contaminated sites is an example of context-specific landscape design that considers bioenergy. Salix (willow) feedstock crops are cultivated in both the United States and Europe to minimize emissions from nutrient-rich waste streams such as biosolids and wastewater, as well as to use certain nutrients to fertilize the biomass crop (Berndes et al., 2008). These systems have a number of advantages, including reduced GHG emissions and waste disposal costs, as well as increased crop yields and lower nutrient input costs.

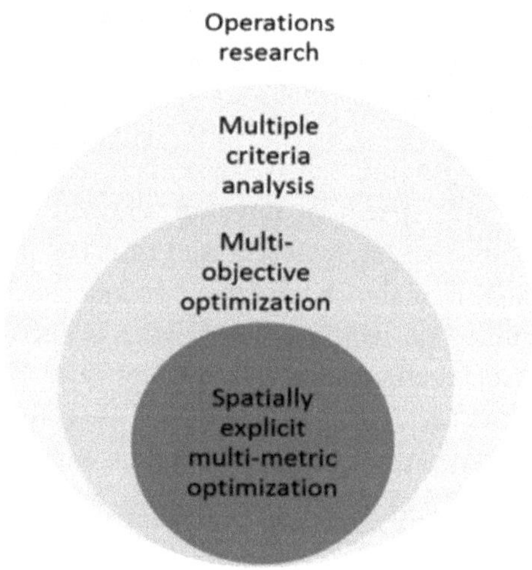

FIGURE 3.1 Hierarchy used in the optimization of landscape design.
Source: Reprinted with permission from Dale et al. (2016). © Elsevier.

3.2 LANDSCAPE DESIGN, DEVELOPMENT, AND IMPLEMENTATION

A landscape design approach followed for bioenergy landscaping includes six steps: (i) develop design objectives in light of the "context; (ii) identify constraints and opportunities; (iii) consider feedstock" appropriate for the context; (iv) analyze and deploy design; (v) track outcomes; and (vi) change as required (Figure 3.2). These actions should be taken in light of the previously discussed concepts of service conservation, context specificity, and adaptive management. This method necessitates stakeholder involvement at all levels of the project and is based on adaptive management principles, under which the initial plan is updated as new information becomes available. The spatial deliberations, magnitude, involvement of various participants, and up-front thinking about various uses and resources over time are all unique features of landscape design. Similar measures for preparing bioenergy feedstock plantation locations were proposed by McCormick et al. (2009), including: (i) screening; (ii) evaluation and consultation at national or regional levels; (iii) comprehensive site-level planning; and (iv) application of accountable land management practices (McCormick, 2009). Sayer et al. (2013) suggested how a landscape viewpoint can help farmers, conservationists, and other land users to coexist. In a similar way, our approach helps us to understand the entire bioenergy supply chain.

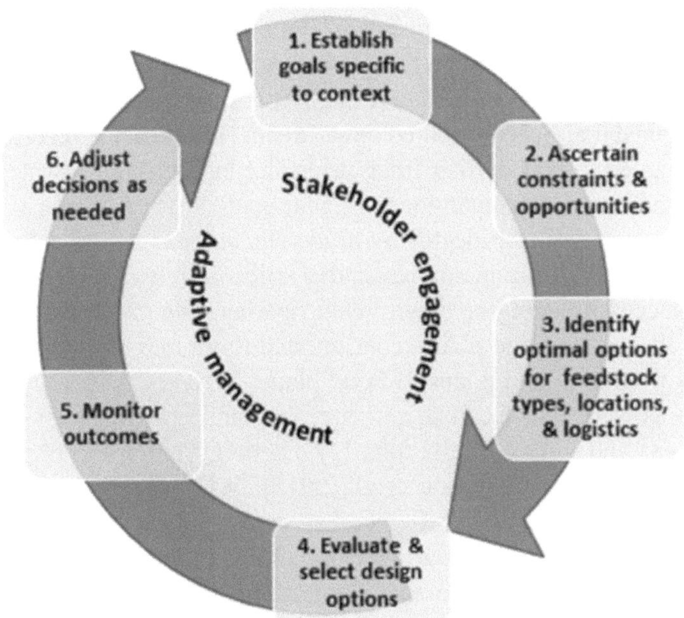

FIGURE 3.2 Steps in landscape design.
Source: Reprinted with permission from Dale et al. (2016). © Elsevier.

3.3 WATERSHED-SCALE PLACEMENT DECISIONS

Both agricultural (Ssegane et al., 2015; Helmers et al., 2012) and forested (Sheridan et al., 1999; Lowrance et al., 1997) production systems require spatial decisions about the location of buffers and filter strips in a watershed. These will help to cool rivers, reduce floods, and prevent erosion and nutrient loss through runoff, as well as provide income to farmers if harvested. Harvested willow (Ssegane et al., 2015) and corn stover are the best two examples of this spatial decision planning (Brandes et al., 2016). The benefits of upstream versus downstream buffer placement from watershed-scale optimization studies are discussed below.

3.3.1 CONDITIONS FOR UPSTREAM PLACEMENT

Water-quality improvement steps implemented upstream spread downstream, which resulted in multiple effects on its performance (Tomer et al., 2009; Dodds et al., 2008). Where the goal is to reduce water temperature, for example, shading is more successful in streams with smaller headwater than in larger mainstems downstream (Wu and Skelton-Groth, 2002). Similarly, filter strips installed on fields with upslope topology having tributary streams show a greater cumulative impact on water quality enhancement in a watershed than when installed downstream (Figure 3.3). Several studies have suggested that putting filter strips and buffers in headwater watershed source areas is a smart idea (Ssegane et al., 2015; Tomer et al., 2009). Since most runoff is collected by headwater streams before entering to a bigger channel, contour, and vegetative strips in several headwaters were more successful than one downstream riparian strip (Tomer et al., 2009). In the absence of flood plains, concentrated flows may occur when runoff moves from highland plains to flood plains or rivers. This causes strong erosive powers, which can be mitigated by planting switchgrass (Panicum virgatum) and willow (Salix alba L.) bioenergy buffers to obstruct and restrict flow paths (Ssegane et al., 2015). While the buffers could be maintained for biomass feedstocks harvesting, this is a great choice for achieving simultaneous environmental and biomass benefits. The lesser cost of installing buffers to additional catchments indicated that placing buffers strategically in a watershed was a cost-effective strategy (Wu and Skelton-Groth, 2002). "Water quality at drainage sources is influenced by spatial decisions made at the watershed scale." Jha et al. (2010) and Dar et al. (2020) estimated decrease in nitrate concentrations to drainage water due to conversion of fields with row crops to grasslands and rice-wheat cropping pattern using a "calibrated soil and water assessment tool (SWAT) and DRAINMOD model." When grasslands substituted crops in erosion prone areas, the reduction was higher (47%) than when crops were converted to grassland in upstream sections of the watershed (16%) or downstream floodplains (8%) (Jha et al., 2010). Furthermore, as opposed to adding buffers in the mainstem, Ha and Wu (2015) found that adding buffers in tributary streams resulted in a greater reduction in the amount of nitrate than in sediment or total phosphorus (TP). This was valid both at the outlet as well as at the basin. The SWAT model was recently used to assess filter strips as mitigation measures for two rivers under a bioenergy scenario expected for 2040 (Efroymson et al., 2016). Simulations were conducted for a 30-m and 50-m buffer strips in Iowa River for a corn-dominated agricultural field (Jager et al., 2017). According to the report, adding a 50-meter buffer to the entire Iowa River Basin would raise coverage to 11.3% of the land, reducing nitrate concentration by 10.8% and sediment load by 80% (Jager et al., 2017). The study estimated that corn stover, willow, and miscanthus would produce 121,000 metric dry tons of feedstock in the Iowa River basin (Jager et al., 2017). Furthermore, harvesting half of the switchgrass grown along a 125-kilometer stretch of stream could yield an additional 121 million dry metric tons of biomass (Jager et al., 2017). These findings matched with those of a smaller sub-basin, of the South Fork Iowa River basin (Ha and Wu, 2015).

3.3.2 HINDRANCES IN IMPLEMENTING DESIGN OF LANDSCAPES

In practice, implementing the landscape design approach suggested here is difficult, particularly at large scales where stakeholders' diversity makes it difficult to agree on shared goals and concerns, let alone indicator baselines and targets. However, in most cases, the bioenergy method only uses a limited portion of the land area as a fuel shed and only meets a portion of the region's energy needs. Consequently, even though not all stakeholders agree, the principles can be generalized to the bioenergy environment. Even at smaller scales, stakeholders may be uninterested in bioenergy and would rather direct the process toward other perceived needs such as health, education, or housing, which are more difficult to link to bioenergy supply chains. Furthermore, certain stakeholder groups can have very specific special interests and are reluctant to compromise. Some indicators may be inadequate due to shortcomings of baseline data and local knowledge. The bigger the geographic area under consideration, the more difficult it would be to gather stakeholders, reach a consensus on goals, and complete a landscape design plan. Bioenergy system sustainability is difficult to measure and define because the processes are complex, the underlying science is still being built, and the inherently context-specific enterprise is difficult to generalize (Dale et al., 2013). Another stumbling block to designing and deploying landscape design is when the landscape is divided in terms of ownership, and the priorities of the owners or local goals do not align with regional, state, global, or international goals. Diverse stakeholders also have conflicting interests, which can lead to roadblocks if not addressed. Landscape design implementation necessitates upfront preparation and scheduling, which can result in higher initial costs or be hampered by insufficient or missing supply chain contact. However, making a strategic investment in supply chain synchronization will result in lower final costs and improve the successful implementation bioenergy system for water quality improvement. If preemptive landscape design processes were used to detect and resolve problems before they became an issue, more bioenergy projects technologies could be adopted successfully. Despite considerable effort and support, initiatives that lack this degree of engagement and communication are more likely to fail (Upreti, 2004).

3.3.3 DOWNSTREAM PLACEMENT LANDSCAPE CONDITIONS

Filter strips can be used in flood-prone areas to protect crops from flood damage. Perennial filter strips will provide income to farmers in the future if harvested for biomass. For example, Li et al. (2019) discovered that in the United States, excessive rainfall and flood damage has decreased corn yield by 34% over the last four decades. In other terms, wasteland areas inside fields can degrade further as topsoil is lost when annual row crops are planted. Contrary to popular belief, perennial lignocellulosic biomass crops can be grown successfully on these lands (Gopalakrishnan et al., 2011). Integrated flood management approaches have been adopted where, 6 m wide grassed buffer strips over an area of 150 ha were planted along the edges of the field in a watershed (Evrard et al., 2008; Vandaele et al., 2013). Biomass Buffer of crops may help to avoid soil degradation and, in the future, may augment farmer income once lignocellulosic feedstocks become cost-competitive (Wu and Skelton-Groth, 2002; Ha and Wu, 2015). The relationship between impaired water quality and the percentage of land used for agriculture is well understood (Moreno-Mateos et al., 2008). "Since prime agricultural lands are more likely to be found on lower slopes, the best opportunities for buffering stream networks will be found downstream where agriculture is the predominant land use." Replacing more intensively controlled crops (e.g., crops with higher fertilizer inputs, etc.), with switchgrass in downstream areas of an East Tennessee, US watershed was proved to be an effective strategy for enhancing water quality in an optimization process using the SWAT model (Parish et al., 2012). As compared to a baseline landscape, intense planting (> 2% area) of perennial switchgrass at the outlet of an agricultural watershed decreased total nitrogen (TN) by 0.08 mg L^{-1}, TP by 0.02 mg L^{-1}, and total suspended solids (TSS) by 25 mg L^{-1} (Parish et al., 2012; Shah et al., 2020). According to another modeling analysis (Jager et al., 2017), the greatest declines in annual crop acreage occurred in the center of the Iowa River watershed.

FIGURE 3.3 Illustration of a vegetated filter strip design include: (1) upland and upstream tributaries placement; (2) multi-layer buffers; and (3) "placement of wetland and vegetation at the outlet of tile-drains."

Source: Reprinted from Krieg et al. (2019). https://creativecommons.org/licenses/by/4.0/

3.4 DECISIONS THAT MAXIMIZE WATER QUALITY

Filter strip effectiveness in trapping nutrients, sediment, and other pollutants is determined by site-specific factors such as strip distance, slope, vegetation, and influent nutrient loading (Liu et al., 2008). The majority of filter strip designs depend on models to decide the best buffer width (Fischer and Fischenich, 2000; Dosskey et al., 2008). Fischer and Fischenich (2000) suggested a 5-m to 30-m width for safe water quality. However, as shown in a field study on drained lands with no tiles, the optimum width of filter strips can be context-dependent. Tiny conversions (10%) of agricultural fields to filter strips with native prairie grasses which planted along the contour and at the foot slopes decreased sediment, TN, and TP by 95%, 85%, and 90%, respectively, according to this report by (Harris and Iyer, 2014). At the field scale, slope may have a major impact on the efficacy of filter strips in trapping nutrients. Filter strips which are planted along the slope of the contours, for example, are efficient in interrupting surface flow on sloped fields (Dabney et al., 2006). Sahu and Gu (2009) "used a calibrated SWAT model to show that strips planted mid-slope on a contour were much more effective than those planted downstream along the riverbank in reducing nitrate export." Finally, when placing filter strips, keep in mind the type of load and the position of influent. The areas separating runoff-generating areas from pollution-loading areas, according to Bentrup (2008), are the most important to buffer. Filter strips are often recommended near water bodies, in part to avoid interfering with farm operations and, in part, to intercept flow until it reaches downstream receiving waters. Geza et al. (2009) found that planting filter strips along field drains (edge-of field) was less costly than planting them upslope in important sediment source areas for a given amount of sediment reduction. In order to generalize recommendations like the ones above, we need to establish acceptable relationships with spatial covariates. "The filter strip model, VFSMOD" (Muoz-Carpena and Parsons, 2004), and the Riparian ecosystems management model (REMM) (Lowrance et al., 1997) are two commonly used models that incorporate factors like slope and erodibility which influences the filter-strip efficiency. These are based on reasonably straightforward relationships between variable buffer width (W [m], buffer slope (S [percent]), and soil erodibility (E). W = (30.5*S1/2)/E (Brown, 1987) and W = 8 + 0.6*S1/2)/E (Brown, 1987) are two examples (Barling and Moore, 1994). In Illinois, Ssegane et al. (2015) evaluated filter strip efficiency, which was found to be a function of strip width and trapping efficiencies (Figure 3.4). For runoff, wider filter strips (6–10 m) were more effective than narrower strips (2–5 m), but not for sediment trapping. The 6–10 m and broader buffers, on the other hand, had no difference in runoff interception or sediment trapping. Buffer trapping efficiencies for nitrate and pesticides did not increase significantly for widths > 5 m. These findings indicate that widths of 6–20 meters are sufficient to achieve a 70% reduction in nutrient and pesticide transport (Figure 3.4). A switch grass or willow buffer could reduce annual leached NO_3 by 61 or 59%, respectively, and N_2O emission by "5.5 or 10.8%" (Ssegane et al., 2015).

3.5 ECONOMIC FEASIBILITY

The cost of filter strip placement is a significant factor to consider. (i) opportunity and profitably costs between the perennials and row crops planted; (ii) income generated by the harvesting of biomass along filter strips; (iii) flood damage to crops that are flood intolerant; (iv) inefficient equipment paths caused by filter strip placement; (v) interfering of filter strips with the existing drainage systems; and (vi) effort and cost of harvesting biomass along filter strips are all factors that influence economic outcomes. While a complete techno-economic study of various placement decisions is not feasible at this time, we discussed a few of these decisions below based on our review. Filter strips are selected by the farmers based on whether they can be implanted in areas where non-bioenergy crops prove unprofitable and whether the perennials used in the strips can be harvested for a profit. These concerns are justified if row crops are more profitable than biomass crops in these regions. However, if the biomass in filter strips can be harvested and sold at a lower cost of management, then planting them will be beneficial financially. The billion-ton 2016 vol 1 study (2016) looked at potential possibilities that took shipping and logistics costs into account when calculating final feedstock prices. "Feedstocks were presumed to be processed into secondary products at a regional facility before being transformed" to "biofuel, biopower, or bioproducts at a biorefinery in this study." They envision a potential feedstock supply system that involves a preprocessing depot that can accumulate multiple feedstock sources and multiple ends uses, transforming raw biomass into a reliable product that can be transported and handled over long distances using existing infrastructure. The POLYSYS model was used to run economic scenario estimates, and based on the previous assumptions, an estimated 217 million tons of biomass could be made available for less than $84 per ton delivered (assuming a roadside/farm gate price of $60 per ton). Farmers could prefer not to plant riparian buffers if they are obstructing the field operations or take up land that could be better used for food crops. Riparian buffers can be used for biofuels generation without affecting the water quality if they are far enough away from water bodies. According to one study in Illinois, planting a 0.32 ha saturated willow buffer on a 2.63 ha corn field would minimize nitrate loss by up to 61% annually and cost the same as other best management practices (Cacho et al., 2018), while also providing another source of livelihood for the farmer. When the filter strip of switchgrass, little bluestem, big bluestem, prairie cordgrass, and cup plant (Silphium perfoliatum) was harvested as an energy feedstock, farmers' income increased by 43% on average, compared to when they only received income from water quality incentives under the USDA's Farm Services Agency Conservation Reserve Program (CRP) (Zilverberg et al., 2016). Based on a 15-year average, Tyndall et al. (2013) projected an annual costs of contour prairie strips to be at $590 to $865/ha. Site planning, maintenance, lost revenue due to land-use shift, establishment were all included in these costs. If the strips were registered in the CRP, the costs were reduced by 85%. Farmers' bottom lines profit from government initiatives that compensate them to preserve buffers as part of a wildlife-friendly landscape. Buffers had major water quality benefits in SWAT simulations of the Iowa river, but only minor impacts on corn and soy residue harvest (Jager et al., 2017), which is interesting to note in an economic sense. In a related analysis, 30-m buffers were simulated around rivers in the South Fork watershed, Iowa, by the SWAT model. "The buffers covered 1,508 hectares and would yield 12,442 dry metric tons of feedstock if harvested (Ha and Wu, 2017)." The theoretical fuel production of 2.7 ML can be estimated using an 80 gallon per dry ton conversion (Ha and Wu, 2017). Climate extremes have recently triggered delays in corn and soy planting in the central United States, and economic trade-offs can be different when a filter strip is placed to intercept moderate runoff volumes versus high-storm flows. "It is more important to locate buffer strips at the edges of fields on fields with low slopes so they do not interfere with field drainage, which can reduce agricultural productivity (Dabney et al., 2006)." Perennial plants that are better suited to wet soils than annual crops are used as biomass feedstocks. In the Prairie Pothole area of the United States, for example, biomass production by a mix of native species was compared (Zilverberg et al., 2016). "Several forbs, C3 grasses, and three C4 grass species [prairie cordgrass (*Spartina pectinate*), big bluestem (*Andropogon gerardii*), and little bluestem (*Schizachyrium scoparium*)] were included in the crop mixture." C4 species performed well at the extremes of a gradient from xeric to mesic conditions, and thus produced harvestable biomass more in consistency than switchgrass grown in a monoculture system (Zilverberg et al., 2016), particularly at the extremes of switchgrass topographic range.

3.6 MULTIPLE OBJECTIVE DECISIONS

Riparian buffers that are multi-tiered (e.g., three zones) may be built to serve different purposes (Figure 3.4) (Bongard and Wyatt, 2010). Streambanks are rehabilitated closest to rivers by growing trees and understory shrubs with fast-growing root system that can withstand wet conditions (Figure 3.3). Buffers should be at least 10-m high and tree height should be at least ½ the width of the stream to provide shade. When trees on the streambank fall, they leave behind large woody debris that aids in the formation of pool-riffle sequences, allochthonous energy sources, and aquatic habitat. The second tier is larger and made up of trees that can withstand flooding. Lowering floodwater, extracting nutrients, and degrading pesticides are all functions of this tier. This tier can also be handled for biomass by cultivating willow, cottonwoods (*Populus* L.), or oil-producing trees (e.g., hazelnut). A third tier should be at least 7 meters high and planted in switchgrass or other warm-season grasses, furthest from the stream and along the field's edge (Bongard and Wyatt, 2010). This tier can also be planted with forbs, which will provide additional pollinator benefits (Kohler et al., 2007; Kutt et al., 2018). Residents, farmers, and academics in central Illinois (USA) were polled on their preferences for 2-tier versus 3-tier buffers by Sullivan et al. (2004). Local non-farmer residents and researchers favored 3-tier buffers over farmers, according to the results of a survey. Economic losses, increased maintenance, and the possibility of troublesome wildlife were among the concerns of farmers. It is necessary to conduct research analysis in order to comprehend and answer these concerns.

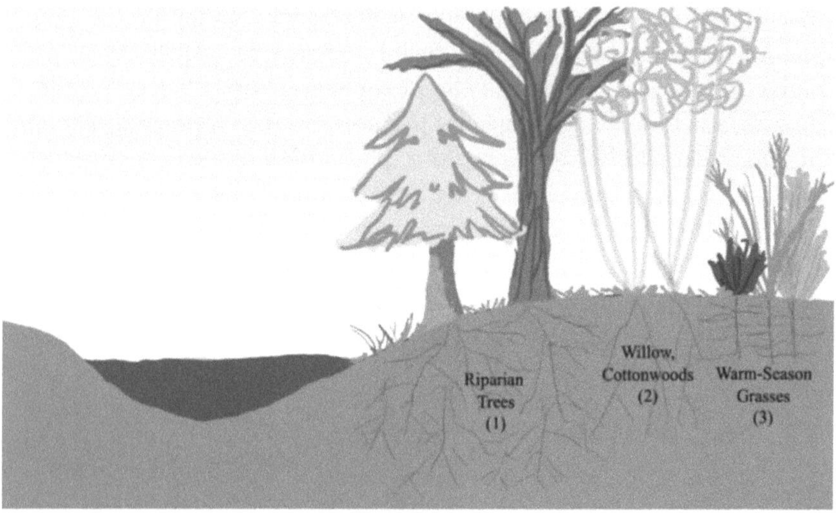

FIGURE 3.4 Multi-tiered riparian buffer.
Source: Reprinted from Krieg et al. (2019). https://creativecommons.org/licenses/by/4.0/

3.7 MODELS FOR SIMULATING BIOENERGY LANDSCAPES

With the growing interest in bioenergy growth, scientists are using science-based field and watershed-scale models to evaluate a variety of important sustainability issues, such as enumerating the nonpoint source pollutant exports from bioenergy landscapes and their actual source areas, and forecasting the potential impact of climate variability and land use change on water quality and the ecosystems. "These models have been used to create the theoretical foundation for bioenergy production management and policy decisions." For simulating the bioenergy environment, field-or watershed-scale models must include the following components: nutrient transport and fate, vegetation growth cycle, hydrology, and the sediment load. Some of these models have features for simulating the best management practices, like as the riparian buffer simulation. Several authors published a comprehensive study of different types of watershed and water quality assessment models (Borah and Bera, 2003). All of the models discussed have a relatively broader user capabilities with the intended applications, ensuring that they continuously undergo modifications for their better applicability. Some of these models, including their more specific rationales, their key capabilities, and attributes relative to the bioenergy landscape are summarized in Table 3.1; this information may be helpful in determining the problems, situations, or conditions for which the models are most suitable; their full potential uses and limitations; and the directions for their enhancements. Here, we focused on public-domain models that have been used to simulate the bioenergy landscape. Three models were described as powerful bioenergy landscape modeling tools: "The agricultural policy/environmental extender (APEX) model, the riparian ecosystem management model (REMM), and the SWAT." The models were also chosen to reflect simulation capabilities at a variety of scales in the landscape. These three models were thought to have advantages in terms of simulating bioenergy feedstocks, management methods, and water quality impact. These models have similar components and capabilities, but the spatial scales they simulate are vastly different.

TABLE 3.1 A Summary of SWAT, APEX, and REMM Models

Model	APEX	REMM	SWAT
Spatial scale	Field/small watershed	Field	Watershed
Computational unit	Subarea	Zone	Hydrologic response unit (HRU)
Runoff/infiltration	Curve number, green-Ampt	Green-Ampt	Curve number, green-Ampt
Subsurface flow	Partitioning of excess soil layer water between quick return flow to channel and subsurface lateral flow	Darcy's equation	Kinematic storage and groundwater flow
Groundwater	Computed as a function of groundwater storage	None	Empirical relations
Runoff in channel	Complete flood routing method with and daily and short time interval	None	Variable storage or Muskingum with Manning's equation
Overland sediment	Universal soil loss equation (USLE), MUSLE, RUSLE	USLE	Modified universal soil loss equation (MUSLE)
Channel sediment	Bagnold's stream power equation with deposition and resuspension allowed	None	Bagnold's stream power equation with deposition and resuspension allowed

TABLE 3.1 *(Continued)*

Model	APEX	REMM	SWAT
Soil nutrient cycles	Phosphorous cycles, carbon, and nitrogen cycles	Phosphorous cycles, carbon, and nitrogen cycles	Phosphorous cycles, carbon, and nitrogen cycles
Channel water quality	Organic nitrogen, phosphorous, algae, CBOD, DO, and pesticides are simulated	None	Organic nitrogen, phosphorous, algae, CBOD, DO, and pesticides are simulated
Plant growth	Heat unit approach	Heat unit approach	Heat unit approach
BMP	Planting, harvest, irrigation, fertilization, pesticide, tillage, grazing, mowing	Planting, harvest, irrigation, fertilization	Planting, harvest, irrigation, fertilization, pesticide, tillage, grazing

Source: Reprinted with permission Zhang and Wu (2015). © Springer Nature.

3.8 CONCLUSIONS

Traditional starch-based crops, which include the oil seeds, agricultural residue corn stover, the forest wood residue, perennial grasses, the short rotation woody crops as well as algae, can all be used to grow biofuels. "The United States Department of Energy (DOE) has made an intensive effort to ensure the environmental sustainability of bioenergy feedstock supply for the bioenergy and bioproducts industries. ." The capacity for biomass feedstock in the United States was estimated in great detail in a biomass resource assessment started by DOE (U.S. DOE, 2011). The report focused on the country's capabilities to produce a billion dry tons of biomass resources for energy purposes/year without negotiating other important farm and forest products like feed, food, and fiber crops. The analysis includes assessments of existing U.S. feedstock capability and the potential for growing crops and agricultural products for renewable energy applications, as well as county-level data for policymakers, industry, and the agricultural community. Following that, studies were undertaken to assess the effects of expected future development on the regional water quality scenario and hydrology in the Mississippi River basin's tributaries (Baskaran et al., 2010; Demissie et al., 2012; Jager et al., 2015; Schweizer et al., 2014). Conservation practices and land use decisions are being used to incorporate landscape design and management principles into bioenergy feedstock development. "The US DOE has set a target of validating landscape design approaches for two bioenergy systems by 2022," with the goal of increasing land use productivity while maintaining ecosystem and social benefits (U.S. DOE Multi-Year Program Plan, 2015). Bioenergy feedstock development allows society to create multi-functional ecosystems that generate food and energy while also promoting environmental quality and ecosystem services using this approach. Along lakes, riparian buffers with fast-growing trees and seasonal grasslands, for example, could be cultivated. These buffers have the potential to minimize surface runoff into streams, improve water quality, and provide wildlife corridors between forest patches. Using a SWAT model, Ha and Wu (2015) investigated the impact of riparian buffers and transforming low productivity land to turn grass on nutrients, the suspended sediments (SS), and the hydrology in the South Fork Iowa River watershed in Iowa. SWAT accurately reflects field buffer and riparian buffer with its respective sub-modules, according to the report. "Switchgrass buffer area coverage had a noticeable impact on nitrogen, phosphorus, and sediment loadings" at the watershed, according to simulation findings. A 15.2% increase in low-productivity land conversion to switch grass could result in reductions of "69.5, 55.8, 46.7, and 13.8% in suspended sediment, TN, TP, and nitrate loadings, respectively." No-till systems could be used instead of conventional tillage, and cover crops could be planted more often to help predatory insects and spiders that manage pests, reducing soil erosion and improving soil quality (Kumar et al., 2016; Bhat et al., 2017, 2019). In the long run, establishing sustainable bioenergy landscapes could boost agricultural productivity by facilitating "crop pollination and natural control of pests," as well as a number of other services that are valuable beyond development. The numerous benefits humans derive from the tools and processes provided by natural and managed ecosystems are referred to as ecosystem services. "As a result, meeting various and often overlapping demands for agricultural products, water quality, and ecosystem service for the benefit of every human being is a major challenge of bioenergy and natural resource use and management." Identifying contaminants of concern (Shah et al., 2020) and the relative roles of the point and nonpoint pollutant sources, strategizing emission reduction and control strategies, and monitoring progress and making changes toward achieving overall watershed targets are all important steps in addressing this obstacle. In this sense, the United States is concentrating on the watershed strategy in conjunction with the US Environmental Protection Agency's (EPA) total maximum daily load (TMDL) program. A watershed scale strategy also identifies the cost-effective ways to think about how to improve the overall health of the watershed environment and conserve biodiversity, rather than only reducing contaminated runoff, water quality emissions and protecting water supplies.

KEYWORDS

- **advanced wastewater treatment**
- **bioenergy**
- **conservation reserve program**
- **landscape design**
- **riparian ecosystems management model**
- **soil and water assessment tool**
- **water quality**

REFERENCES

Barling, R. D., & Moore, I. D., (1994). Role of buffer strips in management of waterway pollution: A review. *Environ. Manag., 18*, 543–558.

Baskaran, L. M., Jager, H. I., Schweizer, P. E., & Srinivasan, R., (2010). Progress toward evaluating the sustainability of switchgrass production at a regional scale. *Trans. ASABE, 53*(5), 1547–1556.

Bentrup, G., (2008). *Conservation Buffers: Design Guidelines for Buffers, Corridors, and Greenways* (p. 110). Department of Agriculture, Forest Service, Southern Research Station, Asheville, NC.

Berndes, G., Borjesson, P., Ostwald, M., & Palm, M., (2008). Multifunctional biomass production systems-an overview with presentation of specific applications in India and Sweden. *Biofuels Bioprod Biorefin., 2*, 16–25.

Beveridge, C. E., & Rocheleau, P., (1995). *Frederick Law Olmsted: Designing the American Landscape.* New York: Rizzoli International.

Bhat, S., Dar, M. U., & Meena, R. M., (2019). Soil erosion and management strategies. *Sustainable Management of Soil and Environment.* doi: 10.1007/978-981-13-8832-3_3.
Bhat, S., et al., (2017). Soil erosion modeling using RUSLE & GIS on micro watershed of J&K. *Journal of Pharmacognosy and Phytochemistry, 6*(5), 838–842.
Bongard, P., Wyatt, G., & U. O. M. Extension (2010). *Benefits of Riparian Forest Buffers* (p. 2).
Borah, D. K., & Bera, M., (2003). Watershed-scale hydrologic and nonpoint-source pollution models: Review of mathematical bases. *Trans. ASABE, 46*(6), 1553–1566.
Brandes, E., McNunn, G., Schulte, I., Bonner, I., Muth, D., Babcock, B., Sharma, B., & Heaton, E., (2016). Subfield profitability analysis reveals an economic case for cropland diversification. *Environ. Res. Lett., 11*, 1–13.
Cacho, J. F., Negri, M. C., Zumpf, C. R., & Campbell, P., (2018). *Introducing Perennial Biomass Crops into Agricultural Landscapes to Address Water Quality Challenges and Provide Other Environmental Services* (Vol. 7). Wiley Interdisciplinary Reviews Energy and Environment.
Crow, T., Berry, T., & De Young, R., (2006). The riverside and Berwyn experience: Contrasts in landscape structure, perceptions of the urban landscape, and their effects on people. *Landsc. Urban Plan, 75*, 282–299.
Dabney, S., Moore, M., & Locke, M., (2006). Integrated management of in-field, edge-of-field, and after-field buffers. *J. Am. Water Resour. Assoc.* Paper No. 04201.
Dale, V. H., Efroymson, R. A., Kline, K. L., & Davitt, M., (2015). A framework for selecting indicators of bioenergy sustainability. *Biofuels Bioprod. Biorefin., 9*(4), 435–446.
Dale, V. H., Efroymson, R. A., Kline, K. L., Langholtz, M. H., et al., (2013). Indicators for assessing socioeconomic sustainability of bioenergy systems: A short list of practical measures. *Ecol. Indic., 26*, 87–102.
Dale, V. H., Kline, K. L., Buford, M. A., Volk, T. A., Smith, C. T., & Stupak, I., (2016). Incorporating bioenergy into sustainable landscape designs. *Renewable and Sustainable Energy Reviews, 56,* 1158–1171.
Dale, V. H., Kline, K. L., Perla, D., & Lucier, A., (2013). Communicating about bioenergy sustainability. *Environ. Manag., 51*, 279–290.
Dar, M. U. D., Shah, A. I., Ali, S., & Bhat, S. A., (2020). *Woodchip Bioreactors for Nitrate Removal in Agricultural Land Drainage: Agricultural Waste Book* (pp. 99–118). Taylor & Franc.
Dar, M. U. D., Singh, J. P., & Ali, S. R., (2020). Watertable behavior under subsurface drainage system in Thehri Muktsar District of Punjab. *Journal of Soil Salinity and Water Quality, 12*(2), 241–249.
Darshini, D., Dwivedi, P., & Glen, K., (2013). Capturing stakeholders' views on oil palm-based biofuel and biomass utilization in Malaysia. *Energy Policy, 62*, 1128–1137.
Demissie, Y., Yan, E., & Wu, M., (2012). Assessing regional hydrology and water quality implications of large-scale biofuel feedstock production in the upper Mississippi River basin. *Environ. Sci. Technol., 46,* 9174–9182.
Dodds, W. K., & Oakes, R. M., (2008). Headwater influences on downstream water quality. *Environ. Manag., 41,* 367–377.
Dosskey, M. G., Eisenhauer, D. E., & Helmers, M. J., (2005). Establishing conservation buffers using precision information. *J. Soil Water Conserv., 60*, 349–354.

Duvenage, I., Langston, C., Stringer, L. C., et al., (2013). Grappling with biofuels in Zimbabwe: Depriving or sustaining societal and environmental integrity? *J. Clean Prod., 42*, 132–140.

Efroymson, R. A., Dale, V. H., Kline, K. L., et al., (2013). Environmental indicators of biofuel sustainability: What about context? *Environ. Manag., 51*, 291–306.

Efroymson, R. A., Langholtz, M. H., Johnson, K., Stokes, B., Brandt, C. C., Davis, M. R., & Dunn, J., (2017). 2016-billion-ton report: Advancing domestic resources for a thriving bioeconomy. *Environmental Sustainability Effects of Select Scenarios* (Vol. 1, 2). Oak Ridge National Laboratory (ORNL), Oak Ridge, TN (United States).

Evrard, O., Vandaele, K., Van, W. B., & Bielders, C. L., (2008). A grassed waterway and earthen dams to control muddy floods from a cultivated catchment of the Belgian loess belt. *Geomorphology, 100*, 419–428.

Fischer, R. A., & Fischenich, J. C., (2000). *Design Recommendations for Riparian Corridors and Vegetated Buffer Strips.* DTIC Document.

Geza, M., Barfield, B., Huhnke, R., Stoecker, A., Storm, D., & Stevens, E., (2009). Comparison of targeted replacement and vegetative filter strips for sediment control and cost-effectiveness. *J. Water Resour. Plan. Manag., 135*, 406–409.

Gopalakrishnan, G., Negri, M. C., & Snyder, S. W., (2011). A novel framework to classify marginal land for sustainable biomass feedstock production. *J. Environ. Qual., 40*, 1593–1600.

Ha, M., & Wu, M., (2015). Simulating and evaluating best management practices for integrated landscape management scenarios in biofuel feedstock production. *Biofuels Bioprod. Biorefining, 9*, 709–721.

Ha, M., & Wu, M., (2015). Simulating riparian buffer in integrated landscape management scenarios for biofuel feedstock production. *BioFPR*. doi: 10.1002/bbb.1579.

Harris, M. A., & Iyer, G., (2014). *Small Changes, Big Impacts: Prairie Conservation Strips* (pp. 1–4). Natural Resource Ecology and Management Publications and Papers.

Hazelton, J. A., Tiwari, S., & Amezaga, J. M., (2013). Stakeholder dynamics in bioenergy feedstock production; the case of *Jatropha curcas*, L for biofuel in Chhattisgarh State, India. *Biomass Bioenergy, 59*, 16–32.

Helmers, M. J., Zhou, X., Asbjornsen, H., Kolka, R., Tomer, M. D., & Cruse, R. M., (2012). Sediment removal by prairie filter strips in row-cropped ephemeral watersheds. *J. Environ. Qual. 41*, 1531–1539.

Holmgren, P., (2013). *On Landscapes – Part 1: Why are Landscapes Important?* Available at: Http://blog.cifor.org/19702/on-landscapes-part-1-why-are-landscapes-important (Center for International Forestry Research) (accessed on 7 July 2022).

Jager, H. I., Baskaran, L. M., Schweizer, P. E., Turhollow, A., Brandt, C. C., & Srinivasan, R., (2015). Forecasting changes in water quality in rivers associated with growing biofuels in the Arkansas-White-Red river drainage, USA. *Glob. Chang. Biol: Bioenergy, 7*(4), 774–784.

Jager, H. I., Wu, M., Ha, M., Baskaran, L., & Kreig, J., (2017). Water quality responses to simulated management practices on agricultural lands producing biomass feedstocks in two tributary basins of the Mississippi River. In: Efroymson, R., Johnson, K., & Langholtz, M., (eds.), *2010 Billion-Ton Report (BT16): Environmental Sustainability Effects of Select Scenarios* (Vol. 1, 2, pp. 139–178). Oak Ridge National Laboratory and the Department of Energy Bioenergy Technologies Office, Oak Ridge, TN.

Jha, M., Schilling, K., Gassman, P., & Wolter, C., (2010). Targeting land-use change for nitrate nitrogen load reductions in an agricultural watershed. *J. Soil Water Conserv., 65*, 342–352.

Jones, K. B., Zurlini, G., Kienast, F., et al., (2012). Informing landscape planning and design for sustaining ecosystem services from existing spatial patterns and knowledge. *Landsc. Ecol., 28*, 1175–1192.

Koh, L. P., & Ghazoul, J., (2010). Spatially explicit scenario analysis for reconciling agricultural expansion, forest protection, and carbon conservation in Indonesia. *Proc. Natl. Acad. Sci., 107*, 11140–11144.

Kohler, F., Verhulst, J., Van, K. R., & Kleijn, D., (2007). At what spatial scale do high-quality habitats enhance the diversity of forbs and pollinators in intensively farmed landscapes? *J. Appl. Ecol., 45*, 753–762.

Kreig, J. A., Ssegane, H., Chaubey, I., Negri, M. C., & Jager, H. I., (2019). Designing bioenergy landscapes to protect water quality. *Biomass and Bioenergy, 128*, 105327.

Kumar, R., Kumar, M., Shah, A. I., Bhat, S. A., Wani, M. A., & Ram, D., (2016). Modeling of soil loss using USLE through remote sensing and geographical information system in micro-watershed of Kashmir valley, India. *Journal of Soil and Water Conservation, 15*(1), 40–45.

Kutt, L., Paal, T., Lohmus, K., Rammi, I. J., Zobel, K., & Liira, J., (2018). Multi-user quality of floral services along a gradient of margin habitats between semi-natural grasslands and forests. *Appl. Veg. Sci., 21*, 363–372.

Li, Y., Guan, K., Schnitkey, G. D., DeLucia, E., & Peng, B., (2019). Excessive rainfall leads to maize yield loss of a comparable magnitude to extreme drought in the United States. *Glob. Chang. Biol., 25*, 2325–2337.

Liu, X., Zhang, X., & Zhang, M., (2008). Major factors influencing the efficacy of vegetated buffers on sediment trapping: A review and analysis. *J. Environ. Qual., 37*, 1667–1674.

Lowrance, R., Altier, L. S., Newbold, J. D., Schnabel, R. R., Groffman, P. M., Denver, J. M., Correll, D. L., et al., (1997). Water quality functions of riparian forest buffers in Chesapeake Bay watersheds. *Environ. Manag., 21*(5), 68–712.

McBride, A., Dale, V. H., Baskaran, L., et al., (2011). Indicators to support environmental sustainability of bioenergy systems. *Ecol. Indic., 11*, 1277–1289.

McCormick, N., (2009). Towards a responsible biofuels development process, discussion paper prepared for UNEP, Oeȩko institute and roundtable on sustainable biofuels. In: *Proceedings of the 2nd Joint International Workshop on Bioenergy, Biodiversity Mapping and Degraded Lands*. Paris, France.

Moreno-Mateos, D., Mander, U., Comin, F. A., Pedrocchi, C., & Uuemaa, E., (2008). Relationships between landscape pattern, wetland characteristics, and water quality in agricultural catchments. *J. Environ. Qual., 37*, 2170–2180.

Muoz-Carpena, R., & Parsons, J. E., (2004). A design procedure for vegetative filter strips using VFSMOD-W. *Trans. ASAE, 47*, 1933–1941.

Nassauer, J. I., & Opdam, P., (2008). Design in science: Extending the landscape ecology paradigm. *Landsc. Ecol., 23*, 633–644.

Parish, E. S., Hilliard, M. R., Baskaran, L. M., Dale, V. H., Griffiths, N. A., Mulholland, P. J., Sorokine, A., et al., (2012). Multimetric spatial optimization of switchgrass plantings across a watershed. *Biofuels Bioprod. Biorefining, 6*, 5872.

Perlack, R. D., & Stokes, B. J., (2011). *U.S. DOE: U.S. Billion-Ton Update: Biomass Supply for a Bioenergy and Bioproducts Industry*. (Leads), ORNL/TM-2011/224. 2011.
Sahu, M., & Gu, R. R., (2009). Modeling the effects of riparian buffer zone and contour strips on stream water quality. *Ecol. Eng., 35*, 1167–1177.
Sayer, J., Sunderland, T., Ghazoul, J., et al., (2013). Ten principles for a landscape approach to reconciling agriculture, conservation, and other competing land uses. *Proc. Natl. Acad. USA, 110*, 8349–8356.
Schweizer, P., & Jager, H. I., (2011). Modeling fish diversity in the Arkansas red-white River Basin. *Trans. Am. Fish Soc., 140*(5), 1227–1239.
Shah, A. I., Dar, M. U. D., Bhat, R. A., Singh, J. P., Singh, K., & Bhat, S. A., (2020). Prospectives and challenges of wastewater treatment technologies to combat contaminants of emerging concerns. *Ecological Engineering, 152,* 105882.
Sheridan, J. M., Lowrance, R., & Bosch, D. D., (1999). Management effects on runoff and sediment transport in riparian forest buffers. *Am. Soc. Agric. Eng., 42*(1), 55–64.
Sinclair, P., Cohen, B., Hansen, Y., Basson, L., & Clift, R., (2015). Stakeholder engagement with the sustainability assessment of bioenergy: Case studies in heat, power and perennial and annual crops in the UK. *Biomass Bioenergy, 73*, 11–22.
Ssegane, H., Negri, M. C., Quinn, J., & Urgun-Demirtas, M., (2015). Multifunctional landscapes: Site characterization and field-scale design to incorporate biomass production into an agricultural system. *Biomass Bioenergy, 80*, 179–190.
Sullivan, W. C., Anderson, O. M., & Lovell, S. T., (2004). Agricultural buffers at the rural-urban fringe: An examination of approval by farmers, residents, and academics in the Midwestern United States. *Landsc. Urban Plan, 69*, 299–313.
Tomer, M. D., Dosskey, M. G., Burkart, M. R., James, D. E., Helmers, M. J., & Eisenhauer, D. E., (2009). Methods to prioritize placement of riparian buffers for improved water quality. *Agrofor. Syst., 75*, 17–25.
Turner, M. G., Gardner, R. H., & O'Nei, R. V., (2001). *Landscape Ecology in Theory and Practice*. New York: Springer-Verlag.
Tyndall, J. C., Schulte, L. A., Liebman, M., & Helmers, M., (2013). Field-level financial assessment of contour prairie strips for enhancement of environmental quality. *Environ. Manag., 52*, 736–747.
U.S. Department of Energy, (2016). Billion-ton report: Advancing domestic resources for a thriving bioeconomy, In: Langholtz, M. H., Stokes, B. J., & Eaton, L. M., (eds.), *Economic Availability of Feedstocks* (Vol. 1). Oak Ridge National Laboratory, Oak Ridge, TN. https://doi.org/10.2172/1271651.
U.S. DOE. (2015). *Multi-year Program Plan.* http://energy.gov/sites/prod/files/2015/04/f22/mypp_beto_march2015.pdf (accessed on 7 July 2022).
Upreti, B. R., (2004). Conflict over biomass energy development in the United Kingdom: Some observations and lessons from England and Wales. *Energy Policy, 32*, 785–800.
Vandaele, K., Lammens, J., Priemen, P., & Evrard, E., (2013). *How to Control Muddy Floods from Cultivated Catchments.* Lessons from the Melsterbeek catchment in Flandersm (Belgium), Samenkering Land en water, St-Truiden, Belgium.
Wu, J., & Skelton-Groth, K., (2002). Targeting conservation efforts in the presence of threshold effects and ecosystem linkages. *Ecol. Econ., 42*, 313–331.

Zhang, Z., & Wu, M., (2015). Progress and challenges in quantifying water quality and ecosystem responses from agricultural, forestry, and bioenergy landscapes. *Current Sustainable/Renewable Energy Reports, 2*(4), 128–135.

Zilverberg, C. J., Teoh, K., Boe, A., Johnson, W. C., & Owens, V., (2016). Strategic use of native species on environmental gradients increases diversity and biomass relative to switchgrass monocultures. *Agric. Ecosyst. Environ., 215*, 110–121.

CHAPTER 4

Water Quality Enhancement with the Use of Eco-Roofs

RAJAT MISHRA,[1] SUSANTA DAS,[2] VISHNU JI AWASTHI,[1] AAMIR SHAH ISHAQ,[2] and ROHIT PRATAP OJHA[1]

[1]*Govind Ballabh Pant University of Agriculture and Technology, Pantnagar, Uttarakhand, India*

[2]*Punjab Agricultural University, Ludhiana, Punjab, India, E-mail: shahaamir2613@gmail.com (A. S. Ishaq)*

ABSTRACT

Eco-roofs mitigate the negative effects of pollution by employing vegetation in the roofs which reduce the runoff, sustain the water quality by different means and provide ample time for clean water harvesting. Eco rooftops mean to develop vegetation on the roofs, which helps to mitigate extreme weather conditions and save energy, provide foods, water, and greenery. This review includes a distributed examination on how eco rooftops can help relieve contamination of water, how green rooftop materials impact the various parameter of runoff, and recommends future exploration bearings. The conversation focuses on how green rooftops build and impact the surrounding entities, carbon sequestration, the water nature of stormwater spillover. Ideas for upcoming headings to investigation incorporate vegetation choice, urban agriculture, advancement of substrates, the nature of the eco-roof overland flow, the supplemental water system, the utilization of graywater, the air contamination, impacts

Advanced Technologies for Water Quality Treatment and Management. Mehraj U. Din Dar, PhD, Aamir Ishaq Shah, MTech, Shakeel Ahmad Bhat, PhD & Syed Rouhullah Ali, PhD (Eds.)

on human wellbeing, and consolidating green rooftops with integrally related advancements, financial issues, and development issues.

4.1 INTRODUCTION

The speedy urbanization culminated in the soil cover changes increase in the impermeable surface area and diminishing of vegetation, forests, grasslands, and croplands these factors also have a significant consequence on the rise in the amount of overland runoff which maximizes the threat of flood occurrence (Zolch et al., 2017). On account of intensified stormwater runoff and expand the impenetrable area, which reduced the soil infiltration rate and thereby enhanced the risk/danger of flood frequency and severity in urban zones. Furthermore, the prevalence of global warming increases the earth's surface temperature result more periodic recurring of extreme precipitation occurrence in urban inundation. As a result, to resolve these issues while still environmental protection and conserving soil and water supplies as urbanization grows, the idea of "eco roofs" can prove to be an innovative and feasible solution as well as an advanced strategy for addressing soil and land-related environmental and agricultural sustainability.

Intense urbanization and industrialization lead to increase population and population density in some major parts. These areas have minimal resources throughout some seasons during the year. That is led to the drastic use of alternative resources which is not a scientific way to sustain these resources. If we use these resources at this alarming rate, they will exhaust after a certain time that will make a critical situation. So proper use strategy of these resources is required for environmental valance and sustainability. This intense urbanization leads to deforestation, pollution; the conversion of green land into habitation creates a huge imbalance in the demand and supply chain of local environmental resources. As food, Energy, Space, Water demand is increased in urban areas. Eco roof is the best solution for Food, Energy, Space, Water resource (Tzoulas et al., 2007). Eco roof is the best solution for the growing negative impacts of urbanization and increase in gray infrastructures (Tzoulas et al., 2007). Eco roofs are the combination of vegetation, the growing medium (i.e., soil or soilless media), support medium (as roof, wall, trailers, etc.). Hundreds of years ago, the Eco rooftops were utilized for seclusion purposes in Nordic countries (Figure 4.1). Eco roofs are being widely adopted and accepted as a suitable water management practice. In Germany, the eco-roofs top development begun in the last part of the 1970s welcoming exploration of biodiversity, rooftop development, innovation, and substrates (Köhler, 2003) adding to the advancement of the cutting edge eco rooftops and plan rules (FLL, 2002). An ever-increasing number of eco rooftops are set up.

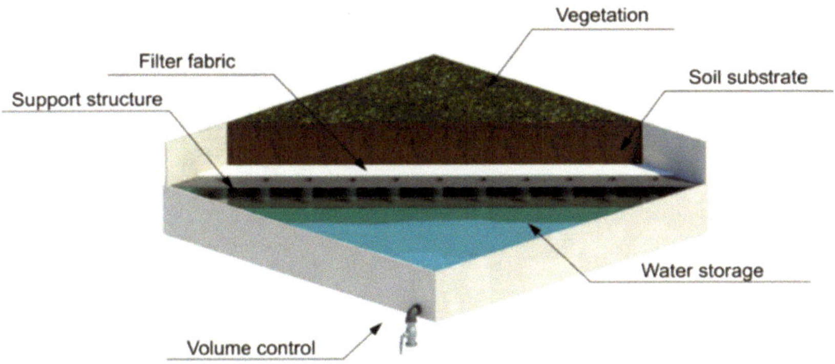

FIGURE 4.1 A multilayer green roof for water harvesting.
Source: Reprinted from Cristiano et al. (2021). https://creativecommons.org/licenses/by-nc-nd/4.0/

Eco roofs introduced in urban areas can prove to be a substantial alternative in solving the stormwater runoff problem which encompasses minimizing runoff volume and reducing the flood risk (Vijayraghavan and Joshi, 2015). The eco-roofing system is more common in rural areas in the Asian continental as they grew their vegetables in their mud/soil roofs. This concept is taking to urban areas for their valuable benefits as food for the population, energy-saving as their cooling and heat resistance provides water for different use, and balancing the ecosystem via greenery. The utilization of Eco roof technology is emerging in urban zones as it contributes to green cover creation, aggrandizing the interception, boosting water retention and evapotranspiration, mitigating flood chances, and provide innovative solutions for sagacious rainwater harvesting and management (Viola et al., 2017; Gwak et al., 2017). The eco-rooftops refer to a sort of drainage layer which supposed to discharge adequate water and acts as a substrate layer of vegetation that supports water and soil conservation (Mentens et al., 2006). In a nutshell, an eco-roof structure comprises three essential layers is growth substrate layer, vegetation layer, and drainage layer (Vijayaraghavan, 2016). The eco-roofs system mainly focuses on enhancing vegetation (Green cover) and water retention and provides ecological benefits without affecting environmental sustainability.

4.2 COMPONENTS OF AN ECO-ROOF

The eco-roof structure has basic integral components aimed at supporting the vegetation and then draining the delayed flow. An eco-roof has four integral components: (1) roof/ceiling layer; (2) drainage layer; (3) growing media layer; and (4) greenery/vegetation layer. This system can be further categorized into different classes based on the specification of these components (Figure 4.2).

4.2.1 VEGETATION/PLANT LAYER

The topmost part of the eco-roof system is the vegetation layer which reduces its kinematic energy of precipitation by interception of the precipitation; this layer also intercepts the excessive radiation which heats the structure that reduces the energy consumption for temperature control of the structure. In this layer different vegetables, flowers, and ornamental, medicinal plants, cultural plants can grow. This layer is further categorized in different aspects like type of plants, different rooting systems, plant height, etc.

4.2.2 GROWING MEDIA LAYER

This is the second and most important layer in the Eco roof system. It dictates the other component with its characteristics. Generally, the soil is used for the growing media, but now a day, soilless cultures like coconut husk, coconut chips; wood chips, clay balls, etc., are used for a clean growing environment. Now a day an intermediate layer is used. This is installed in between of drainage layer and growing media layer; to hold the growing medium and reducing the blockage of drain pores (Vijayaraghavan, 2016).

4.2.3 THE DRAINAGE/FILTER LAYER

This filter layer is a key feature of the Eco rooftops which will provide the water passage from growing media to the outlet point and provide an adequate environment for vegetation. The filter is preventing the loss of the soil substrate, soil particles, and plants. Generally, gravels, coarse sand is used for the drainage layer and there are different materials available for drainage.

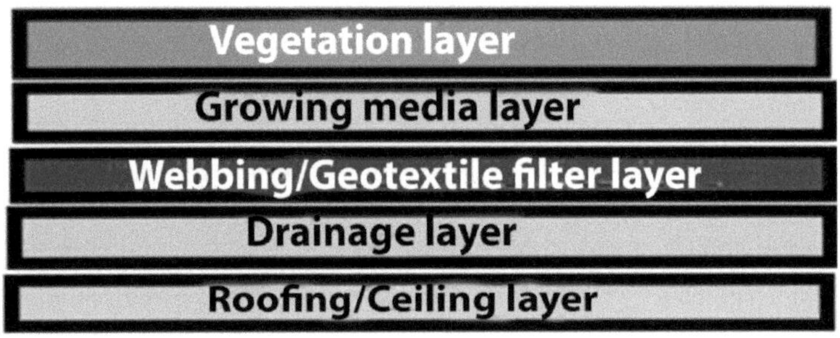

FIGURE 4.2 The main components of the eco-roof.
Source: Berndtsson et al. (2009).

4.2.4 ROOFING/CEILING LAYER

This is a lower layer of the system which isolates the roof from the eco-roof component, and provides a drainage flow path to the outlet, and provides protection against root penetration. There are two main categories of Eco roof technology; the first one is intensive and the second is extensive. The deep soil layer which supports the larger plants, bushes, and ornamentals are categorized in intensive Eco roofs; its maintenance as in the form of fertilizing, watering, and weeding is typically required. The second category of Eco roofs is a thin growing media layer that is best suited for the smaller plants. These smaller plants are supposed to provide full coverage to Eco roofs in their final stage and it is called extensive Eco roofs. This type of Eco roof is maintenance-free and for commercial products, some manuring is often recommended. The third type of eco-roofs which are cultivated with ground covering plants and lawns are categorized in semi-intensive eco-roofs. Regular care like watering, cutting, and fertilization is required in the semi-intensive type of eco-roofs (FLL, 2002) semi-intensive type eco-roofs have a 150 mm substrate layer, which will install in a 10° slope.

4.3 THE CORRELATION BETWEEN PRECIPITATION-RUNOFF AND STORAGE CAPACITY

Many experiments reports agree that the eco-roofs have reduced the storm runoff and the positive effect on water harvesting. Major variables which affect the runoff and water harvesting are initial moisture status, the thickness of the growing media layer, duration, distribution, the intensity of precipitation event, eco-roof's age, the slope for drainage, vegetation cover, etc. Grace et al. (2015; Dar et al., 2020) experimented with over nine months and reported that in unplanted growing media 40% reduction in the runoff, and for the planted area, there was a 60% reduction in runoff. Intensive eco-roofs exposed annual runoff reduction which is about 85–65% of annual precipitation and 81–27% for the extensive roofs found in German conditions from 1987 to 2003 (Mentens et al., 2006). Many researchers have analyzed the eco-roof water retention ability and storm period on an intensity basis and found a negative connection between the eco-roof retention capacity and the depth of precipitation. Trials revealed that 88% retention of stormwater for the low intensity (less than 25 mm), for medium intensity (25–76 mm) more than 55% retention capacity, and 48% retention found for the high-intensity storm(>76.2 mm) (Carter et al., 2006). Simmons et al. (2008) reported all retention of precipitation in the eco-roof of small intensity events(<10 mm); 88–26% retention for 12 mm intensity storm, 43–8% in 28 mm intensity, and 44–13% retention in 49 mm intensity; retention capacity mainly variates through layers width and material used. The moisture-holding capacity depends on not only the precipitation amount but also undoubtedly precipitation duration. Villarreal and Bengtsson (2005) reported for slopes 2% and 14%, and precipitation intensity 0.4 mm/min, 62% and 39% water holding capacity is found for the synchronized storm events, respectively. A linear relation is reported in concrete roofs and stone roofs, but in the case of eco-roofs, the quadratic factor is considered due to extreme events in higher annual precipitation, which leads to lower retention (Bengtsson et al., 2005; Mentens et al., 2006). Peak discharge for minor events was considerably lower from the eco-roofs than a traditional roof, but this influence was much lower for major storms (Carter and Rasmussen, 2006). The eco-roof intercepts the precipitation and absorbs some part of the precipitated water which leads to a time lag for generating runoff that phenomena cannot occur in conventional roofs so they produce a large amount of runoff in a short time which will create the handling of runoff which lead to a flood. VanWoert et al. (2005); and Simmons et al. (2008) reported that the 10 min time lag in 56% of peaks on the planted roof than conventional roofs. The time-delayed between the peaks was observed only 1 minute, according to Villarreal (2007), for synchronized rain events of varying intensity.

4.4 SUBSTRATE SOIL MOISTURE CHARACTERISTICS

The eco-roof enhances the water retention and reduced the precipitated runoff potential which significantly influences soil moisture characteristics. Bengtsson (2005) reported that based on field experiments and laboratory analysis, the eco-roof runoff has a positive correlation with soil characteristics. A 30 mm soil substrata give the 9 mm storage the difference between field capacity and wilting point. DeNardo et al. (2005) estimated the eco roof's soil field capacity to be 33.6%; this assessed field capacity is discussed as the eco roof's substrates water retention capacity (34%). While Bengtsson (2005) defines eco-roof water storage capacities as plant available water (the difference between field capacity and wilting point). DeNardo et al. (2005) studied that the field capacity of the eco-roof relates to roof water retention capacity. Hilten et al. (2008) determined plant available water for optimized the eco-roof medium to be 0.11 m^3/m^3 and 0.08 m^3/m^3, respectively, which are significantly lower than DeNardo et al. (2005); and Bengtsson (2005).

4.5 THE ECO ROOF AGE EFFECT ON OTHER COMPONENTS

The eco rooftop age is mainly affecting the physical and chemical qualities of rooftop material. With time dissolvable substances are washed off or decomposed, the natural substance may swell; soil particles may be washed off, voids ratio might be changed due to the aging of rooftop material. It would thus be able not out of the ordinary that the period of green rooftops would impact spillover elements. Still, not many investigations address the progressions of water-powered execution of maturing eco rooftops. Getter et al. (2007) analyzed the actual condition of the soil and natural matter substance, following five years on the eco rooftop. They discovered that natural matter constituent and pore space multiplied in that span (2%–4% and 41%–81%, separately). The water retention limit expanded from 18% to 68%. Using data from German studies, Mentens et al. (2006) discovered, the phase of the eco rooftop does not necessarily lead to annual overflow.

4.5.1 ROLE OF VEGETATION

Numerous examinations concur that it is the profundity and type of substrate that has a significant effect on eco-roof drainage water, maintenance limit, and not the plant type and cover (Monterusso et al., 2004; Van Woert et al., 2005; Dunnett et al., 2008). In any case, it is likewise tracked down that the vegetation assumes a part in water maintenance and it is, for the most part, articulated in stages with low water accessibility and extreme temperatures and immaterial in winter (Steusloff, 1998; Dunnett et al., 2008; Lundholm, 2008). Van Woert et al. (2005) tracked down that the plant cover played a less significant part in supporting water maintenance when contrasted and the media. They concluded that planted rooftops held 60% precipitation; the media – just rooftops held 50% precipitation and the rock stabilizer rooftop held 27.2% precipitation; notwithstanding, the media-just and the vegetated medicines were not fundamentally extraordinary concerning water maintenance. Similarly, Monterusso et al. (2004), guessed that the distinction in water maintenance among vegetation types was expected to contrasts in substrate profundity and not the specific plant type. Steusloff (1998) affirmed the job of planting in water maintenance during the vegetation time frame: for example, in August the broad furthermore, the semi-concentrated eco rooftop with vegetative layer hold water 30%, separately, more than the broad and the semi-intensive rooftop without plants. During October–December, the vegetative layer did not show any effect on the leachate, and the outcomes for the rooftop model with plantation and the one without were similar. Dunnett et al. (2008) appeared, out of the blue also, inverse to Steusloff (1998) for the uncovered soil substrate delivered the minimum spillover contrasting and planted substrates. The distinctions between various types of vegetation and exposed soil substrate were minor if any throughout the colder time of year time yet huge in summer. The unplanted soil substrate was more susceptible to severe condition of erosion beside this the planted soil substrate mitigate thermal as well as erosion hazards (Kumar et al., 2016; Bhat et al., 2017, 2019; Ul-Zaman et al., 2018). Dunnett et al. (2008), in light of the investigation results, reach a determination that substrate and its profundity is a key factor contrasted with vegetation type in connection with decreasing overflow. Wolf et al. (2008) tracked down that the plantation is to be sure essential to water misfortune in eco rooftop frameworks over and over the water maintenance by the substrate and its resulting vanishing yet, just when water accessibility was extremely low. Mander (2007) saw that the eco-rooftop with two passages on inverse sides appeared, for every single examined occasion, that considerably less spillover happened on one side. The solitary distinction was that the area with minimum spillover had impenetrable vegetation cover. Teemusk et al. (2007) recommend that it was planted in the media layer that retains water and eased back water discharge.

4.6 THE ECO ROOF EFFECT ON THE OVERLAND FLOW WATER CHARACTERISTICS

The Eco roof as a green infrastructure technology has been found to significantly reduce the pollutant load of the runoff water. Physical, chemical, and biological water parameters are significantly affected. Some of the parameters are discussed below.

4.6.1 NITROGEN (N)

Several sources of dangerous nitrogen are a drain in a low amount from eco-roofs compared to conventional roofs, according to some researchers, while others record a large discharge of nitrogen in a different form from eco-roofs. Material characteristics such as the form of soil, drain material, grown vegetation, age of eco-roof, fertilizer use, and maintenance are significant sources of nitrogen release from eco-roofs. Czemiel et al. (2006) experimented in broad vegetated rooftops in southern Sweden and reported that NO_3 is mostly retained by the vegetation and soil. If the nitrogen concentrations in smelling salts structure are compared to precipitation and rooftop spillover, the latter is considerably less. In any case, all-out nitrogen fixations in the downpour and vegetated rooftop spillover water were nearly identical. Natural nitrogen may be delivered from vegetated rooftops, according to Czemiel et al. (2006). Furthermore, Teemusk et al. (2007) discovered that for heavy precipitation, eco rooftops did not increase absolute nitrogen centralization, which was close to precipitation and eco rooftop overflow. According to Grace et al. (2015), nitrogen concentrations in eco-roof runoff were initially >60 mg/L, with concentrations declining over nine months by 10 mg/L. In any case, Czemiel et al. (2009) discovered that an eco-rooftop (built with inorganic lightweight soil in Japan) contributed to a substantial reduction in total nitrogen (TN) in overflow. Moran et al. (2005) observed a large amount of aggregate nitrogen arriving from proposed eco rooftops. Monterusso et al. (2004) found that NO_3 concentrations fluctuate in the wide range between seepage sites.

Furthermore, soil frameworks with the slenderest one tends to have the best distribution (somewhere in the range of 0.22 ppm and 22 ppm). According to Monterusso et al. (2004), there was a rise in nitrate-nitrogen release between the first and second inspections, which took place 140 and 314 days after preparation, respectively. Monterusso et al. (2004) stated that in planted rooftops, NO_3 leachate was based on growing plants, and minimum for rooftops with indigenous plants, which was lower than areas with sedum plugs; the most important delivery was observed for sedum seed areas. According to Köhler et al. (2002), there was a decrease in NO_3 masses in spillover from eco rooftops, decrease in the heap was proportional to the decrease in leachate amount.

4.6.2 PHOSPHORUS (P)

Precipitation water for the most part contains phosphorus in minuscule focuses. The urban overflow might be tainted by phosphorus beginning in, for example, compost utilized in metropolitan cultivating, bird's droppings, and creature's excreta. Eco rooftops with supplemented media (as fertilizer or the expansion of fake compost) and treated eco rooftops are usually a phosphorus source (Moran et al., 2005; Czemiel et al., 2006; Teemusk et al., 2007; Bliss et al., 2009). According to some investigations, majorly, all supplied phosphorus is in the form of phosphates, and there are cases of eco rooftops that do not reveal the arrival of phosphorus (Czemiel et al., 2006, 2009). Others also discovered that total phosphorus (TP) groupings in eco rooftop runoff are significantly higher than phosphate centralizations (Teemusk et al., 2007). According to Köhler et al. (2002), the decrease in phosphates in the heap is linked to a decrease in water volume within the eco rooftop. They also note that the decrease in phosphate load progresses over time. Köhler et al. (2002) observed yearly decreases in phosphate phosphorus load of 26%, 61%, 64%, and 80% over the following four years. According to Grace et al. TP concentrations in eco-roof runoff were initially >30 mg/L, with declining over 9 months to 5 mg/L. This is attributed to plant development, but it may also be linked to the age of the rooftop and the resulting annual depletion of phosphorus from the soil. As a result, the age and preparation schedules of an eco-rooftop can be linked to phosphate spillage.

4.6.3 HEAVY METALS (CU, ZN, PB, CD, CR, ETC.)

The accessible investigations demonstrate insignificant delivery of heavy metals from general eco rooftops. For the most part, the dainty soil of general eco rooftops shows no effect on spillover quality and the focuses found in overland flow are equivalent to fixations in rainfall. Regardless, if less runoff volume is considered, the eco rooftops will, on the whole, reduce heavy metal loads in urban runoff. Heavy metal concentrations in eco rooftop spillover are so much lower than the urban spillover from hard surfaces (Czemiel et al., 2009). Steusloff (1998) investigated the maintenance of heavy metals in the eco-rooftop models and stated that it is primarily dependent on the ability of eco rooftops to reduce spillover. The semi-extensive structures with low height grass family kept 99% of the heap of Cu, Pb, Zn, and also 98% of Cd in the late spring months. Around 96% Zn, 97% Cu, 99% Pb, and 92% Cd were also included in the comprehensive frameworks with vegetation. The semi-escalated rooftop with vegetation held 68% Cu, 92% Zn, 88% Cd, and 94% Pb during the cold months, while the deep rooftop with vegetation held 44% Cu, 72% Zn, 62% Cd, and 91% Pb. Czemiel et al. (2006) discovered that in overflow from eco rooftops in fixations that would contribute to tolerably polluted characteristics water, a few metals turn up in lower concentrations than usually found in metropolitan spillover. The capacity of a planted eco rooftop to store 61% of chromium (Cr), 93% of Pb, 24% of manganese (Mn) and 8% of zinc (loads of precipitation = 100%) is demonstrated when the annual heaps of toxins are measured with the thought to annual water maintenance within one of the eco rooftops considered. The heap of the following hefty metals extended in spillover from the planted eco rooftop: Cu many times, and iron (Fe) one time. Cd concentrations were under-recognition limits.

4.6.4 SUSPENDED SOLIDS (SS) AND TSS

Suspended solids (SS) are particles that cannot move through a 2-m tube. Mud, sand, fine common litter, and other particulate matter can all be included in total SS. A high level of SS can exacerbate problems, such as causing rapid warming of a waterway and reducing photosynthesis in amphibian plants by reducing the amount of light that can pass through the water (Li et al., 2014). Turbidity is a measurement of the amount of light that can pass through water without being scattered by objects. TSS, as well as turbidity, Saadatian et al. have a strong positive relationship (2011). Turbidity and TSS do not appear to be a problem in eco-roof floods since most modern eco rooftops are fixed with geotextile material/webbing that contains the finest particles in the growing media (Li et al., 2014). Morgan et al. (2011) discovered that the first watering occasion resulted in higher TSS levels and turbidity from eco rooftops than the second watering occasion. The investigation's findings revealed that turbidity levels and TSS in vegetated and unvegetated plots changed over a half-year period for four construction media (arkalyte, bottom ash, haydite, and lava). Morgan et al. (2011) studied that in leachate water, the turbidity was decreased by 28–70% and TSS was deceased up to 55–70%, these values were greatly influenced by the age of eco-roofs.

4.6.5 FERTILIZATION

Several scientists have expressed their belief that there is a connection between the arrival of supplements from eco rooftops and the use of manure (Berndtsson et al., 2006; Temsukh et al., 2007; Emilsson et al., 2007; Bliss et al., 2009; Li et al., 2014). Higher supplements fixation found in spillover water from using regular manures than using controlled discharge compost. Additionally, compost should not be used during the rainy season or shortly before precipitation. Monteusso et al. (2004) conducted an exploratory study and discovered that the centralization of phosphorus distributed in the overflow was dependent on the use of specific types of vegetated systems, such as the sedum seed the framework, which resulted in the highest phosphorus arrival. Furthermore, Emilsson et al. (2007) stated that vegetation mats decreased nitrate influx and aggregate nitrogen when compared to surfaces with developing vegetative shoots and un-vegetated substrate; they also concluded that there were no significant differences in total phosphate and potassium overflow for various vegetated frameworks. Following that, supplement loss could be reduced based on knowledge about the specific plant-supplement requirements and using that information to determine the proper preparation procedure that would be suitable for the various phonological stages. Most plants, for example, need more N during their rapid growth times, while P is needed during the plant's foundation cycle. Chen et al. (2011) conducted a promising study in which they used precision water systems and appropriate N deliveries to yields to reduce supplement misfortune to almost nil, compared to 127 kg N/ha misfortune in average cultivation practice (Czemiel et al., 2010; Li et al., 2014).

4.6.6 PH

The pH of leachate water is influenced by eco rooftops, which raises it from around 5 to 6 in rainstorm water, to around 7 to 8 in eco rooftop overflow water (Bliss et al., 2009). This is a significant capability that contributes to lowering the level of fermentation among common water users. Moderately corrosive rainstorms are moderated by the eco rooftops.

4.7 INITIAL RUNOFF OF THE MONSOON SEASON

Normally, the underlying spillover from impenetrable surfaces after a dry span is more polluted than the ensuing overflow. Debris like bird droppings, trash leaves, vegetation, soil particles, and barometrical particles are present in the runoff after a dry spell between storm events that directly affect the rooftops. These are washed out with the primary downpour, and it is perceived that initial runoff start overflow has ordinarily inferior class. For eco rooftops, the principal flush impact would not be normal with deference to the contaminations beginning from the vegetated rooftop itself, as the same source would give contaminations during the whole downpour spillover occasion. Nonetheless, higher centralizations of various contemplated segments were found in the initial overflow tests than in quite a while taken at higher spillover profundities. This is deciphered as an event of a first flush impact. K and broken-down natural C do not show any initial flush impact (Czemiel et al., 2008). Delight et al. (2009) estimated the accompanying eco of top overflow quality boundaries: sulfate, phosphorus, nitrogen, pH, COD, and TSS. They discovered no indications of a first runoff impact on water quality.

4.8 RELATIONSHIP BETWEEN TREATMENT, SPILLOVER CONTAMINATION, AND VEGETATION IMPROVEMENT

The influx of some organic components from eco rooftops can be straightforwardly connected to the application of compost in an Eco rooftop foundation/support stated by Emilsson et al. (2007). The utilization of regular compost causes higher supplement fixations in leachate than the utilization of controlled delivery compost. In the nursery, Emilsson et al. (2007) experimented that no huge contrasts in the all-out spillover of phosphate and potassium between various vegetated frameworks, and not any contrast between unvegetated also, new growth set up frameworks. The vegetation mats show a reduction in nitrate and absolute nitrogen leachate than the barren substrate. Monterusso et al. (2004) studied P fixations in examples taken 140 days later treatment was somewhat greater than in examples engaged 314 days subsequently treatment. They discovered contrasts in the arrival of phosphorus between various vegetated frameworks. The influx of P was most elevated from the sedum seed framework. The eco rooftops are regularly maintained by planting required plant cover. Endurance, what is more, the foundation of various kinds of plants on a model broad planted rooftop with shifting substrate and preparation schedule studied by Rowe et al. (2006). Development was discovered to be straightforwardly identified with the substance of natural material in the substrate with non-delicious locals requiring further substrates, the higher substance of the natural matter, or auxiliary water system. Diffusum stone crop developed at 51 g/m^2 manure snooze slacked after the snoozes of 100 g/m^2 and 150 g/m^2 until the third year when it accomplished equivalent development record rating to the sophisticated manure naps. Imperial pink stonecrop treated with 51 g/m^2 manure did not accomplish a similar development for the 100 and 150 g/m^2 compost medicines. The shortfall of manure affected the two stonecrops; visual evaluations were higher at prepared yields contrasting and the control.

4.9 CONCLUSIONS

The eco-roof is the best technique to overcome the different urban water needs through water harvesting and fresh vegetable requirement. Overflow decrease of more than 61% for storms less than 5 cm has appeared and further help the capacity of eco rooftops to diminish metropolitan stormwater. Nonetheless, the groupings of TN and TP leachate from new eco rooftop media are a worry for the worthiness of water, as abundance supplement heaps can build eutrophication hazard for lakes also, waterways. A significant supplement load was seen in overflow exuding from media tried, likely because of the overabundance of supplements in the fabricated media. Likewise, the organics disintegrated in the eco rooftop spillover can increase the all-out organic oxygen interest in the water. This type of effect will not be seen on a watershed scale from the single rooftop, however, with urban strategies empowering eco rooftops to be consolidated into metropolitan designs, all impacts, and both positive furthermore, negative, from this execution should be thought of. Adjusting the measure of natural matter, sort of natural substance or potentially compost utilized might all prompt a "greener" eco rooftop. On the other hand, in territories where supplement loads are especially tricky, an option media without leachable supplements in any case, with adequate water holding limit might have the option to give a significant part of similar designing advantages without the danger of low-quality overflow.

For leachate quality, it gets obvious that the significant impact on additional parts in spillover is due to the soil material (example – manure) and added fertilizers. On the off chance that and how much treatment is required will be one of the key plan issues. A few investigations show that preparation might be supplanted by watering in the dry period for a similar decent stylish outcome. For certain species, preparation shows even adverse outcomes. The audit demonstrates plainly that there is a need for more exploration into eco rooftop execution in metropolitan climate.

The distinctions estimated by not many existing investigations between the early year's execution of Eco rooftops and the future time show a requirement for long-period observing of Eco rooftops.

KEYWORDS

- **chromium**
- **eco-roofs**
- **iron**
- **manganese**
- **nitrogen**
- **urban agriculture**
- **water quality**

REFERENCES

Arnell, N. W., (1999). The effect of climate change on hydrological regimes in Europe: A continental perspective. *Global Environ. Change, 9*, 5–23.

Bates, B. C., Kundzewicz, Z. W., Wu, S., & Palutikof, J. P., (2008). *Climate Change and Water* (p. 210). Technical paper of the intergovernmental panel on climate change. IPCC Secretariat, Geneva.

Bengtsson, L., (2005). Peak flows from thin sedum-moss roof. *Nordic. Hydrol., 36*(3), 269–280.

Bengtsson, L., Grahn, L., & Olsson, J., (2005). Hydrological function of a thin extensive green roof in southern Sweden. *Nordic. Hydrol., 36*(3), 259–268.

Berndtsson, J. C., Emilsson, T., & Bengtsson, L., (2006). The influence of extensive vegetated roofs on runoff water quality. *Sci. Total Environ., 355*, 48–63.

Berndtsson, J. C., Bengtsson, L., & Jinno, K. (2009). Runoff water quality from intensive and extensive vegetated roofs. *Ecological engineering, 35*(3), 369–380.

Bhat, S., Dar, M. U., & Meena, R. M., (2019). Soil erosion and management strategies. *Sustainable Management of Soil and Environment*. doi: 10.1007/978-981-13-8832-3_3.

Bhat, S., et al., (2017). Soil erosion modeling using RUSLE & GIS on micro watershed of J&K. *Journal of Pharmacognosy and Phytochemistry, 6*(5), 838–842.

Bliss, D. J., Neufeld, R. D., & Ries, R. J., (2009). Stormwater runoff mitigation using a green roof. *Environ. Eng. Sci., 26*(2), 407–417.

Bliss, D. J., Neufeld, R. D., & Ries, R. J., (2009). Stormwater runoff mitigation using a green roof. *Environ. Eng. Sci., 26*, 407–418.

Brenneisen, S., (2003). The benefits of biodiversity from green roofs-key design consequences. In: *Conference Proceedings Greening Rooftops for Sustainable Communities, 2003*. Chicago.

Brenneisen, S., (2006). Space for urban wildlife: Designing green roofs as habitats in Switzerland. *Urban Habitats, 4*(1), 27–36. www.urbanhabitats.org (accessed on 7 July 2022).
Carter, T. L., & Rasmussen, T. C., (2006). Hydrologic behavior of vegetated roofs. *J. Am. Water Resour. Assoc., 42*(5), 1261–1274.
Carter, T., & Jackson, C. R., (2007). Vegetated roofs for stormwater management at multiple spatial scales. *Landscape Urban Plan, 80*, 84–94.
Chen, X. P., Cui, Z. L., Vitousek, P. M., Cassman, K. G., Matson, P. A., Bai, J. S., et al., (2011). Integrated soil–crop system management for food security. *Proc. Natl. Acad. Sci., 108*, 6399–6404.
Cristiano, E., Deidda, R., & Viola, F. (2021). The role of green roofs in urban Water-Energy-Food-Ecosystem nexus: A review. *Science of the Total Environment*, 756, 143876.
Currie, B. A., & Bass, B., (2008). Estimates of air pollution mitigation with green plants and green roofs using the UFORE model. *Urban Ecosyst., 11*, 409–422.
Czemiel, B. J., (2010). Green roof performance towards management of runoff water quantity and quality: A review. *Ecol. Eng., 36*, 351–360.
Czemiel, B. J., Bengtsson, L., & Jinno, K., (2008). First, flush effect from vegetated roofs during simulated rain events. *Hydrol. Res., 39*(3), 171–179.
Czemiel, B. J., Bengtsson, L., & Jinno, K., (2009). Runoff water quality from intensive and extensive vegetated roofs. *Ecol. Eng., 30,* 271–277.
Czemiel, B. J., Emilsson, T., & Bengtsson, L., (2006). The influence of extensive vegetated roofs on runoff quality. *Sci. Total Environ., 355*(1–3), 48–63.
Dar, M. U. D., Bhat, S. A., Ali, & Shah, A. I., (2017). Modeling climate change impact; a study on different procedures and strategies: A review. *Int. J. Pure Appl. Biosci., 5*, 183–200.
Dar, M. U. D., Bhat, S. A., Meena, R. S., & Shah, A. I., (2019). Carbon footprint in eroded soils and its impact on soil health. *Soil Health Restoration and Management,* 1.
Dar, M. U. D., Shah, A. I., Ali, S., & Bhat, S. A., (2020). *Woodchip Bioreactors for Nitrate Removal in Agricultural Land Drainage: Agricultural Waste Book* (pp. 99–118). Taylor & Franc.
DeNardo, J. C., Jarrett, A. R., Manbeck, H. B., Beattie, D. J., & Berghage, R. D., (2005). Stormwater mitigation and surface temperature reduction by green roofs. *Trans. ASAE, 48*(4), 1491–1496.
Dunnett, N., Nagase, A., & Hallam, A., (2008a). The dynamics of planted and colonizing species on a green roof over six growing seasons 2001–2006: Influence of substrate depth. *Urban Ecosyst., 11,* 373–384.
Dunnett, N., Nagase, A., Booth, R., & Grime, P., (2008b). Influence of vegetation composition on runoff in two simulated green roof experiments. *Urban Ecosyst., 11*, 385–398.
Earth Pledge, (2005). *Green Roofs: Ecological Design and Construction.* A Shiffer Design Book, Shiffer Publishing Ltd.
Elena, C., Roberto, D., & Francesco, V., (2020). The role of green roofs in urban water-energy-food-ecosystem nexus: A review. *Science of the Total Environment, 756,* 143876 https://doi.org/10.1016/j.scitotenv.2020.143876.
Emilsson, T. U., Czemiel, B. J., Mattson, J. E., & Rolf, K., (2007). Effect of using conventional and controlled release fertilizer on nutrient runoff from various vegetated roof systems. *Ecol. Eng., 29*, 260–271.

English Nature, (2003). *Green Roofs: Their Existing Status and Potential for Conserving Biodiversity in Urban Areas*. English Nature Research Reports, Report no 498, English Nature, Northminster House, Peterborough, UK.

Fang, C. F., (2008). Evaluating the thermal reduction effect of plant layers on rooftops. *Energy Build., 40,* 1048–1052.

FLL, (2002). *Richtlinie für die Planung, Ausführung und Pflege von Dachbegrünungen (in German) (Guidelines for Planning Execution and Upkeep of Green Roof Sites).*

Getter, K. L., Rowe, D. B., & Andresen, J. A., (2007). Quantifying the effect of slope on extensive green roof stormwater retention. *Ecol. Eng., 31*, 225–231.

Graham, P., & Kim, M., (2005). Evaluating the stormwater management benefits of green roofs through water balance modeling. In: *Green Roofs for Healthy Cities Conference.* Washington, DC.

Gwak, J. H., Lee, B. K., Lee, W. K., & Sohn, S. Y., (2017). Optimal location selection for the installation of urban green roofs considering honeybee habitats along with socioeconomic and environmental effects. *J. Environ. Manage, 189*, 125–133.

Hilten, R. N., Lawrence, T. M., & Tollner, E. W., (2008). Modeling stormwater runoff from green roofs with HYDRUS-1D. *J. Hydrol., 358*, 288–293.

Köhler, M., (2003). Plant survival research and biodiversity: Lessons from Europe. In: *Conference Proceedings Greening Rooftops for Sustainable Communities, 2003.* Chicago.

Köhler, M., Schmidt, M., Grimme, F. W., Laar, M., De Assunc¸ ão, P. V. L., & Tavares, S., (2002). Green roofs in temperate climates and in the hot-humid tropics – far beyond the aesthetics. *Environ. Manage. Health, 13*(4), 382–391.

Kosareo, L., & Ries, R., (2007). Comparative environmental life cycle assessment of green roofs. *Build. Environ., 42*, 2606–2613.

Kumar, R., Kumar, M., Shah, A. I., Bhat, S. A., Wani, M. A., & Ram, D., (2016). Modeling of soil loss using USLE through remote sensing and geographical information system in micro- watershed of Kashmir valley, India. *Journal of Soil and Water Conservation, 15*(1), 40–45.

Li, Y., & Babcock, Jr. R. W., (2014). Green roofs against pollution and climate change. A review. *Agron. Sustain. Dev., 2014*, 1–11.

Mentens, J., Raes, D., & Hermy, M., (2006). Green roofs as a tool for solving the rainwater runoff problem in the urbanized 21st century. *Landscape Urban Plan., 77*, 217–226.

Monterusso, M. A., Rowe, D. B., Rugh, C. L., & Russell, D. K., (2004). Runoff water quantity and quality from green roof systems. *Acta Hort., 639*, 369–376.

Moran, A., Hunt, B., & Smith, J., (2005). Hydrological and water quality performance from green roofs in Goldsboro and Raleigh, North Carolina. In: *Green Roofs for Healthy Cities Conference*. Washington, DC.

Morgan, S., Alyaseri, I., & Retzlaff, W., (2011). Suspended solids in and turbidity of runoff from green roofs. *Int. JP Hytoremediat., 13*, 179–193.

Osmundson, T., (1999). *Roof Gardens: History, Design, and Construction*. W.W. Norton and Company Ltd., London, UK.

Rowe, D. B., (2011). Green roofs as a means of pollution abatement. *Environ Pollut., 159*, 2100–2110.

Rowe, D. B., Monterusso, M. A., & Rugh, C. L., (2006). Assessment of heat-expanded slate and fertility requirements in green roof substrates. *Hort. Technology, 16*(3), 471–477.

Saadatian, O., Sopian, K., Salleh, E., Lim, C. H., Riffat, S., Saadatian, E., et al., (2013). A review of energy aspects of green roofs. *Renew Sustain. Energy Rev., 23*, 155–168.

Simmons, M. T., Gardiner, B., Windhager, S., & Tinsley, J., (2008). Green roofs are not created equal: The hydrologic and thermal performance of six different extensive green roofs and reflective and non-reflective roofs in a sub-tropical climate. *Urban Ecosyst., 11*, 339–348.

Steusloff, S., (1998). Input and output of airborne aggressive substances on green roofs in Karlsruhe. In: Breuste, J., Feldmann, H., & Uhlmann, O., (eds.), *Urban Ecology*. Springer-Verlag, Berlin, Heidelberg, Germany.

Takebayashi, H., & Moriyama, M., (2007). Surface heat budget on green roof and high reflection roof for mitigation of urban heat island. *Build. Environ., 42*, 2971–2979.

Teemusk, A., & Mander, Ü., (2007). Rainwater runoff quantity and quality performance from a green roof: The effects of short-term events. *Ecol. Eng., 30*, 271–277.

Teresa, Z., Johannes, M., Christine, W., Christine, W., & Stephan, P., (2016). Using green infrastructure for urban climate-proofing: An evaluation of heat mitigation measures at the micro-scale. *Urban Forestry & Urban Greening, 20*. doi: 10.1016/j.ufug.2016.09.011.

Tzoulas, K., Korpela, K., Venn, S., Yli-Pelkonen, V., Kazmierczak, A., Niemela, J., & James, P., (2007). Promoting ecosystem and human health in urban areas using green infrastructure: A literature review. *Landsc. Urban Plann., 81*, 167–178.

Ul Zaman, M., Bhat, S., Sharma, S., & Bhat, O., (2018). Methods to control soil erosion-a review. *Int. J. Pure Appl. Biosci., 6*(2), 1114–1121. 10.18782/2320-7051. 6462.

Van, R. T., & Booteldooren, D., (2009). Reducing the acoustical facade load from traffic with green roofs. *Build. Environ., 44*, 1081–1087.

VanWoert, N. D., Rowe, D. B., Andresen, J. A., Rugh, C. L., Fernandez, R. T., & Xiao, L., (2005). Green roofs stormwater retention: Effects of roof surface, slope, and media depth. *J. Environ. Qual., 34*, 1036–1044.

Vijayaraghavan, K., & Joshi, U. M., (2015). Application of seaweed as substrate additive in green roofs: Enhancement of water retention and sorption capacity. *Landscape Urban Plann., 143*, 25–32.

Villarreal, E. L., & Bengtsson, L., (2005). Response of a sedum green-roof to individual rain events. *Ecol. Eng., 25*, 1–7.

Villarreal, E., (2007). Runoff detention effect of a sedum green roof. *Nordic Hydrol., 38*(1), 99–105.

Wolf, D., & Lundholm, J. T., (2008). Water uptake in green roof microcosms: Effects of plant species and water availability. *Ecol. Eng., 33*, 179–186.

Wong, N. H., Chen, Y., Ong, C. L., & Sia, A., (2003). Investigation of thermal benefits of rooftop gardens in the tropical environment. *Build. Environ., 38*, 261–270.

Wong, N. H., Tan, P. Y., & Chen, Y., (2007). Study of thermal performance of extensive rooftop greenery systems in the tropical climate. *Build. Environ., 42*, 25–54.

Yang, J., Yu, Q., & Gong, P., (2008). Quantifying air pollution removal by green roofs in Chicago. *Atmos. Environ., 42*, 7266–7273.

CHAPTER 5

Water Quality Management Models and Systems

VALLU TEJASWINI,[1] JARIPITI TRIVIKRAMA RAJU,[2] and
SHAKEEL AHMAD BHAT[3]

*[1]Department of Soil and Water Engineering, Dr. NTR College of Agricultural Engineering, ANGRAU, Bapatla, Andhra Pradesh, India,
E-mail: vallutejaswini@gmail.com*

*[2]Department of Soil and Water Engineering, College of Technology and Engineering, MPUAT, Udaipur, Rajasthan, India,
E-mail: trivikramraju@gmail.com*

*[3]Department of Soil and Water Engineering, College of Agricultural Engineering and Technology, SKUAST–K, Shalimar Campus, Srinagar – 190025, Jammu and Kashmir, India,
E-mail: wakeelbhat@gmail.com*

ABSTRACT

Water is the crucial element on the earth for the survival of living beings. Unfortunately, its demand is increasing and supply is decreasing day by day leaving the gap very wide. The widening of gap between supply and demand of water leads to the present burning issue, i.e., "water scarcity." Water scarcity is the major problem facing by the world today. As there are difficulties involved in the determination of water quality parameters, water quantity estimation and management are given more importance rather than water quality management in the present system. The world is

Advanced Technologies for Water Quality Treatment and Management. Mehraj U. Din Dar, PhD, Aamir Ishaq Shah, MTech, Shakeel Ahmad Bhat, PhD & Syed Rouhullah Ali, PhD (Eds.)

facing a water quality crisis due to increased population growth coupled with industrialization, urbanization, agricultural activities, increased living standards, etc. Nowadays water quality management has become very critical in many countries. Proper monitoring of contaminants in the water bodies coupled with the use of water quality models which will predict the parameters of future water quality may furnish the foremost option. Ultimately, the main goal of wastewater management is to remove the contaminants from wastewater and to facilitate protection without causing any damage to the human health and the environment. Hence, this chapter presents some wastewater treatment technologies and currently available water quality models which support water quality management.

5.1 INTRODUCTION

Water is the foremost natural element for the entire living beings, but unfortunately, it is a scarce resource in the world. Sustainable development in terms of quantity and quality is impossible without conserving freshwater resources as it is very pivotal for all sectors. Agriculture accounts for nearly 72% of all water withdrawals, whereas municipalities and industries account for 16% and 12% respectively. Maintaining the freshwater quantity and quality is the most challenging task for the countries. The factors answerable for affecting the water in terms of quality and quantity in India include rapid population growth coupled with industrialization, rapid urbanization, and development in agriculture. Nowadays, degradation of water quality is the major problem facing by the world today which can lead to water scarcity. Generally, water quality is the word that indicates the water's suitability to sustain numerous uses or processes. Only, a few percentages of people, i.e., nearly less than 10% of all people in the whole developing world have access to collect wastewater and its suitable treatment (Biswas and Tortajada, 2019). Overall, one in eight people are forecasted to be at a high risk of contamination of the BOD water; one in six is at high risk from contamination with nitrogen and one in four from contamination with phosphorus (International Food Policy Research and Veolia, 2015).

The water quality criteria depend upon the purpose of the water that is being used. For example, drinking water should not comprise any chemicals or microorganisms which are precarious to human health. On the other side, the water for agriculture should be low in salt while it should be low in some other inorganic compounds for steam production and industrial purposes. Some of the most important sources of water pollution are pumping of aquifers, releasing hazardous chemicals, water contamination and the long-range atmospheric transfer of contaminants. The natural influences that affect water quality and quantity include geological, hydrological, and climatic factors. Large-scale mining, fossil fuel burning, urbanization, industrial operations, socio-economic activities, and agricultural production are the human activities that cause an impact on water quality.

A range of contaminants, which lead to disturbances in oxygen balance and are typically accompanied by severe pathogenic contaminations, may be found globally in aquatic ecosystems. Excessive nutrients from several origins such as domestic sewage, agricultural runoff, and agro-industrial wastes mainly affects lakes and impounded rivers in the form of eutrophication. Application agriculture fertilizers, pesticides, and chemicals without any environmental safeguards may deteriorate the soil and water habitat and the primary aquifers as well. Although the land available for crop production may increase through irrigation, salinization caused due to over-application of irrigation water deteriorate the previously fertile soils. Metals from mining discharges, smelting, and industrial manufacturing contaminate the water from rivers, lakes, and wetlands, which is a long-lasting phenomenon. Discharge of aerial metallic impurities not only contaminates in the area of industrial provinces but also in remote areas. Similarly, acidification of surface waters, especially in lakes occurs due to a combination of humidity in the atmosphere with some of the fumes formed when fossil fuels are burned. Direct discharge of synthetic organic micropollutants or transport of pollutants through the atmosphere into surface waters contaminates water. Leaching of wastewater from garbage dumps, tailings in mining and industrial plants contaminate groundwater bodies.

Pollutants can be classified as chemical, physical, physiological, and biological. Inorganic and organic pollutants are the classified form of Chemical pollutants. Organic materials provide a big challenge when transforming them into carbon and water, as demonstrated below:

Organics + Microorganisms + Oxygen + Nutrients → CO_2 + H_2O + more microorganisms (Englande et al., 2015).

Toxicity is the prime consideration of inorganic chemicals. A Physical pollutant includes temperature, SS, color, turbidity, foam, and radioactivity pollutants. Conventional, toxic, or nonconventional are the other classification of pollutants. Domestic sewage including BOD, SS, pH, coliform, oil, and grease comes under conventional contaminants. Pollutants that affect public health and cause aquatic toxicity referred to as toxic pollutants. USEPA defined nonconventional pollutants as pollutants that are neither conventional nor toxic and include COD, TOC, and nutrients such as phosphorous and nitrogen. The kind and level of therapy will rely on the categorization of impurities and the delivering waters.

Most of the countries have been concentrated mainly on amount of water management and water sharing matters. But, many parts of the world are facing serious problems due to poor water quality organization. Significant and notable progress has been made by developed countries in regulatory point sources of pollution, but only a little progress has been made in monitoring non-point sources of contamination. Unfortunately, equivalent progress has not been seen in developing countries in controlling both point and non-point sources of pollution. Indication of water quality deterioration can be seen all over the globe, however the vastness, kinds, and quantum of water quality difficulties vary from one country to another or even within the same country. This difference may be because the population density, industrial, agricultural, and human actions as well as climate, the physical, economic, institutional, and environmental factors are heterogeneous. Hence, local government should take initiatives such as educating and involving the public, command, and control, and use of economic and legal devices which are very helpful under specific conditions. When compared to water quantity, technical, and organization capacities are at a very low level in controlling water pollution. But it is also very difficult and challenging to reverse these trends (Biswas, 2008). However, the present situation needs immediate management strategies for improving the quantity and quality of water resources.

5.2 WATER QUALITY MANAGEMENT

The primary objective of managing wastewater is the sustainable progress of natural resources which also includes the conservation of habitat and public health. In terms of water quality management, sustainable development means managing this natural resource to the extent that the present-day and forthcoming utilizes of this natural element are not compromised. The main aim of wastewater management is to concentrate on wastewater treatment, its reclamation, and reuse. To realize the necessity for a certain level of wastewater, the idea of water quality management is necessary. In terms of the engineering field, the organized use of water must be defined, water quality criteria defined for such usage, and handling or other management procedures based on these needs should thereafter be completed.

The steps needed for the preparation of a water quality management plan are as follows (CPCB, 2008):

- **Step 1: Identify Water Quality Goal:** The first step involves setting a water quality goal which includes identifying designated best use and water quality requirements for it.
- **Step 2: Monitoring of Water Quality:** ISO (International Organization for Standardization) defined Monitoring as "the planned process of sampling, measurement, and successive recording or signaling, or both, of different water characteristics, often to assess conformity to specified objectives." The main aim of monitoring water quality is to confirm whether the observed water quality is fit for proposed uses. Monitoring activities may be long-term, short-term, and continuous monitoring programs. Generally, a water quality monitoring program helps water use, managers in water quality management decision making. This will help to acquire knowledge on the existing water quality of the water body.
- **Step 3: Identifying the Nature and Scale of Pollution:** In this step, the water quality data after repeated observations should compare with the required water quality as per the goal set to identify the nature and magnitude of pollution control needed.
- **Step 4: Inventory of Source:** After step 3, in which the nature and magnitude of pollution are identified, the source of such pollution should be identified.
- **Step 5: Water Quantity Information:** In order to get the required water quality, the water quantity information is essential to know how much effluence load needs to be reduced.
- **Step 6: Selection of Wastewater Treatment Technology:** Wastewater treatment technologies are selected based on the type

of water to be treated. For example, simpler technology can be adopted for treating sewage wastewater. Waste stabilization ponds are suitable for the places where land availability is more like small towns and are considered to be economical ones. A series of waste stabilization ponds are proved to be cost-effective in the places where sewage is found to be flow in open drains and not ideal for anaerobic biological treatment. There are numerous options available for sewage treatments that are proved to be cost-effective and environmentally compatible, like waste stabilization ponds, land treatment, constructed wetlands, aerated lagoon, rotating biological contractors (RBC), duck-weed pond, root zone treatment and up-flow anaerobic sludge blanket system.

- **Step 7: Financing Waste Management:** Efficient usage of water by the domestic and industrial sectors water is achieved only through pricing and demand management which are also important instruments for reducing wastewater volumes and loads. Water-saving technologies such as water recycling and reuse systems are being adopted by the urban organizations when water and sewerage fees are induced on them so that the discharge of pollutants into effluent streams can be minimized. Demand management programs should also be given importance along with price-based incentives which comprise educational and practical components, such as assistance to consumers, promotion, distribution, or sale of water-saving strategies and water conservation campaigns, etc. Beneficiaries will contribute to waste management and hence equal importance should also give to the consideration of beneficiaries.
- **Step 8: Maintenance of Sewage Treatment Plants (STPs):** Trained and skilled personals should manage SWPs and an expert should visit the plant once a month to improve its performance. Most of the STPs are run by the personals that don't have an idea to manage the plants but know the only operation of pumps and motors. According to CPCB's survey report, 39% of the plants are not conforming to general standards which show poor maintenance of plants. Hence, sewage treatment plants (STPs) should be maintained properly for the successful operation of the system.
- **Step 9: Pollution from Industrial Sources:** Pollution from industrial sources can be controlled by:
- Controlling pollution at the source;

- Industrial waste recycling and recovery of resources;
- Waste minimization and clean technologies;
- Standards on wastewater discharge and wastewater pollution charges;
- Wherever advantageous, prefer mixing of sewage with industrial waste.
- **Step 10: Pollution from Non-Point Sources:** Control of pollution from non-point causes such as unsewered sanitation, application of chemicals in agriculture and uncollected wastes dumped in urban and industrial areas should be given extreme importance. The problem with these pollutants is likely to rise in the future with the superior application of these commodities. Hence in this regard, it is important to regulate the use of these toxic pollutants by evolving integrated pest management policy and standards. These toxic chemicals can be replaced by ecologically acceptable substitutes.

Multiple approaches in most circumstances are necessary to assure water quality restoration, because the installation of a water treatment plant and its appropriate process alone may not be enough to preserve water quality in a water body.

5.3 CONVENTIONAL WASTEWATER TREATMENT TECHNOLOGIES

In many situations, the multi-processed strategy is necessary to secure the restructuring of water quality since it alone may not be adequate to enable the creation of a water treatment plant and to conserve the quality of water. Wastewater is the water that has been unfavorably affected in quality by human caused influence. Restoring the water quality is best option to bring the wastewater polluted by humans and nature to a desirable quality (Shah et al., 2020). Wastewater contaminants include SS, biodegradable organics, pathogenic bacteria, and nutrients. The main objectives of wastewater treatment comprise reducing organic content, removal/reduction of nutrients and removal/inactivation of pathogenic microbes. Generally, the conventional treatment processes include a combination of chemical, physical, and biological techniques to conserve water quality by removing solids, organic matter, and nutrients from effluent water. The selection of treatment systems will rely on the needed water quality and site-specific considerations, including accessible resources, climate, access to land, economics, etc.

The existing treatment technologies include preliminary, primary, secondary, and tertiary treatments as shown in Table 5.1. Preliminary treatment involves screening and grit removal. Primary treatment includes various physical and chemical processes like coagulation and flocculation, equalization, dissolved air floatation, sedimentation, neutralization, clarification, and precipitation. Secondary treatment consists of activated sludge treatment, biological filtration which includes trickling filters and RBC and anaerobic treatment systems. Tertiary treatment includes granular media filtration, reverse osmosis systems, membrane filtration, activated carbon, ion exchange, ultraviolet disinfection, etc. Due to the rapid growth of population, industrialization activities, agricultural production, domestic waste discharge also increasing rapidly which diminishes the availability of natural resources increases water pollution. This challenges the present conventional treatment process in which contaminants are still identified at some treatment stages. Hence some advanced technologies are needed to provide better alternatives for the shield of public health and the environment. Types of different wastewater treatment processes and example layouts are shown in Figure 5.1.

TABLE 5.1 Classification of Common Wastewater Treatment Processes According to Their Level of Advancement

Primary	Secondary	Tertiary	Advanced
Bar or bow screen	Activated sludge	Nitrification	Chemical treatment
Grit removal	Extended aeration	Denitrification	Reverse osmosis
Primary sedimentation	Aerated lagoon	Chemical precipitation	Electrodialysis
Comminution	Trickling filter	Disinfection	Carbon adsorption
Oil/fat removal	Rotating bio-discs	(Direct)filtration	Selective ion exchange
Flow equalization	Anaerobic treatment/ UASB	Chemical oxidation	Hyperfiltration
pH neutralization	Anaerobic filter	Biological P removal	Oxidation
Imhoff tank	Stabilization ponds	Constructed wetlands	Detoxification
–	Constructed wetlands	Aquaculture	–
–	Aquaculture	–	–

Source: Veenstra et al. (1997).

Source handling is employed to remove the toxic and other unacceptable pollutants from the wastewater and to prevent the wastewater to intermingle with other waste streams. Pre-treatment, primary treatment, and secondary treatment are the conventional treatment processes, wherever necessary tertiary treatment is also employed to remove specific constituents. Pre-treatment comprises techniques for retention of spills, neutralization, the addition of nutrients, toxins, removal of oil and grease, and the removal of solids by flotation, sedimentation, or filtration. Preliminary treatment eliminates the entry of unsettled/floating materials like wood pieces, papers, tree branches, dead animals, heavy settleable inorganic solids, etc., into primary treatment processes. Preliminary treatment is proceeded by primary treatment, which comprises filtering, removal of silt, by physical separation. Primary treatment removes large suspended organic solids and also a considerable portion of oxygen demanding substances. Secondary treatment is used to remove residual organic matter and suspended material. The contaminants which are not removed in the conventional treatment are removed in tertiary treatment. Precipitation, Filtration, air stripping, coagulation, and flocculation, adsorption, ion exchange, membrane processes, nitrification, and denitrification, etc., comes under tertiary treatment.

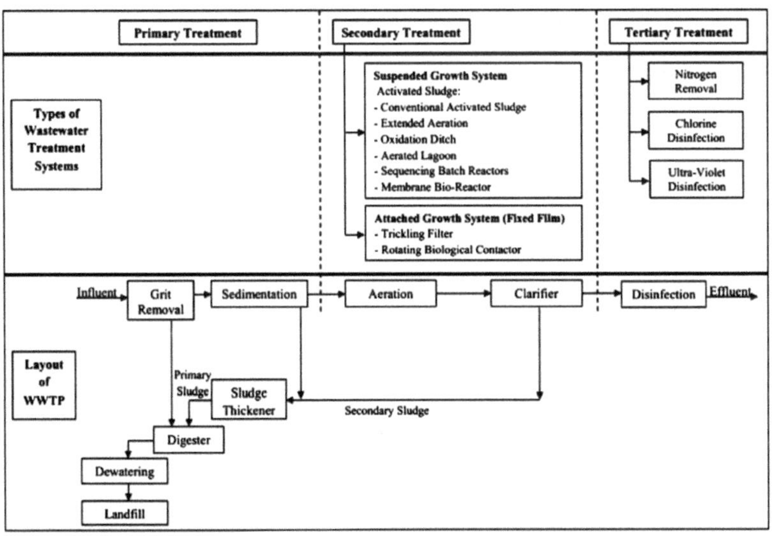

FIGURE 5.1 Types of wastewater treatment processes and layout of WWTP.
Source: Reprinted with permission from Ting and Praveena (2017). © Springer.

Objectives of treating wastewater (Englande et al., 2015).

- Parting of solid from the liquid portion and concentrate the solids composed from the carrier water;
- Eliminate safety elements that have harmful effects on the final disposal of the wastewater and the resulting residue;
- Maximize the reuse capacity of water and residuals from the wastewater treatment plant.

5.4 WASTEWATER TREATMENT PROCESSES

5.4.1 PRETREATMENT OR PRIMARY TREATMENT

Pretreatment techniques are employed to present the wastewater amicable to subsequent treatment processes. A schematic of different pre-treatment technologies is shown in Figure 5.2.

1. **Screening:** The floating materials entangle in the impellers by entering into pumps. The floating materials will choke suction pipes once they enter into them and also it is very challenging to identify their position in the pipeline. Hence, Screening is essential to remove these large, objectionable solid matters and is placed ahead of pumping stations.
2. **Comminuting Devices:** A comminutor is provided with a cutting mechanism to shred the material which is retained and enables the retained material to move along with the sewage. Comminutor generally placed across the flow path to intercept coarse solids and follows grit removal to protect the machinery.
3. **Grit Removal:** Grit, which is composed of minor coarse particles of sand, gravel, or other materials will damage the moving mechanical equipment and pump elements. To prevent abrasion and abnormal wear and tear of the machinery and pump parts, it is important to remove the grit. Constant cleaning of settling tanks and digesters can be greatly reduced by grit removal. An aerated chamber can be used to remove grit, where the air allows the inorganic matter to settle down by keeping the organic matter in suspension. Grit removal can also be done by directing flow velocity through the chamber.
4. **Equalization and Neutralization:** Sometimes, neutralization of huge variations in pH and concentrations of impurities in the

inward flow should be carried out before entering into the treatment plant, especially in industrial waters. Equalization is the most essential procedure to treat industrial wastewater, which helps keep the flow and concentration through the plant generally consistent. In general, neutralization follows equalization, such that acidic and alkaline watercourses in the equalization basin can be somewhat neutralized.

5. **Primary Sedimentation:** After pretreatment, the pre-treated wastewater is introduced into a large rectangular tank in which settleable solids will settle by gravity. In order to diminish the organic load on secondary treatment units, a primary clarifier is placed after screens and grit chambers (Figure 5.3).

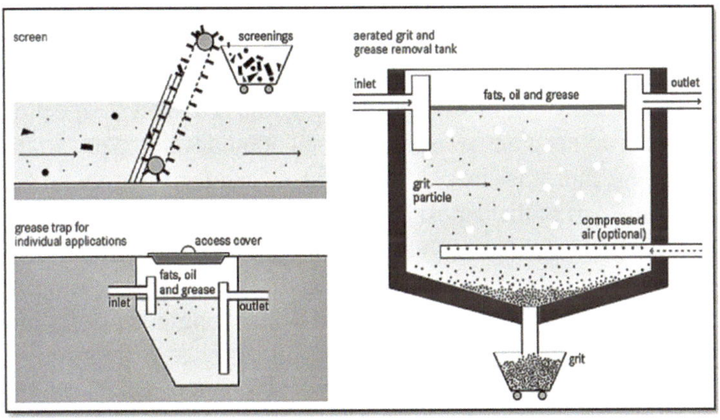

FIGURE 5.2 An outline *of different pretreatment methods.*
Source: Reprinted from Tilley et al. (2014). © Eawag: Swiss Federal Institute of Aquatic Science and Technology, Department Water and Sanitation in Developing Countries (Sandec) Dübendorf, Switzerland, www.sandec.ch

FIGURE 5.3 Grit *chamber in Honduras.*
Source: Reprinted from David and Kristy (2011).

15.4.2 SECONDARY TREATMENT

After pretreatment/primary treatment, secondary treatment is employed where the biological form of treatment is done. There are several methods for biological wastewater treatment; but the most frequently used technologies are activated sludge and trickling filtration processes:

1. **Activated Sludge Process:** In this process, a tank is supplied with oxygen in which an active biological floc is maintained. The entering wastewater and bacteria within the Floc, therefore, get maximum interaction. In a conventional method, the air or pure oxygen is generally employed as bubbles (using diffusers) or by a disturbing agitation of the liquid (using impeller). Then wastewater is introduced into a tank maintained with a concentration of microorganisms. In order to maintain the microbe concentration, some portion of the sludge which moves through the tank is returned into the tank to maintain and is established in a secondary sedimentation basin. A new cell material by synthesis is produced by the activated sludge process. In order to keep active microbe population development, some of the sludge must thus be placed in the tank and feed on organic materials, while part of the deposited material must be discarded.
   Extended aeration is usually designed for full mixed conditions and it is the modified activated sludge treatment process in which aeration operation is given with long detention times. Due to extended sludge ages, heavy solids, or a low ratio of food to microorganisms, sludge breathing will occur and 'burn itself up.' Endogenous breaths are expected to occur. Although some organic residue, inorganics, and solids will always come from the system, occasionally this process is known as total oxidation. As nitrogen is converted into nitrate, due to the long sludge age, the oxygen demand for the effluent is reduced.
2. **Stabilization Ponds and Lagoons:** These are one of the eldest wastewater treatment processes, which are Existing still and can be employed alone or combined with other wastewater treatment process. Factors such as weather, availability of land, objective, and location influence the design and utilization of these ponds. Based on the type of pond and climatic conditions, detention time varies, i.e., from 7 to 180 days. These ponds are classified as facultative, aerated, aerobic, and anaerobic, out of which facultative

pond, which is the most common being used one. To avoid short-circuiting and for greater operational flexibility, it is recommended to use multiple ponds in series.

Oxygen is furnished to aerated lagoons by mechanical or diffused aeration and employed for treatment of non-toxic or non-hazardous waters from food processing, pulp, and paper industries. Aerobic ponds maintain oxygen throughout their depth, whereas anaerobic ponds do not have any aerobic zones which are used as pretreatment units for strong industrial and agricultural wastes.

3. **Filtration Using Trickling Filters:** It is identical to the activated sludge process; the difference is in the former case the organic waste material will be stabilized by the microorganisms which are adhered to a fixed bed, whereas in the latter case they are in suspension. A trickling filter as shown in the Figure 5.4, is an aerobic wastewater treatment process that can be operated easily and also very simple to build. Generally, trickling filters are used for treating effluents from industries and domestic sewage. The wastewater is passed through grown biological film which is formed on a fixed bed with supporting medium. The organic constituents of the effluent will be stabilized by the biological activity of the film. The biofilm is partly in touch with the flow of wastewater and partly exposed to the air for absorption of oxygen. The trickling filters filtration processes are shown in Figure 5.4.

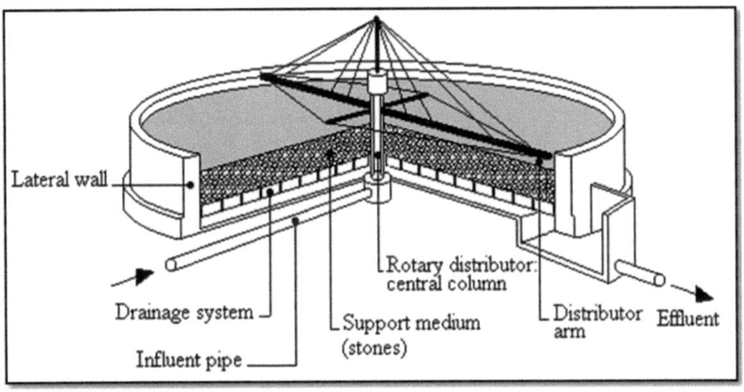

FIGURE 5.4 Classical trickling filter.
Source: Reprinted from Vianna et al., 2021. © 2012 Journal of Urban and Environmental Engineering (JUEE).

5.5 ADVANCED TREATMENT TECHNOLOGIES

Gujarat cleaner production center (GCPC, 2016) reviewed some advanced wastewater treatment technologies such as automated chemostat treatment (ACT), soil biotechnology, reed bed technology, hydrodynamic cavitation, membrane bioreactor, moving bed biological reactor, wet air oxidation, ozonation, electrocoagulation, Fenton, and sequencing batch reactor.

ACT is a flexible new technique which is fully automated and more efficient when compared to current practices. It is used in the treatment of sludge and is the most appropriate solution for treating specific refinery issues. There is no need of adding a complete additional treatment system because it can be employed alone. The output from the ACT is virtually sludge-free and meets the strict disposal standards which require no further handling.

Soil Biotechnology is an environment-friendly waste processing technology and utilizes the principle of trickling filters. In this system, a mix of physical processes such as sedimentation, infiltration, and biochemical procedures are performed to eliminate the pollutants (suspended particles, organic, and inorganic material) from the wastewater. Appropriate mineral elements, cultivation with local micro-flora, and bioindicator plants are the crucial parts of soil biotechnology system. It includes a crude tank, a containment bioreactor, a treatment water tank, pipes, and pumps. Bacteria, earthworms, and mineral additions are used in soil biotechnology for the treatment of solid organic waste as well as for wastewater treatment in a garden-like setting.

RBT dewater solids by using common reed plants in a confined area. In this system, multiple bed shapes can be used based on present land situations and areas. Reed plants are filled in the specially designed ponds and solids are driven into these reed beds. By evaporation, transpiration, and decantation, the dewatering process is carried out. Decanted water enters into underdrains by seeping over the base of the bed and the sand and gravel blanket and moves back for secondary wastewater treatment. During the process of dewatering, the solid cake is produced (solids change from liquid to "cake") which is left in the bed, and the procedure is repetitive.

In Hydrodynamic cavitation, microbubbles are formed, grown, and subsequent collapse in the lowing liquid. In this process, extremely high pressure, shear stresses and temperatures will generate locally. Hydrated lime raises the pH of the wastewater to 10.5, filtered, and then pumped through hydrodynamic cavitation. This is a batch process and to bring the COD to the desired limit, the system is kept under recirculation. Further filtration is done using a filter press and this filtered wastewater is again sent for treatment or can be disposed for can be delivered to the water recycling unit.

MBR technology involves both conventional activated sludge treatment and membrane filtration by an aperture size from 10 nm and 0.4 microns (micro/ultra-filtration), which is 10 suitable for sludge separation. The membrane enables complete disinfection of treated water by retaining all elements, colloids, microorganisms, and germs. The two important process configurations of biomass rejection include IMBR (immersed membrane biomass rejection) and SMBR (Submerged Membrane Biomass Rejection). IMBR is the established configuration in municipal wastewater treatment due to its lower cost of operation. Immersed configuration is less energy-intensive because the module is placed directly into the process tank.

MBBR is originated based on the traditional active sludge process and biofilter process. Moving Bed Biological Reactor is a highly efficient biological treatment process as it is a completely mixed and continuous biological reactor. In this biological reactor, biomass is grown on small carrier elements which are having little lighter density than water. Inside the reactor, these small carrier elements with biomass are kept in movement along with a water stream by aeration or mechanical stirrer. Hence, the choice of the carrier is given utmost importance in MBBR treatment process. This technology is found to be most efficient in treating many effluents, i.e., industrial effluents, wastewater from agricultural industries, municipal wastewater, etc.

Wet air oxidation is one of the technologies accessible which is a hydro thermal process suitable for treating aqueous wastewaters by oxidizing organic and inorganic pollutants of aqueous waste streams. In this technology, in the existence of air or oxygen-containing gas, the waste is oxidized in the liquid phase at higher temperatures(400–573°K) and pressures (0.5–20 MPa). To enhance the reaction rate, methods to improve mass transfer as well as the use of both homogeneous and heterogeneous catalysts are included by this technology. Normally, this system is operated within the super-heated water range, i.e., at $<300C$ and <20 MPa to achieve a higher level of oxidation with higher temperature and for maintaining the liquid state with high-pressure applications. The residence time ranging from seconds to hours and oxygen partial pressure is also vital to the degree of oxidation.

The unstable gas ozone consists of three oxygen atoms, and the gas easily returns to oxygen during a period of from a second to 30 minutes. Ozone is an unstable gas. The water chemistry and the pH and water temperature are the functions of this thermal degradation. Throughout this transformation, a free oxygen atom or free radical form is produced, extremely reactive and short-lived and will last only milliseconds. The colorless smell of Ozone gas resembles air before a huge storm. Ozone has a higher efficacy against germs and viruses as compared to chlorination. Ozone gas oxidizes can also lower iron, manganese, sulfur content and minimize or remove taste and odor issues. Ozone oxides iron and manganese, sulfur, and other elements sulfur into the waters which is removed by post-filtering, to create insoluble metal oxides. Organic particles and pollutants are removed by coagulation or chemical oxidation.

Electrocoagulation removes contaminants from water by passing an electrical current through the water which was proven to be very effective. The technique of electrocoagulation employs a range of geometry of anodes and cathode, e.g., balls, wire mesh, fluidized bed bands and rods. Although the process of electrocoagulation looks to be similar to chemical coagulation, it is in many ways the cationic species that neutralizes the surface loads. In this process, the suspended, emulsified/dissolved pollutants in an aqueous medium are destabilized by the electric current, which is responsible for causing chemical reactions by providing the electromotive force that is introduced into the medium.

Fenton technology is cheaper, easier, and widely used method for treating heavy effluents which are hard to decompose. This technology is one of the AOPs (advanced oxidation process) and can purify various kinds of wastewater. During organic compound removal in the wastewater, the Fenton reaction produces a very effective oxidation agent. Some methods such as UV, ozone, and others are recognized raising the cost of wastewater treatment. Fenton utilizes catalyst ions and hydrogen peroxide, maximum organic compounds easy to oxidize. Fenton technology is not complex equipment, because it does not need high temperature and high pressure. This technology is found to be very efficient in the removal of numerous numbers of dangerous organic impurities from water.

SBRs are the variation of activated-sludge plants. In SBRs, all the treatment and processes are carried out into a single basin or tank, whereas in activated-sludge plants conventional facilities rely on multiple basins.

5.6 WATER QUALITY MODELS

5.6.1 IMPORTANCE OF WATER QUALITY MODELS

Monitoring data is a great way of detecting water with impairment. For many reasons, model predictions are preferred instead of monitoring where monitoring is impossible and water quality assessments in the future are needed. Model predictions may be utilized in combination or rather of monitoring data. Integrated tracking and modeling systems may, ultimately, give greater insights for the same overall costs than any other. Combined models of runoff and water quality connect pressures such as anthropogenic activities to reactions. Optimally, models for water quality management must take account of all five forms of alternative indicators such as differences in the physical ecosystems, shifts in the water's seasonal flow, variations in the system's food base, changes in the encounters of stream biota and release of toxins (Conventional pollutant emissions), which provide for a more integrative manner (Karr, 1990; NRC, 2001).

Models of quality of water are already being used for many types of fresh and salt systems, such as rivers, lakes, estuaries, waterways, and bays. In past decades, water quality models significantly enhanced the reliability and sustainability work of authorities and academics. Now the conditions of water quality may be predicted based on changes in land use, population growth, effluent dumping, and the world climate. Models of water quality often cover data gaps that are a key limitation in the monitoring and treatment of water quality. It aims to detect the causes of contamination and helps understand concurrently numerous complicated biochemical cycles in water bodies that would otherwise be impossible to evaluate in-field surveillance alone. Future events can be accurately predicted. This is meaningful with water quality models that are predictive instruments that depict complicated physical, biological, and chemical phenomena.

Models for the quality of water are normally categorized according to model complexity, water type and water quality metrics (dissolved oxygen

(DO), nutrients, etc.), predicted by the model. The more complicated the model is, the harder and costlier it is to apply to a specific circumstance. Model complexity is based on four criteria, including the quantity and kind of characteristics of water, the degree of geographical information, the degree of temporal detail and the complexity of the analytic water body.

Models for the water quality may be ranked by type of approach (physical, conceptual, empirical), by contaminants (nutrients, sediments, salt, etc.), by application areas (catchment, groundwater, river system, coastal, integrated waters), by nature (deterministic or stochastic), by state analysis (stable or dynamic simulation), by spatial (lumped or distributed), by sizes (1-).(Tsakaris and Alexakis, 2012). The accessibility and consistency of the data are the issues for developing and using water quality management tools. The elaborateness of models utilized for managing water quality must be consistent with the amount and quality of data provided. Static deterministic models generally employ worst-case actual flow estimates to describe the behavior of pollutants under bad realistic conditions. They also require specific estimations of this data. Dynamic models will require flow, temperature, and other variables data from time series. Besides hydraulic data, base-case levels of water quality metrics of relevance require models (DO, mercury, and so on).

Simple problems in water quality can be solved by simple models, but complicated ones may require more sophisticated models or not. The models should connect management options with the significant dependent variable (such as impurity causes and water quality standard constraints). The whole 'chain' from stresses to answers should be incorporated.

5.7 DESCRIPTION OF SOME WATER QUALITY MODELS

5.7.1 DRAINMOD

DRAINMOD is modeled for usage in wetlands for transporting water and reformation of nitrogen in a waterway (Skaggs, 1981). The study technique was originally used to observe the efficiency of water and drainage systems and their impact on the treatment of pollutants, the reaction of crops and their usage of water (Skaggs et al., 1981; Massey et al., 1983; Deal et al., 1986). DRAINMOD has been expanded to forecast nitrogen transport (DRAINMOD-N) and salt flow in shallow soils (DRAINMOD-S).

In DRAINMOD, the daily nitrate leaching for particular cultivation, agriculture, sole, and geo-hydrological conditions may be calculated at a single field scale (El-Sadek, 2007). The information requested in the model to be applied is soil type, land use, climatic, water, and soil profile nitrogen conditions at the initial stage of the simulated period and nitrogen fertilizer practice. El-Sadek (2007) has successfully used several versions of the DRAINMOD model in the Dender catchment basin located in the west of Brussels, to analyze the liquidation of nitrate-nitrogen. DRAINMOD-N and DRAINMOD-S models employ a modified DRAINMOD version to calculate the daily water level and water flows (Tsakiris and Alexakis, 2012). DRAINMOD-N is differentiated by the fact that DRAINMOD-N can forecast nitrogen concentration in-ground and surface and subsurface drainage through functional connections, whereas DRAINMOD-S can estimate the salt drainage water content, soil salinity and salinity impacts for crop production (Skaggs, 1981).

5.7.2 MIKE-11

MIKE-11 is a hydrodynamic 1-D model that analyzes a stable water system flow and can simulate the dynamic flow of water on a DHI river (Danish Hydraulic Institute – Nederland, 1998). The model components are water quality, transport sediments, advection-dispersion, rainfall-runoff, and eutrophication. As either a water model or hydraulic model, MIKE-11 may be utilized both. The MIKE 11 essential module, which models the dynamic fluxes in streams, is the hydrodynamic module that accepts homogenous flow conditions inside this channel. It solves the one-dimensional equations of shallow water SVE (Saint Venant equations) (DHI, 1998).

The advection-dispersion module may mimic the first degree deteriorates of contaminants. In one of six degrees of complexity, the water quality module provides a sophisticated simulation that may be characterized as follows:

- **1st Level:** Dissolved oxygen and BOD degradation, plus temperature and reaeration effects in the first order;
- **2nd Level:** First level + BOD and the sediment exchange reactions;
- **3rd Level:** As nitrate and ammonium balance second level plus without denitrification;

- 4th **Level:** Third Level Plus Denitrifying;
- 5th **Level:** As a result of BODs, the forth level plus oxygen requirement but without nitrogen nitrate and ammonium components; and
- 6th **Level:** Nitrate and ammonium as the fifth level.

Coliform and phosphorus operations can also be introduced to one of the six stages. The quality component of water can be employed with or without sedimentation procedures. It covers toxic metals, iron corrosion, eutrophication, and transfer of nutrients. Several editors in the interface are used to input the advection, hydrodynamic, and water quality characteristics into the system. A water quality file contains the water quality parameters and the default settings. A set of coefficient and water quality constraints including amongst others: degrading of BOD; impact by unit nitrification of ammonia converted into nitrate, ammonium-frequency rate, a continuous nitrification rate (CNR); the demand for nitrification by unit mass of ammonium converted to nitrate.

The MIKE-11 model finalizes the flow simulation before simulating water quality and solvent transportation mechanisms. The model can be used for water quality simulation using current hydrodynamic data. In the Advection Dispersion component, the computation of the mass balance for the water quality parameters is conducted at all intervals in all senses. MIKE-II model depends on Saint-Venant equations as shown below:

$$\frac{\partial Q}{\partial x}+\frac{\partial A}{\partial t}=q$$

$$\frac{\partial Q}{\partial t}+\frac{\partial}{\partial x}\left(\alpha\frac{Q^2}{A}\right)+gA\left(\frac{\partial h}{\partial x}\right)+\frac{n^2 gQ|Q|}{AR^{4/3}}=0$$

where; Q is the discharge; A is the cross-sectional area; q is the lateral inflow; h is the level of water above a reference datum; x is the downstream direction; t is the time; n is the Manning's coefficient; R is the hydraulic radius; g is the gravity acceleration; α is the momentum distribution coefficient.

The result of the MIKE-11 contains time series depths, flow, pollutant, and statistical choices for the description of the findings (Tsakiris and Alexakis, 2012). MIKE-11 constraints include the necessity of large simulated results susceptible to data suitability and the demand for accurate calibration and assessment findings for cross-sectional channel limits at every reach (Cox, 2003). Without experienced assistance, it is difficult to create the model (Olowe et al., 2018).

5.7.3 MIKE SHE

In the model region and different water quality concerns, MIKE SHE replicates the surface and groundwater flow, sediments, nutrient, and pesticide transfer. For a big watershed, MIKESHE can be used (Gayatri et al., 2015). The different processes included overland flow, unsaturated flow, evapotranspiration vegetation, flood water flow, dynamic canal stream and irrigated water quality.

5.7.4 HSPF

The U.S. Environmental Protection Agency (USEPA) created HSPF (Hydrologic Simulation Program Fortran) is a hydrology and water quality model that can evaluate both traditional and hazardous organic contaminants in hydrology and water quality. Model simulations may be carried out by water components of quality such as temperature, fecal coliforms, DO, BOD, total SS, nitrate, orthophosphate, and pH. It includes the inputs of agricultural sediments, nutrients, pesticides, conservation products and fecal coliforms. It is a mix of agricultural runoff control, various anthropogenic runoff and HSP models, based on the aforementioned Model IV of Stanford Watershed (Crawford and Linsley, 1966). The models require factors for example land use patterns and land use management that include temporal history, heat, and radiation from the sun and physical features, land-use patterns. The model limits are as follows: (i) the model applies to infused rivers, reservoirs, and water bodies in one dimension only (Whitehead et al., 1998; Paliwal et al., 2007); (ii) algorithmic problem and procedural; (iii) The precision of the model is restricted by geographical data and by the high enough standard inputs.

5.7.5 INCA MODEL

INCA is a dynamically built model that enables hydrological integration and irrigation quality to be incorporated into the watercourse (Whitehead et al., 2015). It is also frequently utilized in the area of watershed management, in conjunction with land-use changes, climate change, altering pollutant environments (points and diffuse pollution). INCA model is now used for diffuse and point source pathogen stream modeling in catchments. This model measures pollution flows on a day-to-day basis in both surface and aquatic areas of catchments. INCA models enable the geographic character of the watershed to vary the range of reach, the rate, the usage of the land, the flow-speed ratio and the loads from the point, diffuse land origins and diffuse air types of pollutants from input deposits. INCAN-N permits simulation of a single riverbed by tributaries processed as aggregate inputs in a half-distributed way (Whitehead et al., 2015). This model is based on the 4th order Range Kutta approach, a sequence of linked differential equations that may be solved with the numerical integration method.

5.7.6 WASP (WATER QUALITY ANALYSIS SIMULATION PROGRAM) MODEL

Water quality analysis simulation program (WASP) was created by the UEPA in 1983, enabling the simulation of water quality processes in waterways, lakes, reservoirs, and estuaries. The model may be used to mimic water quality. It may be implemented in one-, two- or three-dimensional format and forecasts a range of contaminants such as acid, phosphorus, dissolved, BODs, sediment oxygen demands, algae, periphytons, organics, metals, mercury, pathogens, etc. It is a dynamic model program that can address varied flow conditions through aquatically modeled water environments (Ambrose et al., 1993). Simulation of water quality can be done by DYNHYD and WASP, the computer programs comprised by the model which can be used separately or together. A hydrodynamic software DYNHYD will be employed to model the flow of water over the range of the water system, while the WASP program will simulate the transportation of pollutants inside the column of water. Two broad kinetic components such as Advanced Eutrophication (EUTRO) and Improved Toxicant arrive in a recent version (TOXI). The TOXI model is often used in the simulation of pollutant toxicity by watercourses, while the Advanced Eutrophication Framework addresses the traditional pollution of naturally dissolved phosphate, salinity, solids, and CBOD, DO, BOD, phytoplankton, periphyton, detritus, dissolved organic nitrogen, ammonia, nitrate associated sediment digenetics. WASP may also be associated with transport models of hydrodynamics and sediments which give flows, depths, speeds, temperatures, salinity, and silt flows. The model, however, has several limits, such as the need for large amounts of calibration and verification data (Kannel et al., 2011). The model does not manage plumbing/floating materials or mixing areas or close-to-field impacts. The model employs mass-equilibrium conservation to resolve the governing equations. The system requires certain input data to answer the equations of balance equations. Input, beginning, and limited levels, pollutant origin, coefficient of kinetic parameters, flow characteristics and fluvial design also are significant in solving the system of mass equilibrium. The continuity and momentum equations used by the model were shown below:

$$\frac{dQ}{dt}+\frac{1}{B}\left(\frac{dQ}{dx}\right)=q_s$$

$$\frac{dQ}{dt}+\frac{\partial}{\partial x}\left(\frac{Qw}{A}\right)+gA\left(\frac{\partial z}{\partial x}+\frac{Q_z Q_z}{K^2}\right)=0$$

where; Z is the elevation of water surface; Q is the rate of flow; B is the wetted width; A is the wetted cross-sectional area; t is the time; x is the distance along the channel; K is the conveyance of the channel; g is the gravitational acceleration; q is the discharge per unit channel length.

5.7.7 STREETER PHELPS MODELS

Streeter and Phelps in 1925 designed the first tool for water quality modeling known as S-P model in order to measure water contamination inside the river or stream (Bai et al., 2012). Those models are designed to regulate oxygen and decaying BOD in one order. It is a standard one-dimensional model. Streeter-Phelps equation also known as DO sag is derived on the linear differential equation of the 1st order (Streeter and Phelps, 1925, 1985).

$$\frac{\partial D}{\partial t}=K_1 L_t - K_2 D$$

where; D refers to the saturation deficit; K_1 refers to the deoxygenation rate; K_2 refers to the reaeration rate; and L_t is the oxygen demand remaining at time t.

This concept was initially utilized by sanitary technologist Harold Warner Streeter and Earle Berhand Phelps in the Ohio River in the United States (1876–1953). Like previous models, there is a certain mimicry in this model that is – simplification causes problems in the model due to single input, do sink and DO source (Streeter and Phelps, 1985).

5.7.8 EXPORT COEFFICIENT MODEL (ECM)

The export coefficient model (ECM) was created in 1976 and is a straightforward model (Omernik, 1976). The technique is beneficial for catchment management and evaluation, based on data from publicly available databases and its low data needs (Johnes, 1996; Burt and Johnes, 1997; Johnes and Butterfield, 2002). The total quantity of phosphate and nitrogen provided at a surface water sample location in many nations was frequently utilized for prediction (Bowes et al., 2008; European Commission, 2003; Johnes, 1996; Ierodiaconou et al., 2005).

The ECM technique aims at preaching the loading of nutrients for the water sample at a specified area based on the fertilizer exports from each pollutant in the above site downstream (Johnes, 1996). The ECM forecasts can be presented as the mean yearly concentrate of T-N and T-P in the water body by taking into consideration the total yearly outflow to the samples collected to eliminate the difficulties of foreseeing the accumulation of specific phosphorous fragments and nitrogen species introduced through nutrient cycling processes (Johnes, 1996). Johnes (1996) altered the model to take account of variables of nutritional export like forest and vegetation regions; cultivation of specific plants, which include cereal grains of autumn and fall, rape of oilseeds, field, and vegetable root, momentary, long-lasting, and rough grassland; and human settlement and septic tank systems. The model employs export coefficients chosen from the literature or from field trials for calibration to suit a model for the reported nutrient load of catchment outlets.

The ECM catchment scale simulation model may be separated into three stages: Three data collecting programs such as the investigation of all farmers and owners of land in a catchment, literature surveys, and field measurement device are undertaken in the first stage. In the second phase, the accuracy of the calibration process must be assessed. The validated model is utilized in the predictive stage in the third phase to assess the possibility for water quality improvements and the series of the possible managing options.

5.7.9 MONERIS

The Moneris model accumulates approximately eight different models that replicate the key processes in which suspended particles and nutrients are produced and transported through the river system (Behrendt et al., 2007; Palmeri et al., 2005). That is because a watershed paradigm is pursued in many nations by WFD, the MONERIS concept is designed to promote environmental research in a shared setting (Venohr et al., 2009). MONERIS is a GIS model based on the empirical technique of simply describing complicated connections. It is a conceptual framework for quantifying nutrients emissions in catchment areas greater than 50 km^2 of the point- and non-point emitters (Behrendt et al., 2000). The model contains a scenario manager to calculate the impact of measures on the nutrient emissions of various geographical units and routes. The principal processes and pathways of the MONERIS model are: (a) groundwater, (b) erosion, (c) runoff (dissolved nutrients), (d) tiles drainage, (e) atmospheric deposition on water surface areas, (f) urban regions and (g) point sources (e.g., wastewater treatment plants).

5.7.10 SIMCAT

The model was created in the United Kingdom by Anglian Water (Warn, 1987) and used for predicting conservation and non-conservative contaminants in the stream. It is a stochastic model which uses Monte Carlo simulation methods to assess the effect of pollution emissions on bodies of water (Warn, 2007). SIMCAT (Catchment Simulation), symbolizes the destiny and carriage of soluble compounds in a waterway and inputs to point-source wastewater oxygen concentration that are portrayed by the temperature, reaeration, and BOD decay, the decaying toxins which are non-conservative, and the conservative compounds which are supposed to be non-decaying. The deterrent model (SIMCAT) is used to predict the metrics of river quality and flow characteristics of water (Cox, 2003).

To increase the effectiveness of water by helping to plan actions, it has been created for Britain (Warn, 2010). The major river in either model is separated into user-defined ranges that are the distances between sites of interest or sources. Close to 600 reaches may be modeled and the model periodically picks quality and stream values from the ranges provided for all variables for every run. SIMCAT results are overview data for every approach and each contaminant (90th or 95th percentile and means).

In its operation, it is restricted to use the SIMCAT model simulation software to simulate breathing, photosynthesis, the need for sediment oxygen and reaeration rate. It is an excellent tool for managing the quality of the water, but cannot reflect temporal variability and is not specifically suited for complicated scenarios. SIMCAT is yet a basic, adaptable, and flexible model that may be used to analyze water quality recognition in river systems.

The model utilized its operational idea of mass balancing and depicts the extent of the river as the sequence of continually stirred tank reactors (CSTRS). The contaminants that do not depend on sediment reactions in freshwater can be predicted by SIMCAT. The flow speed is determined using the velocity flow connection and the size distribution for each area is calculated. It implies that pollutants are well mixed all over the river (Cox, 2003).

5.7.11 TOMCAT (TEMPORAL/OVERALL MODEL FOR CATCHMENTS)

The TOMCAT model was created to help to evaluate effluent quality standards on samples using an investigation by Monte Carlo to satisfy surface water quality conservation objectives (Bowden and Brown, 1984). With diurnal and seasonal influences in the measurement data and the quality of water captured, TOMCAT enables more complicated temporal correlations that reproducing these influences in the simulation results. The TOMCAT River system is characterized by a variety of pollution incidents on the rivers, observation stations, abstraction, and effluent dumps. The TOMCAT model requires multiple input types with fixed values (usually physical limitations that describe the response rates), quality, and flow data.

The TOMCAT system is characterized according to the "seasons" contained inside the model, the model is executed, sub-seasons are reproduced, and monthly air temperatures are averaged. Several access variables are provided for each person's reach which contains both BOD and NH_4 concentrations and decomposition rates, oxygen exchange rate, runoff scale, thermal balance rate constant, diffuses drainage basin rate estimation number, runoff scale, reach duration and depth component, and mean sectional area cross-sectional area (Cox, 2003). The model can mimic the effect of overflowing of stormwater. By resolving the operational formulae, TOMCAT estimates the purity and flow at each level. The data collected from surveillance sites are correlated to generated stream flow. In addition, the generated streams are synchronized with the actual data automatically. This model may be used simply to simulate current flow and water quality situations in the stream (e.g., The TOMCAT may be used to forecast the amendments necessary for discharge in order to conform with the river statutory criteria). This model is meant to be straightforward to set up and the output produced may be compared with the values provided in the relevant law. The TOPCAT model has humidity shops such as the roots of the unsaturated region store, the saturated event store, and the old surface store (Quinn et al., 2008).

In short, a percentage of hydrological effective rainfall (HER) rainwater might circumvent the event subsurface and reach the old subsurface storage. A SPLIT parameter defines the fraction of HER that is entering the event surface. In the event of surface storage, SBAR represents the situation humidity condition.

5.7.12 TOPCAT-NP

Model TOPCAT is a MIR-version of TOPMODEL that utilizes floor flux calculations and the same moisture storage for soil (Quinn and Beven, 1993). TOPCAT-NP is a Nitrate and Phosphate model (N-MIR) (P-MIR). This model can estimate nitrate and phosphorus losses every day or every day (Quinn et al., 1999). They constantly try to maintain a certain physical significance for the employed characteristics. Along with two surface runoff factors, TOPCAT has also an additional basic weather flow element. The model utilizes a basic moisture root zone storage to collect possible evapotranspiration supplies and actual rainfall (Quinn et al., 2008). Hydrologically effective rainfall (HER) is another vital component of the model which represents the surplus rain that can penetrate down into the soil. The reported root zone thickness value is normally greater than the actual "anti-drying" (SRMAX) parameter value, which initiates HER release.

5.7.13 QUALS

The United States Environmental Protection Agency (EPA) has provided several QUAL templates such as QUAL2E, QUAL2EUNCAS, QUAL2K and QUAL2Kw, which simulate up to 15 water quality factors (DO, BOD, temperature, algae as chlorophyll a, organic nitrogen N, ammonia as N, nitrites as N, organic phosphorus as P, dissolved phosphorus P, coliform bacteria and three conservative component solutes) in any combo that has been selected by the user. The heat and mass inputs of both QUAL2K and QUAL2E imitate loads and non-point points, abstractions.

5.7.13.1 QUAL2K

QUAL2K is a standard one-dimensional quality of water as well as in the flow model. The model would be neither stochastic nor dynamic. The QUAL2K model can replicate up to 16 determinants of quality of river water and its affluents (Brown and Barnwell, 1987). The stretch of the river is split into many sub-strips equally long. The model has these assumptions:

- The average flow is the basis of the advective transport;
- In the cross-section, water quality indicators are fully mixed;
- The disseminative transport is linked to the gradient.

The model enables the user to evaluate the following variables in any combined effect:

- Dissolved oxygen;
- Temperature;

- Phosphorous;
- Nitrate, nitrite, ammonium, and organic nitrogen;
- Chlorophyll-a;
- Up to three conservative solutes;
- One non-conservative constituent solute;
- Coliform bacteria.

In addition, nitrate, phosphate, and DO have additional details; most drivers have been modeled as degradation of the first order. The QUAL2K is designed to represent contaminants in freshwater which rely on relationships between sediment, in particular as a tank of inorganic and organic compounds. The first stage of the QUAL2K standard model is to split the river up to 50 spans, and each of them would then be subdivided into computer parts of similar length (up to 20 per reach).

The flow and water quality data needs of the model require individual values for each variable. The river reaches are the major feature of such a model. The required data by the model with each river reach are as follows:

- Flow data and hydraulic terms;
- Initial conditions;
- Response amount coefficients;
- Local climatological data for heat equilibrium calculations; and
- Rate constraints for all of the organic and chemical responses.

The model output contains answers to the mass balance of pollutants and the flow for every reach. The major benefit of QUAL2K is its capacity to simulate algae (Chlorophyll-a), its code and the theory of its rich literature. You may get QUALK2K from its website free of charge. The model requires a minimal quantity of sediment data and just incomplete hydraulic conditions.

5.7.13.2 QUAL2E

QUAL2E, an equilibrium model for the prediction of pollutants in extending streams and mixed lakes, was created by US Environment Protection Agency (USEPA) (Cox, 2003; Brown and Barnwell, 1987). It was considered that the river is a one-dimensional model which forecasts substantial nutrient cycle reactions, sediments, algae development, negative impacts on the environment and its impact on the DO content in the body (Birgand, 2004). It helps to forecast geographical and temporal fluctuations within the water column of certain characteristics, such as temperature, nutrients, BOD, and DO levels (Kannel et al., 2011). QUAL2E may be used to know the influence of multiple pollutant sources discharged into water systems and their effect on water quality. In addition, the impact of algal expansion and mortality rate may be anticipated using the model on DO in the water body. An implicit technique of finite-difference calculates the differential equation of the QUAL2E model.

The QUAL2E model mimics the actual elements and components related to the amount of DO. Biochemical oxygen demands, algae processes and nutrients responses are the important sinks and/or sources in the equation. Furthermore, QUAL2E may demonstrate the reaeration that occurs when water pours over dams. The QUAL2E is created to mimic several aspects of water quality, including oxygen dissolved in water sources, nutrients, and conservative contaminants (Cox, 2003). A model QUAL2E cannot be used on a watercourse with a temporal fluctuation. Macrophytes, mobility of the sediments and denitrification mechanisms cannot be modeled. Because of its constant presumption of state, the model cannot represent the changeable flow conditions.

5.7.14 AQUATOX

USEPA has created an AQUATOX model that forecasts the destiny and effect of various pollutants in aquatic environments (Park et al., 2008; Sharma and Kansal, 2013). The model aims to anticipate environmental stressors and associated ecosystem consequences. It is an ecological mechanistic model and predicts different aspects of water quality, including nutrients, sediments, and hazardous substances. Furthermore, the model simulates their influence on aquatic creatures and plants (Park and Clough, 2004; Bingli et al., 2008). The model incorporates its methods from the clean model used for the living aquatic ecosystem and concurrently predicts almost 20 factors inside the water environment, making the model is among the finest WQM. It is highly advanced and developed to predict the fate of contaminants in the aquatic environment as a spatially and zeitally determined mechanistic model (Sharma and Kansal, 2013). It considers that the river consists of many well-mixed sections and uses mean flow values for its functioning each time. The Runge-Kutta integration techniques in the fourth and fifth order were implemented to overcome the model problem. The model may therefore be used for understanding the influence on aquatic ecosystems of different water quality factors. The AQUATOX model can identify several environmental stresses causing ecological damage. It forecasts the ecosystem effects of contaminants. The model has assessed the impacts on watercourses of numerous environmental factors, including nutrients, silt, organic waste, hazardous chemicals, temperature, periphyton, phytoplankton, and macroscopic materials. The limits of this model have included the inability while modeling a variation in nutrient, chemicals, and Sediment levels in the watercourse and the lack of a representation for interior nutrients in the algae bioenergy sector, to model elements and connect hydrodynamic models. It is, nevertheless, superior to water quality models since it is open to the public and may be integrated with certain watershed models.

5.7.15 CE-QUAL-RIV1

A single-dimensional hydrodynamic water quality model was created in the US Army Engineers waterways experiment station (WES) in 1991 for simulation of water quality parameters in tributaries and island creeks (Sharma and Kansal, 2013). Most models have been created under stable flow circumstances, but the CE QUAL-RIV1 model replicates extremely unstable river system flows (Martin et al., 2002). CE QUAL-RIV1 comprises two codes, such as RIV1Q and hydrodynamic code, program code (RIV1H). At the very beginning, the river hydraulics are calculated by using RIV1H code for the water column with geometrical river characteristics and boundary conditions. For its modeling of water quality, RIV1Q leverages the output generated from RIV1H. The RIV1H code can be used to measure the flow rate, depth, speeds, and breadth of the rivers in the four-point implicit solution procedure of the St. Venant equation. Whether the model is applicable in simulated characteristics in the water body, determines the character of the river flow.

For its modeling of water quality, RIV1Q leverages the result generated from RIV1H. The RIV1H code can be used to measure the flow rate, depth, speeds, and breadth of the rivers in the four-point implicit analytic solution of the St. Venant equation. Whether the model is applicable in simulated characteristics in the body of water, determines the character of the stream flow. The equations governing the model were continuity and momentum equations. The 1-D advection and dispersion equation united with the sink and source was shown below:

$$\frac{\partial \alpha}{\partial t}+U\frac{\partial \alpha}{\partial x}=D\frac{\partial^2 \alpha}{\partial x^2}+\frac{q}{A}(k-a)-K_s\alpha+\sin k_s$$

where; Q is the flow rate, (m^3/s); U is the mean velocity, (m/min); A is the channel area, (m^2); h is the depth, (m); g is the gravitational acceleration; S_o and S_f are the bottom slope and friction slope, respectively; x is the length (m); t is the time; α is the parameters; q is the lateral flow rate (m^3/s); D is the dispersion coefficient (m^2/min).

The problem was computationally resolved since the variable was nonlinear, hyperbolic, and partial. CE-QUAL-RIV1 may be used to mimic river processes of chemistry, biology, and physics. With this model, it is possible to predict the lakes and estuaries reaction to pollutants. It might be utilized for the prediction and estimation of the hydraulic and geometrical characteristics of a watershed with several hydraulic structures like dams. In addition, several aspects of the quality of water, including thermal stratification, algal development and macrophytes, may be simulated. It is, therefore, a one-dimensional model that requires considerable training sessions to efficiently use it. In addition, the model cannot be utilized for simulating activities of sediment movement in the systems of the river.

5.7.16 CE-QUAL-W2

The CE QUAL-W2 model for stream, estuarine, lakes, reservoirs, and basin river systems are a two-dimensional hydrodynamic (longitudinal and vertical) model of water quality. It has been created at the Waterway Research Institute of the United States Army Corps (Cole and Buchak, 1995). The model takes the form of horizontal uniformity that is especially suitable for water applications with low horizontal water quality fluctuation. Total organic carbon, DO, biological oxygen demand (BOD), algae, and *Escherichia coli* may also be simulated on this model. Around 17 types of factors in the quality of water affect concentration.

$$\frac{\partial BC}{\partial t}+\frac{\partial UBC}{\partial x}+\frac{\partial WBC}{\partial z}-\frac{\partial(BD_x\frac{\partial C}{\partial x}}{\partial x}$$

$$\frac{\partial\left(BD_z\frac{\partial C}{\partial z}\right)}{\partial x}=C_qB+SB$$

Here, the layer of time-space change was designated as *B*; The horizontal component of the normal concentration as *C*; *U* and *W* are the lateral average velocities in the x-direction and y-direction, respectively; The diffusion coefficients of temperature and composition as D_x and D_z respectively; The elements of input and output flow of material flow rate as C_q, and Sources and sinks as S.

5.7.17 SWAT

SWAT is a USDA Agricultural Research Service (USDA-ARS), a hydrological and river basin model created for water management (Neitsch et al., 2002; Tolson and Shoemaker, 2007). This program is publicly accessible and is used to estimate the long-term influence on the vast and complicated watershed, land use and management situations of freshwater, products of sediment and agriculture by agricultural and rural management techniques (Srinivasan et al., 1998). It can mimic the movement of groundwater, nutrients from canals and dams. It may also be used to assess the effects on water resources of nutrient, toxic, and particle adsorption. The SWAT was used to forecast the impact on the aquatic ecosystem of agricultural and rural management strategies.

The SWAT model requires precise foliage, terrain, climate, soil characteristics, and water body type to be functional. The model SWAT works every day. The model simulation comprises a division into a great number of compartments coupled into such a hydrological reaction unit and then further split into (HRU). Each hydrological responsive unit is used for water and pollution simulations to the watershed outflow via the stream channel. It may be utilized to address different management issues effectively and used to research unique pollution transport mechanisms in the vast river basin.

Everyday subsurface flows and transport of nutrients can be simulated by SWAT from canals and lakes and calculate the quality characteristics of flux water (algae, DO, and carbonated BOD). This concept has been used in diverse situations in a variety of applications.

SWAT, though, does not replicate events that take place sub-daily, such as one cyclone and diurnal DO variations in a large body of water. The data types of a 100 whenever the water shift is so huge and separated into hundreds of reaction units, also tough to handle and alter. SWAT does not replicate comprehensive event-based floods and sediments and also has problems simulating lowland and snowmelt erosion in the springtime and fall periods (Shoemaker et al., 2005; Dai et al., 2005).

5.7.18 ELCOM-CAEDYM

The Center for Water Research at the University of Western Australia (COM–CAEDYM–ESL) is a three-dimensional hydrodynamic and ecology model, which has been created by the Center on Water Research at the University of Australia. It is commonly utilized in ponds, dams, and coastlines (Daniel and Imberger, 2012; Yajima and Choi, 2013). This is an environment-friendly model capable of simulating biological three-dimensional phenomena (Hamilton and Schladow, 1997), macroscopic, zooplankton, fish, and bottom invertebrates (Hipsey et al., 2006).

5.7.19 EFDC

A comprehensive system of surface waves, including hydrodynamics, transporting sediments, transporting hazardous contaminants, and water quality-eutrophication elements have been created by Hamrick (1996). EFDC (environmental fluid dynamics code). It has been utilized for various water bodies, streams, lakes, reservoirs, humid areas, estuaries, and coastal areas in ecological monitoring and treatment. It is among the most extensively used water-dynamic models (Xie et al., 2010; Liu and Huang, 2009).

5.8 COMPLEXITIES INVOLVED IN WATER QUALITY MANAGEMENT

There are many complexities involved in water quality maintenance. For example, in the case of data requirements, it requires a huge number of constraints to be estimated and also the data varies with a variety of factors. Unlike water quality management, data requirements and the total number of parameters required for assessing the water quantity are very few.

The field of knowledge has, nevertheless, advanced in recent years, especially in terms of the interconnections between contaminants and human health. There is also an increase of devices' capacity to detect tremendously low levels of water bodies pollution. In contrast to parts per million only a few years ago, it is now feasible to detect extremely low levels of contamination (i.e., parts per milliard). In the context of these technological advancements, currently, newer, and emerging contaminants are monitored because of their real or imagined detrimental health effects. In recent decades, therefore, a quantity and monitoring of the number of water quality criteria and parameters estimated in the drinking water system have substantially risen in certain large cities in the industrialized nations. However, it is costing to monitor a large number of water quality related parameters, i.e., nearly 350 parameters, since it involves the collection of pure data volumes, advanced analytical laboratory and technically skilled data collection, analysis, and effective management employees. It can be conceivable, in some urban centers of the industrialized nations, to enable such complex data gathering, processing, data interpretation and management systems. But it is difficult for many poor countries to create a viable and useful system that can manage even around 25 water quality criteria routinely. This is simply one illustration of the complexity of an effective water quality management system as compared to a water quantity system (Biswas, 2007).

5.9 CONCLUSION

This chapter provides various conventional, advanced wastewater treatment technologies along with some of the currently available water quality models which are crucial for managing water quality.

From this chapter, it can be concluded that the future water crisis may arise due to water quality deterioration. Even at present, most of the highly populated countries like China and India are facing water scarcity due to polluted water in several parts of their countries. Hence, immediate measures should be taken to control water pollution which will otherwise constitute a dire threat to the entire life in the future. Even though Management of water quality is much more challenging than management of water quantities, judicious use of water along with some treating technologies, monitoring programs, people's participation, government decisions to control water pollution and using water quality models can bring down water quality related issues to a great extent. Along with some conventional treatment technologies, some advanced treatment technologies are now available some of them may be costlier or some may be cost-effective. Now-a days, low-cost technologies which make use of low-cost materials for filtration, such as natural fibers, sand, gravel, etc., are also proven to be very effective in treating pollutants from different sectors. Due to the growing water quality significance, more, and more elements should be included in the model which is again a challenging task. Technologies to be selected mainly depend upon the extent and source of pollution. Combined water quality monitoring and water quality predictions using models may provide the best solution to combat water quality problems.

KEYWORDS

- **automated chemostat treatment**
- **DRAINMOD**
- **MIKE SHE**
- **quality assessment**
- **simulation models**
- **wastewater treatment**
- **water management**
- **water quality**

REFERENCES

Ambrose, R. B., Wool, T. A., & Martin, J. L., (1993). The water quality analysis simulation program, WASP5, part A: Model documentation. In: *Environmental Research Laboratory, US Environmental Protection Agency,* Athens, GA.

Bai, J. H., Gao, H. F., Xiao, R., Wang, J. J., & Huang, C., (2012). A review of soil nitrogen mineralization in coastal wetlands: Issues and methods. *CLEAN—Soil, Air, Water, 40*(10), 1099–1105.

Behrendt, H., Huber, P., Kornmilch, M., Opitz, D., Schmoll, O., Scholz, G., & Uebe, R., (2000). *Nutrient Emissions into River Basins of Germany.* UBA-FB.

Behrendt, H., Venohr, M., Hirt, U., Hofmann, J., Opitz, D., & Gericke, A., (2007). *The Model System Moneris Version 2.0 User's Manual.* Leibniz Institute of Freshwater Ecology and Inland Fisheries in the Forschungsverbund Berline. V., Müggelseedamm 310, D-12587 Berlin, Germany.

Bingli, L., Huang, S., Min, Q., Tianyun, L., & Zijian, W., (2008). Prediction of the environmental fate and aquatic ecological impact of nitrobenzene in the Songhua River using the modified AQUATOX model. *J. Environ. Sci., 20*, 769–777.

Birgand, F., (2004). Evaluation of QUAL2E. Agricultural non-point source water quality models: Their use and application. *Southern Cooperative Series Bulletin, 398,* 99–107.

Biswas, A. K., & Tortajada, C., (2019). Water quality management: A globally neglected issue. *Int. J. Water Resour. Dev., 35*(6), 913–916.

Biswas, A. K., (2007). Water as a human right in the MENA region: Challenges and opportunities. *Int. J. Water Resour. Dev., 23*(2), 209–225.

Biswas, A. K., (2008). Asian water development outlook: Achieving water security for Asia. *Int. J.Water Resour. Dev., 24*(1), 145–176.

Bowden, K., & Brown, S. R., (1984). Relating effluent control parameters to river quality objectives using a generalized catchment simulation model. *Water Sci. Technol., 16,* 197–205.

Bowes, M. J., Smith, J. T., Jarvie, H. P., & Neal, C., (2008). Modeling of phosphorus inputs to rivers from diffuse and point sources. *Sci. Total Environ., 395*, 125–138.

Brown, L. C., & Barnwell, T. O., (1987). *The Enhanced Stream Water Quality Models QUAL2E and QUAL2E-UNCAS: Documentation and User Manual.* Athens, Georgia: US EPA.

Burt, T. P., & Johnes, P. J., (1997). Managing water quality in agricultural catchments. *Trans. Inst. Br. Geogr., 22*(1), 61–69.

Cole, T. M., & Buchak, E. M., (1995). *CE-QUAL-W2: A Two-Dimensional, Laterally Averaged, Hydrodynamic and Water Quality Model.* User manual (Version 2.0). Army Engineer Waterways Experimental Station Vicksburg MS Environmental Lab.

Cox, B. A., (2003). A review of currently available in-stream water quality models and their applicability for simulating dissolved oxygen in lowland rivers. *Sci. Environ., 314–316*, 335–377.

CPCB (Central Pollution Control Board), (2008). Central pollution control board guidelines for water quality management. *Int. Environ. Law Res. Cent.*

Crawford, N. H., & Linsley, R. K., (1966). *Digital Simulation in Hydrology Stanford Watershed Model 4.*

Dai, T., Koenig, J., Shoemaker, L., Tech, T., & Hantush, M., (2005). *TMDL model Evaluation and Research Needs.* National Risk Management Research Laboratory, Office of Research and Development.

Daniel, A. M., & Imberger, J., (2012). Managing wastewater effluent to enhance aquatic receiving ecosystem productivity: A coastal lagoon in Western Australia. *J. Environ. Manag., 99*, 52–60.

David & Kristy, (2014). *A Small Scale Grit Chamber in Honduras.* https://sswm.info/factsheet/pre-treatment-technologies (accessed on 7 July 2022).

Deal, S. C., Gilliam, J. W., Skaggs, R. W., & Konyha, K. D., (1986). Prediction of nitrogen and phosphorus losses as related to agricultural drainage system design. *Agri. Ecosyst. Environ., 18*, 37–51.

DHI (Danish Hydraulic Institute), (1998). *MIKE 11: A Microcomputer Based Modeling System for Rivers and Channels.* Reference manual of the Danish Hydraulic Institute, Hoersholm, Denmark.

El-Sadek, A., (2007). Upscaling field scale hydrology and water quality modeling to catchment scale. *Water Resour. Manage., 21*, 149–169.

Englande, A. J., Krenkel, P., & Shamas, J., (2015). Wastewater treatment & water reclamation. *Reference Module in Earth Syst. Environ. Sci.*, 1–32.

European Commission Common Implementation Strategy for the Water Framework Directive. (2003). *Guidance Document No. 7, Monitoring Under the Water Framework Directive*. The Directorate General Environment of the European Commission, Brussels.

GayatriDevi, K., Ganasri, D. P., & Dwarakish, G. S., (2015). Review on hydrological models. International conference on water resources, costal and water engineering. *Science Report, Aquatic Procedia, 4*, 1001–1007.

GCPC (Gujarat Cleaner Production Centre)-ENVIS Centre, (2016). *Advanced Wastewater Treatment Technologies,* 1–97.

Hamilton, D. P., & Schladow, S. G., (1997). Prediction of water quality in lake sand reservoirs. Part I : Model description. *Ecol. Modeling, 96*(1), 91–110.

Hamrick, J. M., (1996). *User's Manual for the Environmental Fluid Dynamics Computer Code*. Department of Physical Sciences, School of Marine Science, Virginia Institute of Marine Science, College of William and Mary.

Hipsey, M. R., Romero, J. R., & Antenucci, J. P., (2006). *Computational Aquatic Ecosystem Dynamics Model: CAEDYM v2* (p. 90). Contract Research Group, Centre for Water Research, University of Western Australia.

Ierodiaconou, D., Laurenson, L., & Leblanc, M., (2005). The consequences of land use change on nutrient exports: A regional scale assessment in south-west Victoria, Australia. *J. Environ. Manag., 74*, 305–316.

International Food Policy Research Institute & Veolia, (2015). *The Murky Future of Global Water Quality: New Global Study Projects Rapid Deterioration in Water Quality* . Washington, DC:IFPRI and Veolia Water North America.

Johnes, P. J., & Butterfield, D., (2002). Landscape, regional and global estimates of nitrogen flux from land to sea: Errors and uncertainties. *Biogeochem, 57, 58*, 429–476.

Johnes, P. J., (1996). Evaluation and management of the impact of land use on the nitrogen and phosphorus load delivered to surface waters: The export coefficient modeling approach. *J. Hydrol., 183*, 323–349.

Jordão, E. P., & Pessoa, C. A., (2009). *Tratamento de Esgotos Domésticos* (5th edn.). Rio de Janeiro: *ABES.*

Kannel, P. R., Kanel, S. R., Lee, S., Lee, Y. S., & Gan, T. Y., (2011). A review of public domain water quality models for simulating dissolved oxygen in rivers and streams. *Environ. Modeling Assess.*, 16, 183–204.

Karr, J. R., (1990). Bioassessment and non-point source pollution: An overview. In: *Second National Symposium on Water Quality Assessment* (pp. 4–18). Washington, D.C, EPA Office of Water.

Liu, X., & Huang, W., (2009). Modeling sediment resuspension and transport induced by storm wind in Apalachicola Bay, USA. *Environmental Modeling & Software, 24*(11), 1302–1313.

Martin, J., Wool, T., & Olson, R., (2002). *A Dynamic One-Dimensional Model of Hydrodynamics and Water Quality (EPD-RIV1), Version 1.0.* Model Documentation and User Manual, Atlanta, GA: Georgia Environmental Protection Division.

Massey, F. C., Skaggs, R. W., & Sneed, R. E., (1983). Energy and water requirements for subirrigation vs. sprinkler irrigation. *Transactions of the ASAE, 26*(1), 126–133.

Neitsch, S., Arnold, J., Kiniry, J. E. A., Srinivasan, R., & Williams, J., (2002). *Soil and Water Assessment Tool User's Manual Version 2000* (p. 202). GSWRL Report.

NRC (National Research Council), (2001). *Assessing the TMDL Approach to Water Quality Management Committee to Assess the Scientific Basis of the Total Maximum Daily Load Approach to Water Pollution Reduction.* Washington, D.C., Water Science and Technology Board, National Academy Press.

Olowe, K. O., & Kumaraswamy, M. V., (2018). Assessment of some existing water quality models. *Nat. Environ. Pollut. Tech., 17*(3), 939–948.

Omernik, J. M., (1976). *The Influence of Land Use on Stream Nutrient Levels.* USEPA Ecological Research Series, EPA-60013-76-014. US Environmental Protection Agency, Corvallis, OR.

Paliwal, R., Sharma, P., & Kansal, A., (2007). Water quality modeling of the river Yamuna (India) usingQUAL2E- UNCAS. *J. Environ. Manag., 83*(2), 131–144.

Palmeri, L., Bendroricchio, G., & Artioli, Y., (2005). Modeling nutrient emissions from river systems and loads to the coastal zone: Po River case study, Italy. *Ecol. Modeling, 184*, 37–53.

Park, R. A., Clough, J. S., & Wellman, M. C., (2008). AQUATOX: Modeling environmental fate and ecological effects in aquatic ecosystems. *Ecol. Modeling, 213*, 1–15.

Park, R., & Clough, J., (2004). *AQUATOX (Release 2) Modeling Environmental Fate and Ecological Effects in Aquatic Ecosystems* (Vol. 2). Technical Documentation; US Environmental Protection Agency, Washington, DC.

Quinn, P. F., & Beven, K. J., (1993). Spatial and temporal predictions of soil moisture dynamics, runoff, variable source areas and evapotranspiration for Plynlimon, mid-Wales. *Hydrol. Process., 7*, 425–448.

Quinn, P. F., Anthony, S., & Lord, E., (1999). Basin-scale nitrate modeling using a minimum information requirement approach. In: Trudgill, S., Walling, D., & Webb, B., (eds.). *Water Quality: Processes and Policy* (pp. 101–117). Wiley: Chichester.

Quinn, P. F., Hewett, C. J. M., & Dayawansa, N. D. K., (2008). TOPCAT-NP: A minimum information requirement model for simulation of flow and nutrient transport from agricultural systems. *Hydrol. Process., 22*, 2565–2580.

Shah, A. I., Din, D. M. U., Bhat, R. A., Singh, J. P., Singh, K., & Bhat, S. A., (2020). Prospectives and challenges of wastewater treatment technologies to combat contaminants of emerging concerns. *Ecol. Eng., 152*, 105882 https://doi.org/10.1016/j.ecoleng.2020.105882.

Sharma, D., & Kansal, A., (2013). Assessment of river quality models: A review. *Rev. Environ. Sci. Biotechnol., 12*, 285–311.

Shoemaker, C. A., Haith, D. A., & Benaman, J., (2005). Calibration and validation of soil and water assessment tool on an agricultural watershed in upstate New York. *Journal of Hydrologic Engineering, 10*, 363–374.

Skaggs, R. W., (1981). DRAINMOD reference report. *Methods for Design and Evaluation of Drainage-Water Management Systems for Soils with High Water Tables* (pp. 329). USDA- SCS,.

Skaggs, R. W., Fausey, N. R., & Nolte, B. H., (1981). Water management evaluation for Northcentral Ohio. *Transactions of the ASAE, 24*(4), 922–928.

Srinivasin, R., Muttiah, R. S., & Arnold, J. G., (1998). Large area modeling and assessment part I-model development. *J. Am. Water Resour. Assoc.*, (34), 73–89.

Streeter, H. W., & Phelps, E. B., (1925). *A Study of the Pollution and Natural Purification of the Ohio River.* United States Public Health Service, U.S. Department of Health, Education and Welfare.

Streeter, H. W., & Phelps, E. B., (1985). A study of the pollution and natural purification of the Orhio river. III. *Factors Concerned in the Phenomena of Oxidation and Reaeration, Public Health Bulletin no. 146.*

Tilley, E., Ulrich, L., Luethi, C., Reymond, P., & Zurbruegg, C., (2014). *Compendium of Sanitation Systems and Technologies* (2nd revised edn.). Duebendorf, Switzerland: Swiss Federal Institute of Aquatic Science and Technology (Eawag).

Ting, Y. F., & Praveena, S. M., (2017). Sources, mechanisms, and fate of steroid estrogens in wastewater treatment plants: A mini-review. *Environ. Monitoring Assess, 189*, 178.

Tolson, B. A., & Shoemaker, C. A., (2007). Cannonsville reservoir watershed SWAT2000 model development, calibration and validation. *J. Hydrol., 337*, 68–86.

Tsakiris, G., & Alexakis, D., (2012). Water quality models: An overview. *Eur. Water, 37*, 33–46.

Veenstra, S., Alaerts, G. J., & Bijlsma, M., (1997). *Water Pollution Control- A Guide to the Use of Water Quality Management Principles.*

Venohr, M., Hirt, U., Hofmann, J., Opitz, D., Gericke, A., Wetzig, A., Ortelbach, K., et al., (2009). *The Model System MONERIS Version 2.14.1vba Manual.* Leibniz Institute of Freshwater Ecology and Inland Fisheries in the Forschungsverbund Berlin E.V., Müggelseedamm 310, D-12587 Berlin, Germany.

Vianna, M. R., de Melo, G. C. B., & Neto, M. R. V. (2012). Wastewater Treatment In Trickling Filters Using Luffa cyllindrica As Biofilm Supporting Medium. Journal of Urban and Environmental Engineering, 6(2), 57–66. http://www.jstor.org/stable/26203371

Warn, A., (1987). SIMCAT-A catchment simulation model for planning investment for river quality. *Systems Analysis in Water Quality Management* (pp. 211–218). Pergamon Press, New York.

Warn, T., (2007). *SIMCAT 10.1-A Guide and Reference for Users* (p. 210). Environment Agency.

Warn, T., (2010). *SIMCAT 11.5 A Guide and Reference for Users.* Environment Agency.

Whitehead, P. G., Barbour, E., Futter, M. N., Sarkar, S., Rodda, H., Caser, J., Butterfield, D., et al., (2015). *Impact of Climate Change and Socio-Economic Scenarios on Flow and Water Quality of the Ganges, Brahmaputra and Meghna (GBM) River System: Low Flow and Flood Statistics.*

Whitehead, P. G., Wilson, E. J., Butterfield, D., & Seed, K., (1998). A semi – distributed integrated nitrogen model for multiple source assessment in catchments (INCA): Part I model structure and process equations. *Sci. Total Environ.,* 547–558.

Xie, R., Wu, D., Yan, Y., & Zhou, H., (2010). Fine silt particle path line of dredging sediment in the Yangtze River deep water navigation channel based on EFDC model. *J. Hydrodynamics., 22*(6), 760–772.

Yajima, H., & Choi, J., (2013). Changes in phytoplankton biomass due to diversion of an inflow into the Urayama Reservoir. *Ecol. Eng., 58*, 180–191.

RELEVANT WEBSITES

https://aosts.com/types-wastewater-screening/ (accessed on 7 July 2022).

https://water.mecc.edu/courses/ENV195WWI/Lesson2_print.htm (accessed on 7 July 2022).

https://www.hydrofluxindustrial.com.au/epco-australia/package-sewage-plants/configurations/trickling-filters/ (accessed on 7 July 2022).

https://www.waterpathogens.org/book/activated-sludge (accessed on 7 July 2022).

CHAPTER 6

Riparian Wetlands and Water Quality

SYED ROUHULLAH ALI and MAHRUKH

College of Agricultural Engineering and Technology, SKUAST–K, Jammu and Kashmir – 190025, India,
E-mail: agasyedrouhullah@gmail.com (S. R. Ali)

ABSTRACT

Riparian areas exist as strips of vegetative cover adjacent to rivers, streams, wetlands, lakes, reservoirs, and other freshwater water bodies that affect and are impacted by the presence of water. This vegetation contributes to specific ecosystems which perform a wide range of ecological functions. Together the human and biodiversity values are crucial for the benefits of riparian/wetland habitats. The nearness of the river also shows that riparian areas provide a significant effect upon the quality of the water. Riparian vegetation impedes drainage from built lands, the removal of toxins, and the significant improvements in water quality that enters streams which results in improved water quality, specifically, by controlling non-point source contaminants. Except where human activity has occurred, riparian zones can be located next to most of the water sources. Human impacts vary from changes in riparian and river hydrology and geomorphic structural modification for loss of riparian vegetation. That is all why effective efforts ought to be made to preserve or to recover riparian buffers that are present next to streams by comprehensive watershed approaches.

Advanced Technologies for Water Quality Treatment and Management. Mehraj U. Din Dar, PhD, Aamir Ishaq Shah, MTech, Shakeel Ahmad Bhat, PhD & Syed Rouhullah Ali, PhD (Eds.)

6.1 INTRODUCTION

Riparian zones and wetland habitats are vulnerable physical or vegetative environments that are formed through interaction with surface or subsurface water (Leonard et al., 1992). Riparian areas are generally known as the lands that surround streams, ponds, and reservoirs. An essential element of the arrangement and functioning of riparian regions is this interaction with adjacent waters. The riparian habitats are defined by National Water Act (1998) as: "Riparian habitat is described as the physical structure and vegetation of areas adjacent to a watercourse that are typically characterized by alluvial soils that are inundated or flooded to a degree and with a frequency sufficient to support vegetation of species with a composition and physical structure distinct from that of adjacent land areas." These areas are characterized by numerous hydrology, soil, and biota transformations and occupy the gap between land and water. Riparian regions have a high-water table and are prone to erosion and intrusion from the surrounding water body on a regular basis. Certain scientists (Gregory, 1997), defined riparian regions as "ecotones" or the transition zones between aquatic and terrestrial environments through some (Brinson et al., 1981) also have accepted riparian communities as units of landscape that contain a variety of divisions that stretch from water to upland surroundings. Ilhardt et al. (2000) define riparian regions by way of "three-dimensional ecotones of activity that involve terrestrial and aquatic habitats, spreading down into the soil, over the vegetation, over the floodplain, up surrounding slopes that drain into the river, laterally into the terrestrial environment, and along the waterway at variable distance."

Wetlands perform critical biological and environmental roles such as flow attenuation, sediment, and nitrogen preservation, and nature conservation. Wetlands are distinguished by a number of characteristics, the most prominent of which are the appearance of water around or close to the surface, distinct hydromorphic soil, and vegetative cover suited to or resistant to soils that are saturated. At lower elevations, the water table is generally at or above the surface, or the ground is surrounded by shallow water not deeper than 6.6 feet. Wetlands' importance has increasingly been recognized, and numerous wetland management programs have been developed (Junk et al., 2013). Unfortunately, farming operations and urbanization have resulted in the loss of the bulk of wetlands in settlements (Douglas et al., 1994; Dugan, 1992; Johnson, 2013). Wetlands conserve and maintain water resources. They act like giant sponges, holding onto water amid flooding and releasing this throughout dry spells. Wetlands help to mitigate soil erosion and flood destruction by controlling water levels during floods (Kumar et al., 2016; Bhat et al., 2019). They also recharge groundwater supplies and eliminate contaminants out from the water. Since they are natural filters, they aid in the purification of water by capturing a variety of contaminants such as heavy metals, sediments, and disease-causing species. Riparian wetland ecosystems are distinguished by high species diversity, density, and sustainability of both plant and animal species. Through nutrient, energy and organism transfer, the marine, riparian, and adjacent upland areas are constantly interacting (Meehan et al., 1997; Odum, 1979; Cummins, 1980; Franklin et al., 1981; Swanson et al., 1981).

Riparian areas have typically been among the most productive areas on the planet, as well as valuable environmental resources, since they offer a variety of ecological assets, such as a high density of different species of plants as well as wildlife habitat (Bentrup, 1998). Wetlands and riparian areas play important roles in water quality improvement, marine ecosystems, stream filtering, shoreline stability, flood attenuation, and groundwater exchange. Plant communities in riparian regions perform numerous functions in the absence of human alteration, including bank stability due to roots, accumulation of sediments on river banks at periods of overbank flow, interstitial flow of sediments, and abundant wood supplies, all of which have a direct impact on channel formation and ecological features. Substantial nutrient and contaminant retention capacities are characteristics of riparian wetlands (Fisher and Acreman, 2004) that are defined as the transitional zones between the unsaturated zone, groundwater, and surface water (Casey et al., 2001; Zhao et al., 2009). Also, perhaps because of their vicinity to water sources and their physical, biological, and chemical characteristics they are amongst the most susceptible to erosion. Although sediment deposition is normal river processes, heavy floods along with human activities can accelerate the erosion, which could result in severe environmental issues. Soil erosion in streamlines and on rivers will lead, for example, to instability and failure in streamlines (Chu-Agor et al., 2008, 2009; Daly et al., 2015; Bhat et al., 2017; Ul-Zaman et al., 2018) as well as increased load of sediment into the streams (Fox et al., 2016). Nevertheless, utilizing naturally occurring traces of radionuclides,

Wilson et al. (2008) observed that 48% of suspended load was due to the streambank erosion.

Riparian functions and water quality are intrinsic to hydrology because the water distribution canals grow throughout time dependent on ample water or saturated soils, depending largely on the conditions underlying atmosphere, geomorphic location, and stable vegetative growth (Winward, 2000; Rosgen, 2006). Hydrology affects the consistency of water by affecting chemical composition, thermal mass, and aquatic habitats, not just the effect on soil, flooding, or deposition but also the profile, pattern, and size of channels (Bilotta and Brazier, 2009; Van Vliet and Zwolsman, 2008). Diverse studies have identified the relevance of storm events for solvent exportation (Ulen, 1995; Neal et al., 2006; Mitchell et al., 2006; Raymond and Saiers, 2010). Particularly for smaller catchments, there is a considerable annual flux during individual storm events and a change in quality of the surface water within a few hours (Kirchner, 2003). Systematic transfers of fluids during storm flux cycles, such as variations on water distribution pathways, are also attributed to hydrological processes (Christophersen et al., 1982; Davies et al., 1992; Hagedorn et al., 2000; Lyon et al., 2011). The latter may be attributed to fluctuations in the elevation of the riparian wetland water table, resulting in a variety of runoff events (Emmett et al., 1994; Bechtold et al., 2003; Inamdar et al., 2008; Kerr et al., 2008). Prior and Johnes (2002); Hooper (2008); Piatek et al. (2009); and O'Brien et al. (2013) found that processes in riparian wetlands frequently superimposed the effect of the catchment's hill slope region on surface water content. Biogeochemists, in contrast to hydrologists, research wetland processes over weeks or months and, except for shifting groundwater levels, prefer to neglect changing hydrological patterns and boundary fluxes. Those trials concentrate on internal processes and are not related to output streams through the quality of the discharging stream (Knorr et al., 2009; Reiche et al., 2009; Pennington and Watmough, 2015). This will probably affect significantly the quality of water sources by changing biogeochemical processes in riparian ecosystems, e.g., as extreme weather events (extended dry spells, severe rainfall) become increasingly previewed by changes in climate (Emmett et al., 1994; Kull et al., 2008; Szkokan-Emilson, 2013; Watmough and Orlovskaya, 2015; Dar et al., 2017, 2020).

Diverse studies have clarified the interplay of hydrological and biogeochemical processes that have an impact upon the export of solutes from rivers (Piatek et al., 2009; Vidon et al., 2014; Arnold et al., 2015), which comprise of fluctuation in seasonal hydrologic events (Kerr et al., 2008), antecedent moisture conditions (Inamdar et al., 2009; Mitchell et al., 2006), climatic conditions (Kull et al., 2008), factors related to seasons (Muller and Tankéré-Muller, 2012), variations in the stream flow routes (Christophersen et al., 1982; Mitchell et al., 2006; Kerr et al., 2008), accessible solute concentration present in the wetland (Piatek et al., 2009), dilution of solutes (Inamdar et al., 2009), redox processes (Kerr et al., 2008), and the flushing of solutes after drought periods (Emmett et al., 1994; Szkokan-Emilson et al., 2013). Seasonal variations of fluctuations caused by flooding of stream water chemicals using a probabilistic clustering process studied by Aubert et al. (2013). Similarly, the necessity to recognize temporal changes in the end-member chemistry as a function of catchment wetness for improved characterization of catchment flow-paths and mixing responses was highlighted by Inamdar et al. (2013). Knorr (2012) emphasized the importance of both hydrological and redox conditions in the wetlands for solute export in the catchment of this study.

The concerns regarding water quality and ecosystem services reductions are among the many factors that environmental science, management, and decision-makers are seriously concerned about streambank and riparian degradation. The water quality functions are affected by the position of a wetland within the terrain affects. Many instances have been widely analyzed to determine how a specific wetland on a landscape scale would influence the water quality (Whigham et al., 1988). Whigham et al. (1988) consider wetlands that surround first-order streams to be effective in extracting nitrate from groundwater and sediments from the streams. However, for phosphor extraction, they are not shown to be much effective. Because of the lower proportion of water coming into contact with wetlands, the sediments and nutrients were found less successfully to be removed from the stream downstream of the first order streams (Whigham et al., 1988). Also, there are reports that suggest the portion of wetlands or the riparian environments immediately under the source of non-point emissions are the most efficient filters (Lowrance et al., 1983; Cooper et al., 1986; Phillips, 1989). Hydrological modeling based on Watersheds are being more and more developed and implemented to investigate wetland quantity and water quality impacts and agriculture practices (Fossey et al., 2015; Mahdizadeh et al., 2015; Valipour, 2015). Yang et al. (2010) have implemented HEW-based SWAT model to characterize flaws on an isolated Canadian prairie watershed and have shown that conservation of wetlands can provide substantial environmental advantages for reducing sediments and nutrients. Applying wetland modeling to a Michigan watershed, Martineze-Martinezet al. (2014) demonstrated that growing wetlands had higher flood absorption than a rising depth of wetlands. In order to analyze various wetland regeneration scenarios for the removal of sediments in a Michigan, Martineze-Martinez et al. (2015) had created a SWAT pairing and a SUSTAIN field scale model (Alvi et al., 2009). Liu et al. (2008) have specifically established a module on riparian wetland in the SWAT for the evaluation of riparian wetlands and sediments formed from upland drainages and routed through riparian wetlands. These studies have shown the value of modeling wetlands in characterizing wetland systems, quantifying magnitudes of wetland hydrology effects, and considering spatial changes in wetland effects for restoration scenarios.

Intact ecological riparian zones conserving, and recycling nutrients change local microclimate and sustain large networks of food, contributing to a wide variety of fish and other fauna. Typically, the ecological buffers within uplands and surrounding water sources are naturally wetlands and riparian regions. Wetlands and riparian areas can play a crucial role in reducing NPS emissions by surface river intercepting, subsurface flow, and other drainage flows. The aim of their work in improving water quality is to filter, remove, convert, and store contaminant products such as sediments, arsenic, ammonia, and some other heavy metals. Wetlands and riparian areas also shield the environment from the effects of pollutants and discourage toxins from entering water. The importance of the collection, removal, transformation, and preservation of pollutants such as sediments, arsenic, ammonia, and some heavy metals is part of improving the water quality. Wetlands and riparian areas also avoid the entrance into waters of pollutants and safeguard the water from contaminant effects. Riparian woods have also been found to minimize instream emissions during flooding (Karr and Gorman, 1975; Kleiss et al., 1989).

In increasingly advanced urban areas, building, filling, channeling or some other major modificatory measures can almost destroy wetlands and riparian ecosystems. The intervention and the connection of dams inter-basin diversion, and irrigation of hydrological systems affect the biophysical characteristics of riparian ecosystems. For agricultural regions, wetland, and riparian zones might be disturbed by misuse of the pasture field and also by the loss of natural vegetative habitat and substitution with annual or perennial cover. The elimination of the vegetation on the riverside not just eliminates the holding impact of root on the ground, but also reduces bank hydraulic strength and increases flow speeds on the bank (Sedell et al., 1991). Additionally, there may have been significant hydrological alterations to accelerate the drainage of farmland. Other significant impacts can result from a variety of activities like construction of highways, surface mining, dredged material build-up, and harbor and marina excavations. These kinds of practices will minimize or remove wetland and riparian roles that improve the quality of water and can worsen the issues with NPS. Modifications that would degrade its current functionality would therefore inhibit the functioning of wetlands and riparian zones. Degraded wetlands and riparian zones should also be preserved in order to act as potential in reducing the NPS emissions. The erosion process can be accelerated with extreme floods and human disturbance and cause significant environmental issues. Shields et al. (1995), for example, estimated that about 480,000 km of degraded stream bank deposited about 450 million tons of sediment annually in the U.S. The effect of non-point sources of pollutants by catching overland flood nutrients and sediments (Bentrup, 1998) and reducing banks' susceptibility to erosion decreases with riparian plant buffers. Wetland conservation, water quality protection, the protection of threatened or endangered species and flood hazard control are direct concern for the protection and restoration of degraded riparian ecosystems. Hence, steps to restore riparian lands are also urgently necessary.

6.2 IDENTIFICATION AND DELINEATION OF RIPARIAN AREAS

Riparian zones occupy a number of regions in different geomorphologic conditions (Ffolliott et al., 2004). These kinds of areas exist, in the vast regions with massive meandering waterways, mountainous terrain with small and deep ravine or valleys, to lowlands Riparian areas must be described and contrasted by other adjacent land and water components. Riparian areas are described also as "zones of influence" that exist within aquatic and terrestrial ecosystems. Riparian regions usually contain a portion or more of the wetlands, but they also include portions of adjacent upland and aquatic ecosystems. Figures 6.1 and 6.2 illustrates a "generic" informative riparian impact zone.

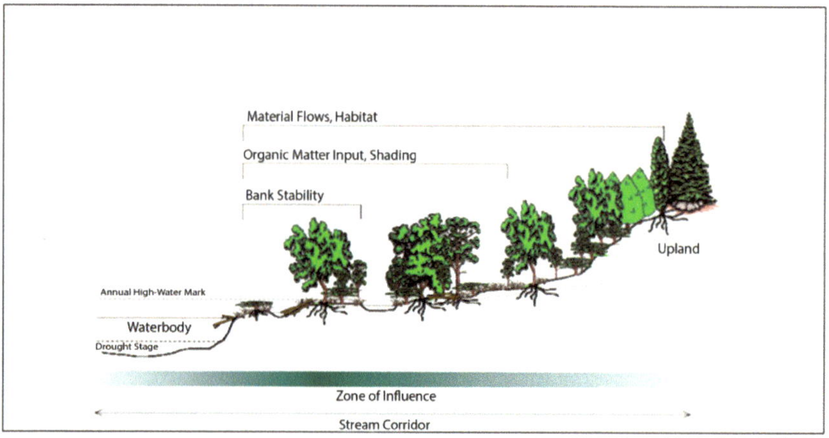

FIGURE 6.1 A characteristic riparian zone diagram that demonstrates influences on aquatic and upland areas.
Source: Adapted from: Ilhardt et al. (2000).

Areas of the riparia are distinguished from aquatic ecosystems, wetlands, and uplands on the basis of the same characteristics—soils, vegetation, and hydrology. Riparian habitats are usually populated by grasslands, trees and herbaceous plants as compared to aquatic environments in which these vegetation species are absent. Given the apparent distinction in the moisture regimes between riparian areas and aquatic habitats and their depth of water, a borderline that is reliable for separation cannot be established. This may define the aquatic boundaries of riparian areas where permanent water begins.

The zonal orientation and location in the landscape are essential characteristics of riparian regions. In contrast with wetlands, riparian zones from some perspectives are much wider and more restrictive. They are larger, and they include not just portions of the wetlands but also parts of areas that are not vegetated. Riparian zones are constrained rather than wetlands, as they are confined to the unique geomorphic conditions of the rivers, lakes, and marine habitats. For example, large peatlands cannot be referred to as riparian areas just because almost all of these deprive of linear landscape features, distinctive zones, and water body adjacency. Forests in riparian areas continue to exist in arid environments, with inadequate moisture to facilitate upland forestry (Brinson, 1990), and with distinctive borders as compared to uplands that receive precipitation, both adjacent uplands and aquatic environments may provide water as their primary or sole water supply to riparian regions. Water takes the course of groundwater discharge, a surface and shallow sub-surface flows from the uplands into riparian areas. Contrary to uplands, riparian floods are important agents for the geomorphic transition, the degradation of soils and sediments from certain riparian sites and their deposit in several other areas to establish new geomorphic landscapes. A basic distinction between riparian and upland areas is the lateral movement of resources from beyond the floodplains to riparian areas. Riparian zones are comparatively shallow where the water table is depleted by some trees and shrubs that drain water directly from flooded areas which typically cannot occur in upland regions (Robinson, 1958). For arid landscape at lower altitudes, riparian can only be the place of willow and cottonwoods and the only place in which mesquite grows intensely (Stromberg et al., 1996).

Riparian areas have their own collection of markers that can be used to delineate them when looking for their existence. The following factors must be considered during the delineation process (Department of Water Affairs and Forestry, 2005):

- Watercourse-related topography;
- Vegetation; and
- Alluvial and sediment deposits.

The final delineation of the riparian zones is supported by a range of information sources, including topographical maps, aerial photos and videos and eco-regions (predictive capability).

Normally, riparian zones exist around a river or a stream flooded area. The edge of a macro channel bank is a good predictor of the riparian outer edge. The riparian region is mostly determined by plant indicators. The external border of a riparian region ought to be adjacent to the waterway and identified by vegetation as the zone where distinctive changes occur in the composition of the species relative to the adjacent terrestrial area and in physical structure such as vigor or strength of growth of species similar to those of adjacent territory (Figure 6.2). These variations between riparian and terrestrial plants are largely due to a higher availability of water for species adjacent to water bodies than for those that grow at a farther distance. Riparian regions have alluvial deposits mostly, though not necessarily. While the occurrence of alluvial soils cannot always be used to reliably demarcate riparian areas as the main predictor the topography and vegetation may also be verified. With alluvial deposits observable from the sky, aerial photographs and videos can help determine the deposits' scale and the plant line that indicates a variation in the distribution of the species or robust growth. The distribution of plant species in riparian areas differs amongst regions depending upon climatic and geological conditions. The identification of species varieties that ought to grow in riparian zones in a specific region of the world is facilitated by using established eco-regional classification schemes.

FIGURE 6.2 Vegetation line between riparian and terrestrial vegetation.
Source: Department of Water Affairs and Forestry (2005).

6.3 IMPORTANCE OF RIPARIAN AREAS

Riparian area fulfills a number of socially valuable purposes, including the conservation and improvement of water supplies and the facilitation of habitats for fauna and flora. These regions have vital connections to aquatic and terrestrial habitats (Elmore, 1992; Osborne and Kovacic, 1993).

Because of their usage qualities, the riparian areas are particularly important (Clary and Booth, 1993). Riparian regions perform several roles, including:

- Storing water and help in reducing floods;
- Stabilizing stream banks;

- Enhancement of water quality by sediments and nutrients trapping;
- Preserve the temperature of water for aquatic life;
- Provide birds and animals with shelter and food;
- Serve as a barrier between aquatic and adjoining land uses, etc.

The higher riparian vegetative density than neighboring uplands reduces the rate of overland or bank-overflow runoff, eliminating sediments and nutrients effectively (Correll, 1997). More density of vegetation leads to increased chemicals and nutrients being assimilated, immobilized, stored, and transformed (Schultz et al., 2000). As riparian regions are slowing down and flood waters disperse through banks, more water penetrates the ground, recharges groundwater, and reaches the underlying streams (Wissmar and Swanson, 1990; Elmore, 1992). A very common feature of riparian vegetation is the dense root density. The high root density improves protection, reduces soil erosion, and improves the rate of infiltration. Increased riparian infiltration rates may contribute to a major reduction in overland flow or off-bank flood volumes (Schultz et al., 2000). Sediments, chemicals, and nutrients that otherwise reach the stream are reduced by the riparian vegetation. Therefore, these areas can enhance water quality by reducing emissions from non-point sources. Wetlands and riparian areas typically grow between the highlands and the neighboring waters as the natural buffers. The depletion of such mechanisms enables NPS toxins to be more directly involved in water receipt. Sediment is removed, nutrients, and other contaminants are removed from the overland and low surface flow of riparian areas, and habitat values are created by streamside, creating coarse and fine particulates and wildlife cover and food.

6.4 EFFECT OF RIPARIAN AREAS ON WATER QUALITY

The collection, removal, transformation, and preservation of substances like sediment, nitrogen, ammonia, and some heavy metals are all part of riparian areas' role in enhancing water quality. As a result, wetland and riparian areas protect water from pollution by preventing pollutants from accessing the water. Wetland and riparian areas are efficient at eliminating dissolved solids, nutrients, and other toxins from upland drainage while also keeping the stream channel temperature stable. Riparian areas of natural plants are critical for maintaining water quality. Grassy and wooded riparian areas have been studied for their role in regulating the transport of sediments, nutrients, chemicals, pollutants, microorganisms, and other pollutants into waterways by using buffers in and along the riparian areas. The capacity of a wetland or riparian area to eliminate NPS contaminants from drainage flowing through the wetland or riparian region is referred to as the abatement feature of NPS contamination. Riparian wetlands perform two important NPS emission control functions: as a phosphorus sink and by denitrification, which converts nitrate into nitrogen gas. Wetlands and riparian areas are more efficient at reducing non-point source contamination because they are components of an integrated land management plan that includes nutrient, sediment, and soil erosion management. Furthermore, few researches suggest that riparian and wetland vegetation serve as a drain for nutrient uptake and storage (Richardson, 1988). This role, though, may be linked to the age of the wetland or riparian forest (Lowrance et al., 1983).

Numerous experiments have looked at the impact of wetlands on water availability and quality. Gilliam (1994) conducted a study on riparian wetlands and water quality for North Carolina Coastal Plain watershed, USA. The existence of vegetated riparian areas between human activities and small streams was found to be the most significant factor affecting the amount of sediment and nutrients entering streams from nonpoint sources in the Eastern and Southern United States. Several studies have shown that sediment and nitrate concentrations in water flowing across riparian areas have decreased by more than 90% (Chescheir et al., 1991; Haycock and Pinay, 1993). Swanson et al. (2017) summarized the interagency riparian proper functioning condition (PFC) evaluation for lotic riparian habitats and drew links between PFC and water quality features. By dissipating energy associated with high water flow, the PFC relationship of hydrology, vegetation, and soils/landforms affects water quality by minimizing vertical instability and lateral erosion while creating floodplains with trapped sediment and nutrients. Another research was undertaken in the Black River basin in southern Ontario, Canada, to establish SWAT-based wetland modeling to analyze the water quality impact of riparian wetland degradation and regeneration scenarios (Yang et al., 2016). The research findings revealed that for 100% riparian wetland depletion, sediment, total nitrogen (TN), and total phosphorus (TP) loadings increased by 251.0%, 260.5%, and 890.9%, respectively, while for 100% riparian wetland regeneration, they decreased by 34.5%, 28.33%, and 37.0%, respectively. The findings also showed that as the depletion of riparian wetlands rises, the resulting environmental damage worsens at a faster pace. The environmental effects of riparian wetland regeneration, on the other hand, rise at a slower pace as the restoration progresses. In the Lehstenbach catchment, an analysis was undertaken to further understand the observed pronounced temporal variability of stream water quality, i.e., to distinguish between biogeochemical and hydrological drivers (Weyer et al., 2018). Around 280 samples were collected in a creek that flowed through a shallow wetland. To distinguish stream water samples, researchers used cluster analysis (CA) based on the results of an Isomap analysis. In addition, the water stream sampler classification was contrasted with the renowned Schollerand Piper water sample classification (Piper, 1944; Schoeller, 1962). The conclusion was that bioactivity, interfering with dynamics of the water table in the riparian wetlands, was a major determinant for the quality of the stream and for solutions export. A thorough understanding of this dynamic is important for evaluation of the effects on water quality, nitrogen exports and carbon sequestration of the riparian wetlands of both climate and land use transition.

The most important physical, chemical, and biological fates and processes for transport linked to riparian areas are infiltration, absorption, filtration, adsorption, decay, and assimilation. Infiltration is a primary method for transfer of water and dissolved chemical substances and particles to the ground. Infiltration is also known as percolation. Infiltrating is one of the processes that can be easily quantified and that influence the success of the riparian area used for chemicals removal. Therefore, several built-in riparian buffers are built to optimize precipitation infiltration (Midwest Plan Service. Livestock Waste Facilities Handbook, 1985). Adsorption of clay and organic matter in soils is more efficient in trapping dissolved materials with positive charges, such as heavy metals, orthophosphorus, and some contaminants, while filtration is more effective in trapping larger soil particles, aggregates, and organic particulates (National Research Council, 2002). When conditions are favorable, that is, when nitrate-laden groundwater flows from areas which are both rich in organic matter and anaerobic, denitrification is very good at separating nitrate from subsurface flow. Under aerobic or unsaturated soil environments, denitrification can also occur in localized soil micro-sites with large concentrations of organic carbon particulate matter (Parkin, 1987). Schade et al. (2001)

recognized the value of denitrification over plant assimilation as a method for extracting nitrate from shallow groundwater. Organic carbon provided by riparian shrubs served as an energy source for denitrification in this case. Slack water ecosystems (i.e., shallow, and slowly flowing portions of surface water channels) and hyporheic zones may experience the same processes mentioned above as subsurface water moving through riparian soils. Denitrification, for example, removes nitrate from running water in hyporheic zones (Hinkle et al., 2001).

The length of time runoff and associated chemicals are stored in the riparian environment, which is largely a hydrological phenomenon, determines the degree to which these processes operate. Where overland flow and shallow subsurface flow are spread evenly across the riparian region, overall pollutant management is more efficient. When one progresses from ephemeral and first and second-order streams to broader, higher-order streams, the importance of biogeochemical transitions in terms of stopping toxins from accessing neighboring waters decreases. Before accessing the channel network, a larger portion of the current flows through riparian areas along low-order waterways, rendering their riparian areas more effective in removing toxins from runoff. If water quality is a top priority, regeneration of usable riparian areas along ephemeral and first and second-order streams should take precedence over broader, higher-order streams (Brinson, 1993; National Research Council, 2002).

6.4.1 EFFECT ON SURFACE WATER QUALITY

Before reaching the stream, drainage water from either agricultural or industrial uplands can travel through the riparian forest. The quality of the water joining the stream along its journey can be influenced by chemical, physical, and biological activities in the riparian environment. Runoff water from developed areas, such as farm fields, also absorbs elevated levels of nitrate-nitrogen (NO_3^-–N) and phosphorus, which are important nutrients for growing crops but toxic to humans and animals in high concentrations. Furthermore, excess nutrients in receiving waters will promote algae formation, reducing the oxygen supply required for fishing and disrupting the aquatic food chain (Evans et al., 1996). When farm runoff water was drained into a wetland instead of released directly into a stream, plant nutrient and sediment concentrations in the water were decreased by 30% to 100% (Chescheir et al., 1991). The smaller the quantities of contaminants that entered the stream when the drainage water remained in the wetland for a longer time. Only a small amount of drained water, as well as all contaminants were left in the wetland. The amount of nutrients and sediments reaching the sea also increased as runoff volume and velocity grew. Researchers conclude that a large amount of nitrate nitrogen has been lost as it crosses the rivers. The nitrate-nitrogen content of the highlands river was determined to be nearly 85% per year, with a shift of 27 pounds of NO_3^-–N per acre on the farm's edge to about 4 pounds per acre at the edge of the wetland near streams (Cooper et al., 1987a). Most nitrate-nitrogen was lost when flow traveled through farm fields and small streams through the thin vegetation bands. Most investigators believe that denitrification and plant intake are the main mechanisms of the removal of NO_3^- in riparian areas. It was noticed that the NO_3/Cl^- ratio was greatly changed because the water poured through the riparian area from the agricultural region. Denitrification was concluded to be primarily responsible for removing NO_3 (Dar et al., 2020).

Riparian wetlands are able to extract a considerable number of sediments and phosphorus from the surface flow. The thick growth of vegetation typical at the edge of riparian areas in the field-forest delays the water flow on the soil and deposits much of the coarse sediments. As the water flows through the wetland into the higher order streams and finally to the flora, the composition of the deposited sediments becomes finer and cleaner. When it enters the floodplain swamp, the sediment layer is very thin and consists mostly of clay-sized fragments. The impact of sediment and phosphorus accumulation on land surface runoff was calculated by a report (Cooper et al., 1987b). The investigators find that much of the sediment had been deposited at the field edge in the riparian region (Figure 6.3). Around 85%–90% (after extrapolation of the data) of the water in wooded areas was trapped (Figure 6.4) and did never reach large streams.

Another research was conducted also to assess the efficacy of riparian filters in separating sediment from surface waters using ^{137}Cs (Lowrance et al., 1988). It was discovered that the riparian forest deposited more soil than the surrounding farm area lost. This disparity was due to deposition from the watershed's upstream sections (Lowrance et al., 1986). This group therefore claimed that riparian environments are effective and essential sediment traps based on this and their previous works. Similarly, it was discovered that 90% of the sediments were depleted in a forested wetland receiving pumped farm runoff (Chescheir et al., 1991). The withdrawal of P from drained agriculture runoff water necessitated sediment co-deposition (Chescheir et al., 1991). However, a study revealed that the soluble amounts in surface water that enters and leaves riparian areas remained unchanged (Peterjohn and Correll, 1984), implying that riparian buffers do a decent job at eliminating P bound to sediments but are largely inefficient at reducing soluble P (Gilliam, 1994).

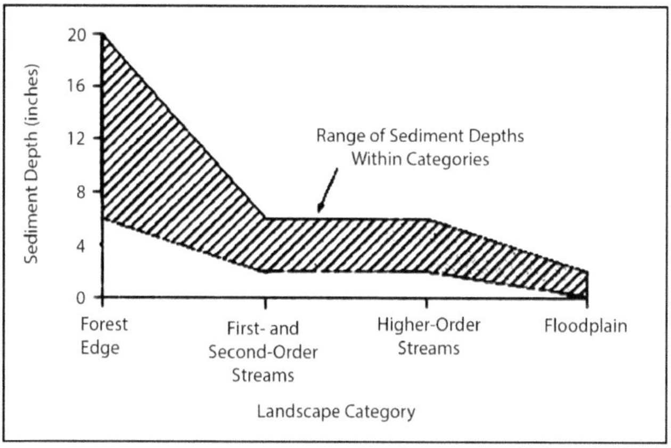

FIGURE 6.3 Depth of sediment accumulation in riparian wetlands.
Source: Cooper et al. (1987a, b).

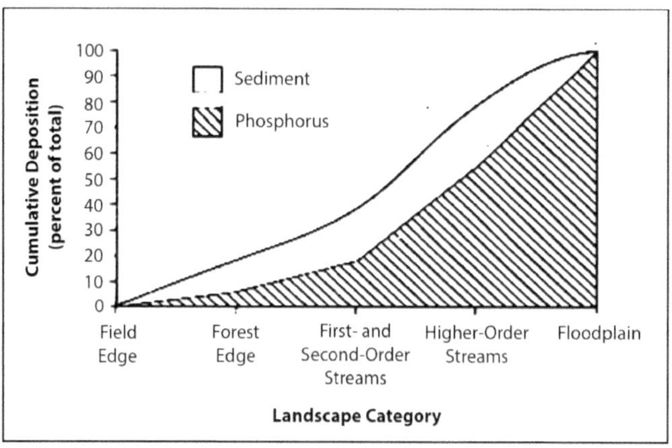

FIGURE 6.4 Cumulative deposition of sediments on base of riparian wetland (Cooper et al., 1987a); and phosphorus deposits and its spread in riparian wetlands (Cooper et al., 1987b).

6.4.2 EFFECT ON GROUNDWATER QUALITY

A high quality, but rarely significant in amount, water percolates from unfertilized wetland soils to existing groundwater supply. Wetlands are damp due to one or both factors: (i) they produce a huge amount of water from surface flow or precipitate or (ii) water is percolated naturally through the soils. Subsequently the amount of groundwater recharge is small when percolation is small (Evans et al., 1996). In the Coastal Plain of Northern Carolina, Jacobs and Gilliam (1985) noted that NO_3^-–N was reduced from approximately 15 mg/L to 1–2 mg/L in shallow waters flowing from agriculture to the small streams shown in Figure 6.5. This decrease occurred on the riparian region of the first 10 to 15 m.

Riparian wetlands, for example, corn fields or interstream receive a significant amount of drainage from steep slope areas. Whereas some contaminants are eliminated and trapped, groundwater has little time for refilling as this water flows rapidly across the soil surface on its way to the outlets. In addition, riparian regions present next to streams directly drain groundwater into them.

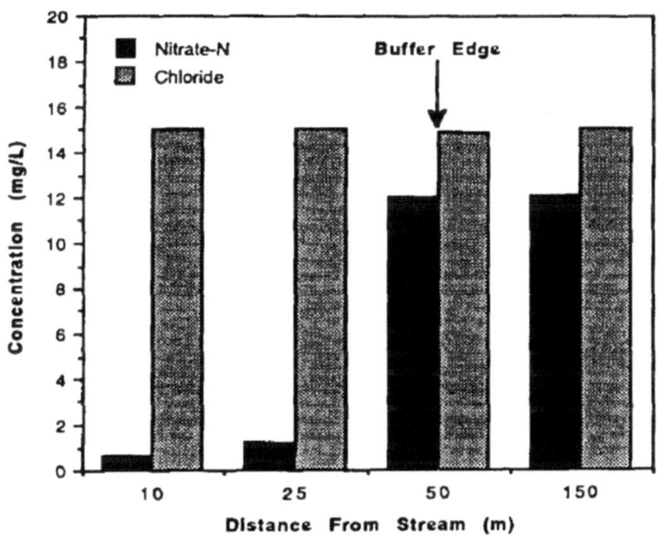

FIGURE 6.5 Concentrations of nitrate–nitrogen and chloride in shallow groundwater flowing from a cultivated coastal plain area to a small stream through a wooded riparian filter.

Source: Adapted from: Jacobs and Gilliam (1985).

6.5 MANAGEMENT OF RIPARIAN WETLANDS

Although wetland regions and riparian areas reduce pollution due to NPS, it is done under certain specific operating conditions. If hydrological modifications or NPS emissions surpass the normal assimilation capability of such systems, areas of the wetlands and riparian areas are disturbed and may be degraded. Deteriorated wetlands are less able to eliminate contaminants from non-point sources and to control flood water peak flows (Richardson and Davis, 1987; Bedford and Preston, 1988). A degraded wetland may also provide the surrounding water bodies with elevated amounts of sediment, nutrients, and other environmental contaminants, which are not a treatment, but a source of non-point contamination (Brinson, 1988). There must also be protection against alterations to wetlands and riparian regions that would disrupt their prevailing roles. Additionally, deteriorated wetland and riparian area shall, where appropriate, be reclaimed to facilitate a reduction in NPS emissions. The conservation projects seek to maintain current roles as part of the NPS for improvements in water quality in wetlands and riparian areas. The solution in general is to establish a series of operations that preserve wetland and riparian functions and avoid adverse effects on areas which have a reduction in the emissions caused by the NPS. A combination of programmatic and structural practices should protect the ecosystem and water quality functions of wetlands and riparian areas serving an NPS pollution abatement function. The ecology and water quality role of wetlands and riparian areas performing an NPS pollution reduction role should be protected by a balance of programmatic and structural activities.

Erosion of the horizontal channel (meanders), typical in the floodplain, which damages soil and riparian vegetation only to be deposited downstream. Floods intensify erosion and deposition, deluge vegetation on the banks of the river, and distort, destroy, ruin, or uproot vegetation. Mass flooding, sediment torrents and organic debris barriers are also related to floods (Swanson et al., 1976; Swanston, 1978, 1980). These activities disrupt riparian vegetation but may also develop new riparian or wetland conditions with the modification of the stream profiles. If unchecked, wildlife species can have major impacts in riparian and wetland habitat concentrated locations. An effective riparian planning strategy for protecting water quality and fishery and biodiversity resources is the maintenance of barriers or leave strips. Forest routes, if built along waterways or over waterways, limit wetland habitat and riparian productivity for many species of wildlife. Road building has the greatest and much more long-lasting effects on riparian areas than any other management or developmental work (Thomas et al., 1979). When grazing in riparian areas and wetlands is intense and persistent, it can diminish, alter, or remove the vegetation which supports the important habitats for wildlife (Kennedy, 1977; Platts, 1981; Storch, 1979).

Protection of uninterrupted riparian areas is extremely important, as conservation on both environmental and economic grounds can require much more than merely avoiding changes caused by humans as a management method. Measures to maintain intact habitats are often reasonably simple to introduce, are equally possible and less expensive than to restore damaged structures (NRC, 1992; Cairns, 1993). Protocols must be formulated in order to promote effective measures in the conservation of productive riparian/wetland habitats in response to the effects of drought as well. Alternatives of riparian or wetland monitoring must incorporate changes of grazing seasons and seasonal restrictions in order to safeguard the needs of flora, vulnerable watersheds, riverine, fauna, and other resources values. Restructuring is also a social and scientific challenge in deteriorated riparian areas. This means that restoration applies to both the recovery and the expected outcome of the degraded riparian areas. A primary goal of riparian restoration is to encourage natural pathways and connections among terrestrial, river, and aquatic environments on a self-sustaining basis. Initiatives such as the Conservation Reserve USDA Program and the Wetland Reserve Program focus on providing ways to restructure and preserve riparian regions and wetlands. The ecosystem restoration of Riparians may include attempts to modify or repair their properties or environments physically, chemically, scientifically, biologically or in the interest of the wetland habitats. Some of the riparian restoration approaches include control over exotic species, riparian-growing, cottonwood protection, proper use of fire, contouring, and stabilizing floodplains, closed-loops, optimized grazing methodologies, better education or even other forms of handling or management to enhance the healthy ecological systems. In addition, implementation of a habitat conservation plan in the riparian/aquatic/floodplain areas may be considered part of management programs.

6.6 CONCLUSION

Wetlands and riparian areas serve significant roles, such as feeding a range of birds, nesting materials sources, habitat for aquatic animals, and wildlife nurseries (Atcheson et al., 1979). Floodwater storage, erosion control, and groundwater recharge are other important purposes of the wetlands and riparian areas. Riparian areas protect water receiving from pollutant impacts by preventing pollutant intrusion into the receiving waters. Usually, wetlands, and riparian areas act as natural buffers between uplands and adjacent water bodies. Decline of such networks enables NPS contaminants to make an increasing contribution to the water supply. Over 50% of suspended particles enter wetlands and riparian zones and are removed.

Wetlands and riparian areas typically remove 60–75% of overall nitrogen contaminants from surface water as well as from groundwater. The removal of phosphorus ranges from 50% to 80% in wetlands and riparian areas. Therefore, a strong effort should be made for preserving wetland acting as a vegetated buffer alongside seasonal and intermittent channels or streams. Building, excavation, channeling, or other major changes may virtually destroy wetland and riparian regions for highly developed urban areas. Overexploitation of the grassland or loss of natural plants and substitution by annual or perennial cultivation may be affecting wetlands and riparian areas in agricultural areas. These are such practices that will degrade or eliminate wetland and riparian's role to enhance the quality of water and may exacerbate NPS problems. Riparian and wetland zones should be protected, as these areas are the most critical areas to preserve water quality and enable control of both the NSP and the corollary advantages of such natural aquatic environments.

KEYWORDS

- **cluster analysis**
- **ecological riparian**
- **hydrological systems**
- **proper functioning condition**
- **total nitrogen**
- **total phosphorus**

REFERENCES

Alvi, K., Zhen, J. X., Paul, S., & Rafi, T., (2009). *SUSTAIN: A Framework for Placement of Best Management Practices in Urban Watersheds to Protect Water Quality*. United States Environmental Protection Agency, Office of Research and Development, National Risk Management Research Laboratory, Water Supply and Water Resources Division, Cincinnati, OH. EPA/600/R-09/095.

Arnold, C., Ghezzehei, T. A., & Berhe, A. A., (2015). Decomposition of distinct organic matter pools is regulated by moisture status in structured wetland soils. *Soil Biol. Biochem., 81*, 28–37.

Atcheson, J., Conrad, E. T., Bailey, S. F. W., & Hughes, Jr. M., (1979). *Analysis of Selected Functional Characteristics of Wetlands*. Prepared for the U.S. Army Coastal Engineering Research Centre.

Aubert, A. H., Tavenard, R., Emonet, R., De Lavenne, A., Malinowski, S., Guyet, T., Quiniou, R., et al., (2013). Clustering flood events from water quality time series using latent Dirichlet allocation model. *Water Resour. Res., 49*(12), 8187–8199.

Bechtold, J. S., Edwards, R. T., & Naiman, R. J., (2003). Biotic versus hydrologic control over seasonal nitrate leaching in a floodplain forest. *Biogeochem., 63*, 53–72.

Bedford, B. L., & Preston, E. M., (1988). Developing the scientific basis for assessing cumulative effects of wetland loss and degradation on landscape functions: Status, perspectives, and prospects. *Environmental Management, 12*(5), 751–771.

Bentrup, G., (1998). *The Practical Streambank Bioengineering Guide*. USDA. Natural Resources Conservation Service.

Bhat, S., Dar, M. U., & Meena, R. M., (2019). Soil erosion and management strategies. *Sustainable Management of Soil and Environment*. doi: 10.1007/978-981-13-8832-3_3.

Bhat, S., et al., (2017). Soil erosion modeling using RUSLE & GIS on micro watershed of J&K. *Journal of Pharmacognosy and Phytochemistry, 6*(5), 838–842.

Bilotta, G., & Brazier, R., (2008). Understanding the influence of suspended solids on water quality and aquatic biota. *Water Research, 42*, 2849–2861.

Brinson, M. M., (1988). Strategies for assessing the cumulative effects of wetland alteration on water quality. *Environmental Management, 12*(5), 655–662.

Brinson, M. M., (1990). Riverine forests. In: Lugo, A. E., Brinson, M., & Brown, S., (eds.), *Forested Wetlands*. New York: Elsevier.

Brinson, M. M., (1993). Changes in the functioning of wetlands along environmental gradients. *Wetlands, 13*, 65–67.

Brinson, M. M., Swift, B. L., Plantico, R. C., & Barclay, J. S., (1981). *Riparian Ecosystems: Their Ecology and Status*. FWS/OBS–81/17. Kearneysville, WV: U.S. Fish and Wildlife Service.

Cairns, Jr. J., (1993). Is restoration ecology practical? *Restoration Ecology, 1*, 3–6.

Casey, R. E., & Klaine, S. J., (2001). Nutrient attenuation by a riparian wetland during natural and artificial runoff events. *J. Environ. Qual., 30*, 1720–1731.

Chescheir, G. M., Gilliam, J. W., Skaggs, R. W., & Broadhead, R. B., (1991). Nutrient and sediment removal in forested wetlands receiving pumped agricultural drainage water. *Wetlands, 11*(1), 87–103.

Christophersen, N., Seip, H. M., & Wright, R. F., (1982). A model for stream water chemistry at Birkenes, Norway. *Water Resour. Res., 18*, 977–996.

Chu-Agor, M., Fox, G. A., & Wilson, G. V., (2009). Empirical sediment transport function predicting seepage erosion undercutting for cohesive bank failure prediction. *J. Hydrol., 377*, 155–164.

Chu-Agor, M., Wilson, G. V., & Fox, G. A., (2008). Numerical modeling of bank instability by seepage erosion undercutting of layered streambanks. *J. Hydrol. Eng., 13*(12), 1133–1145.

Clary, W. P., & Booth, G. D., (1993). Early season utilization of mountain meadow pastures. *Journal of Range Management, 46*, 493–497.

Cooper, J. R., & Gilliam, J. W., (1987a). Phosphorus redistribution from cultivated fields into riparian areas. *Soil Sci. Soc. Am. J., 51*, 1600–1604.

Cooper, J. R., Gilliam, J. W., & Jacobs, T. C., (1986). Riparian areas as a control of nonpoint pollutants. In: Correll, D., (ed.), *Watershed Research Perspectives* (pp. 166–192). Smithsonian Institution Press, Washington, DC.

Cooper, J. R., Gilliam, J. W., Daniels, R. B., & Robarge, W. P., (1987b). Riparian areas as filters for agricultural sediment. *Soil Soc. Am. Proc., 51*, 416–420.

Correll, D. L., (1997). Buffer zones and water quality protection: General principles. In: Hancock, N. E., et al., (eds.), *Buffer Zones: Their Processes and Potential in Water Protection*. Quest Environmental. Hertfordshire, England.

Cummins, K. W., (1980). The multiple linkages of forests and streams. In: Waring, R. H., (ed.), *Forests: Fresh Perspectives from Ecosystem Analysis* (pp. 191–198). Corvallis, OR: Proc. 40th Annual Biol. Colloq; Oregon State University Press.

Daly, E. R., Miller, R. B., & Fox, G. A., (2015). Modeling streambank erosion and failure along protected and unprotected composite streambanks. *Adv. Water Resour., 81*, 114–127.

Dar, M. U. D., Bhat, S. A., Ali, & Shah, A. I., (2017). Modeling climate change impact; a study on different procedures and strategies: A review. *Int. J. Pure Appl. Biosci. 5*, 183–200.

Dar, M. U. D., Bhat, S. A., Meena, R. S., & Shah, A. I., (2019). Carbon footprint in eroded soils and its impact on soil health. *Soil Health Restoration and Management*, 1.

Dar, M. U. D., Shah, A. I., Ali, S., & Bhat, S. A., (2020). *Woodchip Bioreactors for Nitrate Removal in Agricultural Land Drainage: Agricultural Waste B*ook (pp. 99–118). Taylor & Franc.

Davies, T. D., Tranter, M., Wigington, P. J. Jr., Eshleman, K. N. (1992)., 'Acidic episodes' in surface waters in Europe. *J. Hydrol., 132*, 25–69.

Department of Water Affairs and Forestry, South Africa, (2005). *A Practical Field Procedure for Identification and Delineation of Wetlands and Riparian Areas* (p. 49). Document published by the Department of Water Affairs and Forestry.

Douglas, A. J., & Johnson, R. L., (1994). Drainage investment and wetlands loss: An analysis of the national resources inventory data. *J. Environ. Manag., 40*(4), 341–355.

Dugan, P., (1992). *Wetlands in Danger*. Mitchell Beazley London Ltd.

Elmore, W., (1992). Riparian responses to grazing practices. In: Naiman, R. J., (ed.), *Watershed Management* (pp. 442–457). Springer-Verlag. New York, NY.

Emmett, B. A., Hudson, J. A., Coward, P. A., & Reynolds, B., (1994). The impact of a riparian wetland on stream water quality in a recently afforested upland catchment. *J. Hydrol., 162*, 337–353.

Evans, R., Gilliam, J. W., & Lilly, J. P., (1996). *Wetland and Water Quality*. North Carolina Cooperative Extension Service. Publication Number: AG 473-7.

Ffolliott, P. F., Baker, M. B. Jr., DeBano, L. F., & Neary, D. G., (2004). Introduction. In: Baker, M. B., et al., (eds.), *Riparian Areas of the Southwestern United States: Hydrology, Ecology and Management* (pp. 1–9). CRC Press. Boca Raton, FL.

Fisher, J., & Acreman, M. C., (2004). Wetland nutrient removal: A review of the evidence. *Hydrol. Earth Syst. Sci., 8*(4), 673–685.

Fossey, M., Rousseau, A. N., Bensalma, F., Savary, S., & Royer, A., (2015). Integrating isolated and riparian wetland modules in the PHYSITEL/HYDROTEL modeling platform: Model performance and diagnosis. *Hydrol. Process.*, 10.1002/hyp.10534.

Fox, G. A., Sheshukov, A., Cruse, R., Kolar, R. L., & Guertault, L., (2016). Reservoir sedimentation and upstream sediment sources: Perspectives and future research needs on streambank and gully erosion. *Environ. Manag.*, http://dx.doi.org/10.1007/s00267-016-0671-9.

Franklin, J. F., Cromack, K. Jr., Denison, W., et al., (1981). *Ecological Characteristics of Old-Growth Douglas-fir* Forests (p. 48). Gen. Tech. Rep. PNW-118. Portland, OR: U.S. Department of Agriculture, Forest Service, Pacific Northwest Forest and Range Experiment Station.

Gilliam, J., (1994). Riparian wetlands and water quality. *Journal of Environmental Quality, 23*(5), 896–900.

Gregory, S. V., (1997). Riparian management in the 21st century. In: Kohm, K. A., & Franklin, J. F., (eds.), *Creating a Forestry for the 21st Century: The Science of Ecosystem Management* (pp. 69–85). Washington, DC, and Covelo, CA: Island Press.

Hagedorn, F., Schleppi, P., Waldner, P., & Flühler, H., (2000). Export of dissolved organic carbon and nitrogen from Gleysol dominated catchments – the significance of water flow paths. *Biogeochem., 50*, 137–161.

Haycock, N. E., & Pinay, G., (1993). Nitrate retention in grass and poplar vegetated buffer strips during the winter. *J. Environ. Qual., 22*, 273–278.

Hinkle, S. R., Duff, J. H., Triska, F. J., Laenen, A., Gates, E. B., Bencala, K. E., Wentz, D. A., & Silva, S. R., (2001). Linking hyporheic flow and nitrogen cycling near the Willamette River—A large river in Oregon, USA. *J. Hydrol., 244*, 157–180.

Hooper, R. P., (2001). Applying the scientific method to small catchment studies: A review of the Panola Mountain experience. *Hydrol. Process., 15*, 2039–2050.

Ilhardt, B. L., Verry, E. S., & Palik, B. J., (2000). Defining riparian areas. In: Verry, E. S., Hornbeck, J. W., & Dolloff, C. A., (eds.), *Riparian Management in Forests of the Continental Eastern United States* (p. 29). New York: Lewis Publishers.

Inamdar, S., Dhillon, G., Singh, S., Dutta, S., Levia, D., Scott, D., Mitchell, M., Van, S. J., & McHale, P., (2013). Temporal variation in end-member chemistry and its influence on runoff mixing patterns in a forested, piedmont catchment. *Water Resour. Res., 49*(4), 1828–1844.

Inamdar, S., Rupp, J., & Mitchell, M., (2008). Differences in dissolved organic carbon and nitrogen responses to storm-event and ground-water conditions in a forested, glaciated watershed in western New York. *J. Am. Water Resour. Assoc., 44*(6), 1458–1473.

Inamdar, S., Rupp, J., & Mitchell, M., (2009). Groundwater flushing of solutes at wetland and hillslope positions during storm events in a small glaciated catchment in western New York, USA. *Hydrol. Process., 23*, 1912–1926.

Jacobs, T. J., & Gilliam, J. W., (1985). Riparian losses of nitrate from agricultural drainage waters. *J. Environ. Qual., 14*, 472–478.

Johnson, C. A., (2013). Wetland losses due to row crop expansion in the Dakota prairie pothole region. *Wetlands, 33*(1), 175–182.

Junk, W. J., An, S., Finlayson, C. M., Gopal, B., Kvet, J., Mitchell, S. A., & Robarts, R. D., (2013). Current state of knowledge regarding the world's wetlands and their future under global climate change: A synthesis. *Aquat. Sci., 75*(1), 151–167.

Karr, J. R., & Gorman, O. T., (1975). *Effects of Land Treatment on the Aquatic Environment* (pp. 1–18). In U.S. Environmental Protection Agency Non-Point Source Pollution Seminar, Washington, DC. EPA 905/9-75-007.

Kennedy, C. E., (1977). Wildlife conflicts in riparian management: Water. In: Johnson, R. R., & Jones, D. A., (eds.), *Importance, Preservation and Management of Riparian Habitat; A Symposium* (pp. 52–58). Tucson, AZ. Gen. Tech. Rep. RM-43. Ft. Collins, CO: U.S. Department of Agriculture, Forest Service, Rocky Mountain Forest and Range Experiment Station.

Kerr, S. C., Shafer, M. M., Overdier, J., & Armstrong, D. E., (2008). Hydrologic and biogeochemical controls on trace element export from northern Wisconsin wetlands. *Biogeochem., 89*, 273–294.

Kirchner, J. W., (2003). A double paradox in catchment hydrology and geochemistry. *Hydrol. Process., 17*, 871–874.

Kleiss, B. A., Morris, E. E., Nix, J. F., & Barko, J. W., (1989). Modification of riverine water quality by an adjacent bottomland hardwood wetland. In: Fisk, D. W., (ed.), *Proceedings Wetlands: Concerns and Successes* (pp: 429–438). Tampa, FL. American Water Resources Association, Bethesda, MD. TPS 89-3.

Knorr, K. H., (2012). DOC-dynamics in a small headwater catchment as driven by redox fluctuations and hydrological flow paths – are DOC exports mediated by iron reduction/oxidation cycles? *Biogeosci. Discuss., 9*, 1–34.

Knorr, K. H., Lischeid, G., & Blodau, C., (2009). Dynamics of redox processes in a minerotrophic fen exposed to a water table manipulation. *Geoderma., 153*, 379–392.

Kull, A., Kull, A., Jaagus, J., Kuusemets, V., & Mander, U., (2008). The effects of fluctuating climatic conditions and weather events on nutrient dynamics in a narrow mosaic riparian peatland. *Boreal Environ. Res., 13*, 243–263.

Kumar, R., Kumar, M., Shah, A. I., Bhat, S. A., Wani, M. A., & Ram, D., (2016). Modeling of soil loss using USLE through remote sensing and geographical information system in micro-watershed of Kashmir valley, India. *Journal of Soil and Water Conservation, 15*(1), 40–45.

Leonard, S., Staidl, G., Fogg, J., Gebhardt, K., Hagenbuck, W., & Prichard, D., (1992). *Riparian Area Management: Procedures for Ecological Site Inventory with Special Reference to Riparian-Wetland Sites* (p. 135). Bureau of Land Management, BLM/SC/PT-92/004+ 1737, Service Center, CO.

Liu, Y. B., Yang, W., & Wang, X., (2008). Development of a SWAT extension module to simulate riparian wetland hydrologic processes at a watershed scale. *Hydrol. Process., 22*(16), 2901–2915.

Lowrance, R. R., McIntyre, S., & Lance, C., (1988). Erosion and deposition in a field/forest system estimated using cesium-137 activity. *J. Soil Water Conservation, 43*, 195–199.

Lowrance, R. R., Sharpe, J. K., & Sheridan, J. M., (1986). Long-term sediment deposition in the riparian zone of a coastal plain watershed. *J. Soil Water Conservation, 41*, 266–271.

Lowrance, R. R., Todd, R. L., & Asmussen, L. E., (1983). Waterborne nutrient budgets for the riparian zone of an agricultural watershed. *Agriculture, Ecosystems and Environment, 10*, 371–384.

Lyon, S. W., Grabs, T., Laudon, H., Bishop, K. H., & Seibert, J., (2011). Variability of groundwater levels and total organic carbon in the riparian zone of a boreal catchment. *Journal of Geophys. Res. Biogeosci.* doi: 10.1029/2010JG001452.

Mahdizadeh, K. M., Gholami, S. M. A., & Valipour, M., (2015). Simulation of open-and closed-end border irrigation systems using SIRMOD. *Archives Agron. Soil Sci., 61(7)*, 929–941.

Martinez-Martinez, E., Nejadhashemi, A. P., Woznicki, S. A., & Love, B. J., (2014). Modeling the hydrological significance of wetland restoration scenarios. *J. Environ. Manag., 133*, 121134.

Martinez-Martinez, E., Nejadhashemi, A. P., Woznicki, S. A., Adhikari, U., & Giri, S., (2015). Assessing the significance of wetland restoration scenarios on sediment mitigation plan. *Ecol. Eng., 77*, 103–113.

Meehan, W. R., Swanson, F. J., & Sedell, J. R., (1977). Influences of riparian vegetation on aquatic ecosystems with particular reference to salmonid fishes and their food supply. In: Johnson, R. R., & Jones, D. A., (eds.), *Importance, Preservation and Management of Riparian Habitat; A Symposium* (pp. 137–148).Tucson, AZ: Gen. Tech. Rep. RM-43. Ft. Collins, CO: U.S. Department of Agriculture, Forest Service, Rocky Mountain Forest and Range Experiment Station.

Midwest Plan Service, (1985). *Livestock Waste Facilities Handbook.* Ames, IA: Iowa State University.

Mitchell, M. J., Piatek, K. B., Christopher, S., Mayer, B., Kendall, C., & McHale, P., (2006). Solute sources in stream water during consecutive fall storms in a northern hardwood forest watershed: A combined hydrological, chemical and isotopic approach. *Biogeochem., 78*, 217–246.

Muller, F. L. L., & Tankéré-Muller, S. P. C., (2012). Seasonal variations in surface water chemistry at disturbed and pristine peatland sites in the Flow Country of northern Scotland. *Sci. Total Environ., 435, 436*, 351–362.

National Research Council, (2002). *Riparian Areas: Functions and Strategies for Management.* Washington, DC: The National Academies Press.

Neal, C., House, W. A., Jarvie, H. P., Neal, M., Hill, L., & Wickham, H., (2006). The water quality of the River Dun and the Kennet and Avon canal. *J. Hydrol., 330*, 155–170.

NRC (National Research Council), (1992). *Restoration of Aquatic Ecosystems*. Washington, DC: National Academy Press.

O'Brien, H. D., Eimers, M. C., Watmough, S. A., & Casson, N. J., (2013). Spatial and temporal patterns in total phosphorus in south-central Ontario streams: The role of wetlands and past disturbance. *Can. J. Fish. Aquat Sci., 70*(5), 766–774.

Odum, E. P., (1979). Ecological importance of the riparian zone. Strategies for protection and management of floodplain wetlands and other riparian ecosystems. In: Johnson, R. R., & McCormick, J. F., (eds.), *Proceedings of the Symposium – 1978* (pp. 2–4).

Calloway Gardens, GA Gen. Tech. Rep. WO-12. Washington, DC: U.S. Department of Agriculture, Forest Service.

Osborne, L. L., & Kovacic, D. A., (1993). Riparian vegetated buffer strips in water-quality restoration and stream management. *Freshwater Biology, 29*, 243–258.

Parkin, T. B., (1987). Soil microsites as a source of denitrification variability. *Soil Science Society of America Journal., 51*(5), 1194–1199.

Pennington, P. R., & Watmough, S., (2015). The biogeochemistry of metal-contaminated peatlands in Sudbury, Ontario, Canada. *Water Air Soil Poll., 226*(10), 326. doi: 10.1007/s11270-015-2572-6.

Peterjohn, W. T., & Correll, D. T., (1984). Nutrient dynamics in an agricultural watershed: Observations on the role of a riparian forest. *Ecology, 65*, 1466–1475.

Phillips, J. D., (1989). Nonpoint source pollution control effectiveness of riparian forests along a coastal Plain River. *J. Hydrol., 110*(1989), 221–237.

Piatek, K. B., Christopher, S. F., & Mitchell, M. J., (2009). Spatial and temporal dynamics of stream chemistry in a forested watershed. *Hydrol. Earth Syst. Sci., 13*, 423–439.

Piper, A. M., (1944). A graphic procedure in the geochemical interpretation of water analyses. *Transactions – Am. Geophys. Union, 25*, 914–928.

Platts, W. S., (1981). Effects of livestock grazing. In: *Influence of Forest and Rangeland Management on Anadromous Fish Habitat in Western North America* (p. 25). Gen. Tech. Rep. PNW-124, Portland, OR: U.S. Department of Agriculture, Forest Service, Pacific Northwest Forest and Range Experiment Station.

Prior, H., & Johnes, P. J., (2002). Regulation of surface water quality in a cretaceous chalk catchment, UK: An assessment of the relative importance of instream and wetland processes. *Sci. Total Environ., 282*, 159–174.

Raymond, P. A., & Saiers, J. E., (2010). Event controlled DOC export from forested watersheds. *Biogeochem., 100*, 197–209.

Reiche, M., Hädrich, A., Lischeid, G., & Küsel, K., (2009). Impact of manipulated drought and heavy rainfall events on peat mineralization processes and source-sink functions of an acidic fen. *J. Geophys. Res. Biogeosci., 114*, G02021.

Republic of South Africa. (1998). National Water Act. Act No. 36 of 1998. *Government Gazette, 19182.*

Richardson, C. J., & Davis, J. A., (1987). Natural and artificial wetland ecosystems: Ecological opportunities and limitations. In: Reddy, K. H., & Smith, W. H., (eds.), *Aquatic Plants for Water Treatment and Resource Recovery* (pp, 819–854). Magnolia Publishing Inc.

Richardson, C. J., (1988). *Freshwater Wetlands: Transformers, Filters, or Sinks?* (Vol. 11, No. 2, pp. 3–9). FOREM, School of Forestry and Environmental Studies, Duke University.

Robinson, T. W., (1958). *Phreatophytes* (p. 84). U.S. Geological Survey Water-Supply Paper 1423.

Rosgen, D., (2006). *Watershed Assessment of River Stability and Sediment Supply (WARSSS).* Ft. Collins, CO: Wildland Hydrology.

Schade, J. D., Fisher, S. G., Grimm, N. B., & Seddon, J. A., (2001). The influence of a riparian shrub on nitrogen cycling in a Sonoran Desert stream. *Ecology., 82*, 3363–3376.

Schoeller, H., (1962). Groundwater. *Dynamic and Chemical Hydrology, Research, Exploitation and Evaluation of Resources* (p. 642). Masson et Cie, Paris.

Schultz, R. C., Colletti, J. P., Isenhart, T. M., Marquez, C. O., Simpkins, W. W., & Ball, C. J., (2000). Riparian forest buffer practices. In: Garett, H. E., et al., (eds.), *North American Agroforestry: An Integrated Science and Practice* (pp. 189–281). American Society of Agronomy, Inc. Madison, WI.

Sedell, J. R., & Beschta, R. L., (1991). Bringing back the "bio" in bioengineering. *American Fisheries Society Symposium, 10*, 160–175.

Shields, F. D., Bowie, A. J., & Cooper, C. M., (1995). Control of streambank erosion due to bed degradation with vegetation and structure. *Water Resour. Bull., 31*, 475–489.

Storch, R. L., (1979). Livestock/streamside management programs in eastern Oregon. In: Cope, O. B., (ed.), *Proceedings of the Forum -G Razing and Riparian/Stream Ecosystems* (pp. 56–59). 1978 November 3–4; Denver, CO: Trout Unlimited.

Stromberg, J. C., Tiller, R., & Richter, B., (1996). Effects of groundwater decline on riparian vegetation of semiarid regions: The San Pedro, Arizona. *Ecological Applications, 6*, 113–131.

Swanson, F. J., Gregory, S. V., Sedell, J. R., & Campbell, A. S., (1981). Land-water interaction: The riparian zone. In: Edmunds, R. L., (ed.), *The Natural Behavior and Response to Stress of Western Coniferous Forests*. Stroudsburg, PA: Dowden, Hutchinson and Ross, Inc.

Swanson, F. J., Lienkaemper, G. W., & Sedell, J. R., (1976). *History, Physical Effects and Management Implications of Large Organic Debris in Western Oregon Streams* (p. 15). Gen. Tech. Rep. PNW-56. Portland, OR: U.S. Department of Agriculture, Forest Service, Pacific Northwest Forest and Range Experiment Station.

Swanson, S., Kozlowski, D., Hall, R., Heggem, D., & Lin, J., (2017). Riparian proper functioning condition assessment to improve watershed management for water quality. *Journal of Soil and Water Conservation, 72*(2), 168–182.

Swanston, D. N., (1978). Effect of geology on soil mass movement activity in the Pacific Northwest. In: Youngberg, C. T., (ed.), *Forest Soils and Land Use* (pp. 89–115). Ft. Collins, CO: Proc. Fifth North Amer. For. Soils Conf., Colorado State University.

Swanston, D. N., (1980). Impact of natural events. In: Meehan, W. R., (ed.), *Influence of Forest and Rangeland Management on an Anadromous Fish Habitat in Western North America* (p. 27). Gen. Tech. Rep. PNW-104. Portland, OR: U.S. Department of Agriculture, Forest Service, Pacific Northwest Forest and Range Experiment Station.

Szkokan-Emilson, E. J., Kielstra, B., Watmough, S., & Gunn, J., (2013). Drought-induced release of metals from peatlands in watersheds recovering from historical metal and Sulphur deposition. *Biogeochem., 116*(1–3), 131–145.

Thomas, J. W., Maser, C., & Rodiek, J. E., (1979). Riparian zones. In: Thomas, J. W., (ed.), *Wildlife Habitats in Managed Forests-the Blue Mountains of Oregon and Washington; Agric. Handbook 553* (pp. 40–47). Washington, DC: U.S. Department of Agriculture, Forest Service.

Ul Zaman, M., Bhat, S., Sharma, S., & Bhat, O., (2018). Methods to control soil erosion: A review. *Int. J. Pure Appl. Biosci., 6*(2), 1114–1121. 10.18782/2320-7051. 6462.

Ulen, B., (1995). Episodic precipitation and discharge events and their influence on losses of phosphorus and nitrogen from tile-drained arable fields. *Swed. J. Agr. Res., 25*, 25–31.

Valipour, M., (2015). Long-term runoff study using SARIMA and ARIMA models in the United States. *Meteorol. Appl., 22*, 592–598.

Van, V. M., & Zwolsman, J., (2008). Impact of summer droughts on the water quality of the Meuse River. *Journal of Hydrology, 353*, 1–17.

Vidon, P., Jacinthe, P. A., Liu, X., Fisher, K., & Baker, M., (2014). Hydrobiogeochemical controls on riparian nutrient and greenhouse gas dynamics: 10 years pots-restoration. *J. Am. Water Resour. Assoc., 50*(3), 639–652.

Watmough, S. A., & Orlovskaya, L., (2015). Predicting metal release from peatlands in Sudbury, Ontario, in response to drought. *Water Air Soil Poll., 226*(4), 103.

Weyer, C., Peiffer, S., & Lischeid, G., (2018). Stream water quality affected by interacting hydrological and biogeochemical processes in a riparian wetland. *Journal of Hydrology*. doi: Https://doi.org/10.1016/j.jhydrol.2018.05.067.

Whigham, D. F., Chitterling, C., & Palmer, B., (1988). Impacts of freshwater wetlands on water quality: A landscape perspective. *Environmental Management, 12*(5), 663–671.

Wilson, C. G., Kuhnle, R. A., Bosch, D. D., Steiner, J. L., Starks, P. J., Tomer, M. D., & Wilson, G. V., (2008). Quantifying relative contributions from sediment sources in conservation effects assessment project watersheds. *J. Soil Water Conserv., 63*(6), 523–532.

Winward, A. H., (2000). *Monitoring the Vegetation Resources in Riparian Areas*. General Technical Report RMRS-GTR-47. Fort Collins, CO: USDA, Forest Service, Rocky Mountain Research Station.

Wissmar, R. C., & Swanson, F. J., (1990). Landscape disturbance and lotic ecotones. In: Naiman, R. J., & Decamps, H., (eds.), *The Ecology and Management of Aquatic Terrestrial Ecotones* (pp. 65–89). Unesco and the Parthenon Publishing Group. Paris, France.

Yang, W., Liu, Y., Ou, C., & Gabor, S., (2016). Examining water quality effects of riparian wetland loss and restoration scenarios in a southern Ontario watershed. *Journal of Environmental Management, 174*, 26–34.

Yang, W., Wang, X., Liu, Y. B., Gabor, S., Boychuk, L., & Badiou, P., (2010). Simulated environmental effects of wetland restoration scenarios in a typical Canadian prairie watershed. *Well. Ecol. Manag., 18*(3), 269–279.

Zhao, T. Q., Xu, H. S., He, Y. X., Tai, C., Meng, H. Q., Zeng, F. F., & Xing, M. L., (2009). Agricultural non-point nitrogen pollution control function of different vegetation types in riparian wetlands: A case study in the Yellow River wetland in China. *J. Environ. Sci., 21*(7), 933–939.

CHAPTER 7

Role of Remote Sensing in Assessing Quality of Water

POOJA GOYAL, AAMIR ISHAQ SHAH, NEHA SINGHAL, and MEHRAJ U. DIN DAR*

Department of Soil and Water Engineering, Punjab Agricultural University, Ludhiana – 141004, Punjab, India,
**E-mail: mehrajdar417@gmail.com (M. U. D. Dar)*

ABSTRACT

The deteriorating water quality in almost all the water systems of the world is an alarming concern for scientists throughout the world. Therefore, identification of the contaminants, the contribution of different sources of pollution and to keep its track in real-time is a challenging task. Continuous and accurate monitoring of large-scale waterbodies with limited resources can be effectively managed through remote sensing technologies. Satellite imaging can help us develop an integrated water quality monitoring database for effective assessment and management of our precious resources. The water quality parameters help in deciding the suitability of particular water to its best possible use based on its quality. The satellite remote sensing helps to acquire information on the water quality parameters based on the optical properties of water has proved to be a great boon to provide real-time synoptic view for mapping the water quality parameters. With the improvement in technological developments in sensor technology in terms of resolutions and accuracy, i.e., the latest hyperspectral imaging, improved algorithms, availability of data, use of artificial intelligence

Advanced Technologies for Water Quality Treatment and Management. Mehraj U. Din Dar, PhD, Aamir Ishaq Shah, MTech, Shakeel Ahmad Bhat, PhD & Syed Rouhullah Ali, PhD (Eds.)

and neural networks have enabled us in monitoring not only the optically active water quality parameters, but also the distribution of optically inactive water quality parameters. In this chapter will discuss in brief about the principle behind the estimation of different water quality parameters using remote sensing, the type of sensors employed both spaceborne and airborne and different approaches used to estimate various types of water quality parameters.

7.1 INTRODUCTION

Water resources are one of the most precious and important natural resources on earth. Almost all living beings rely on water for their survival including human beings. Our water resources play an important role in sustaining different ecosystems, acting as a source of food, income, and livelihood for many. With a rapidly growing population, socio-economic development, urbanization, and globalization, an increased amount of stress has been placed on our water resources especially in times of changing and uncertain climate change (Hameed et al., 2017; Dar et al., 2017). The water quality has been deteriorating in more than half of the water systems around the world due to anthropogenic pollution (Dai et al., 2017).

Water quality, in general, describes the characteristics of water in terms of chemical, physical, and biological attributes of water. Water quality is rather a relative term than an absolute, usually defined with respect to its suitability for a specific need or purpose. The water quality parameters suitable for irrigating the crop may not be suitable for direct human consumption. So, the criteria, one standard fit all is not applicable in this case and the treatment level required depends on the purpose for what it is being utilized. Most commonly, the standards used to assess the water quality differs with its requirements for human consumption, agriculture, animal husbandry and industry needs. Water quality is affected by pollutants delivered from both point and non-point sources (Dar et al., 2020). Point source refers to a single source of pollution, such as a pipe outlet or ditch. Non-point sources are difficult to trace as they are diffuse and depends on the response of watershed to the water movement, land use management, and human and natural activities in a watershed (Ritchie et al., 2003). Agricultural and urban runoff, treated, and untreated sewage discharge, industrial activities like mining, construction, and manufacturing are all sources of pollution that negatively affects the quality of water. Water pollution not only affects the inland freshwater resources but also the fragile coastal ecosystems like estuaries, which are more sensitive to changing environments due to anthropogenic activities. Therefore, it is vital to develop an integrated water quality monitoring database for efficient and sustainable management of existing water resources (Wang et al., 2017).

Traditionally, the water quality indicators, including physical, chemical, and biological properties, are determined by *in situ* observations by collecting samples from point of interest followed by analyzing them in the laboratory. Although these conventional point sampling measurements are accurate at particular discrete points, they are time consuming, labor intensive and costly. Moreover, they do not provide real-time synoptic view of water quality parameters required for accurate assessment and management of global water systems. Different studies have demonstrated that remote sensing can act as a valuable tool for monitoring water quality in a cost-effective way, especially for large scale monitoring. It provides quick, detailed, and reliable information for systematic water management. Remote sensing has an upper hand over conventional *in situ* measurements because it makes it possible to identify, monitor, and quantify the water quality parameters in a more efficient and effective way (Lindell et al., 1999). Remote sensing can be used successfully in countries that either lack in conventional water quality monitoring programs or they are unsatisfactory (Bonansea et al., 2019).

Many different types of sensors mounted either on a spacecraft or any other platforms are used to measure the radiation reflected from the water surface at different wavelengths for water quality analysis. This reflected radiance has different characteristics because of the presence of different constituents in water. These reflections are used to detect different water quality indicators like chlorophyll concentrations (chl-a), total suspended solids (TSS), turbidity, colored dissolved organic matter (CDOM), Secchi disk depth (SDD), water temperature, salinity, pH, dissolved oxygen (DO), total phosphorus (TP), etc., either directly or indirectly. The essential factors for monitoring and assessment of water quality depend on the spectral signatures of water and its constituents. This chapter will review in brief the different types of sensors used for analyzing water quality and the different water quality indicators that are currently analyzed through remote sensing. Many different types of multispectral and hyperspectral satellites have successfully been used for the estimation of different water quality parameters that are discussed in detail in further sections.

7.2 BASIC PRINCIPLE BEHIND WATER QUALITY MEASUREMENT USING REMOTE SENSING

The remote sensing of water bodies is based on the radiation backscattered from its surface. The water body receives the radiation emitted from the sun. The presence of different substances causes water to backscatter the light differently which is then sensed by the remote sensing sensor. Observing the spectrum of water also called spectral signature helps us to interpret water quality parameters like color, saturation, transparency, etc. The satellite images derived from the sensor needs to be processed before utilizing them for analysis purpose. The images need to be preprocessed for geometric, radiometric/atmospheric corrections and reduce noise. Geometric corrections are done to ensure that each pixel in the image is at its correct position w.r.t. earth surface. Atmospheric or radiometric correction is done to remove the effect of clouds, water vapor, or other gasses present in the atmosphere. Water generally reflects light in the visible range of the EM spectrum however some also reflect light in the near-infrared (NIR) section of the spectrum.

In general, the major water quality indicators that affect water quality can be classified as physical, chemical, biological, and esthetic indicators as given in Figure 7.1. But for estimating water quality parameters through remote sensing, the water quality parameters can be classified into two categories based on their optical properties, i.e., optically active and optically inactive components. These parameters are described in detail in the section. The presence of optically active elements in water interacts with light and changes the energy spectra of reflected solar/thermal radiation through different absorption of the spectrum. The changes in the spectral signature backscattered from water are sensed through remote sensing sensor and related to the parameter of interest using empirical or analytical models. Some of the chemicals and pathogens, which are not optically active, are inferred indirectly from the measurement of detectable water quality parameters having strong correlations with them (Ritchie et al.,

2003). The optimal wavelength corresponding to a water quality parameter depends on the substance being measured, its concentration, and the sensor characteristics.

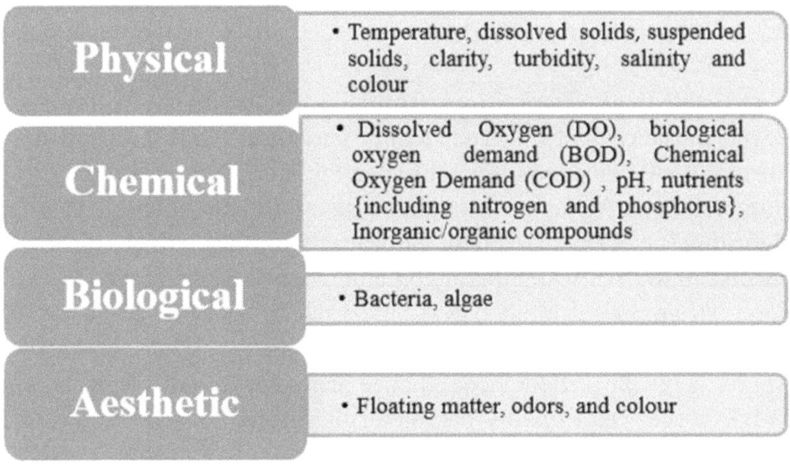

FIGURE 7.1 Classification of water quality indicators.

The development of remote sensing technology for water quality measurement through airborne, space-borne platforms or close-range optical devices like spectrometers and radiometers started in the late 1970s. The initial techniques focused on measuring the spectral and thermal differences in the emitted energy from the water surfaces and empirical relationships were drawn between spectral properties and the water quality parameter of interest. The various approaches used in remote sensing to estimate water quality parameters are discussed in further sections.

7.3 REMOTE SENSING APPROACHES FOR WATER QUALITY ASSESSMENT

Various approaches are used to derive water quality parameters through remote sensing. They are divided into empirical, semi-empirical, analytical, and semi-analytical methods (Wang and Yang, 2019). The various approaches are used and then validated by *in situ* measurements. The images obtained through remote sensing are high in accuracy and can generate spatially explicit water quality maps in a more efficient way even limited *in situ* data (Gholizadeh et al., 2016; Giardino et al., 2010; Wang et al., 2012).

7.3.1 EMPIRICAL METHOD

Here, statistical relationships/regression is established between the *in-situ* measurement of water quality parameters and the characteristic spectral bands or combination of spectral bands, i.e., reflectance values extracted from the image through remote-sensing to obtain the empirical algorithm (Qi et al., 2014). This empirical algorithm is then used to obtain the distribution of water quality parameters in a larger space for a longer time. An empirical method was first used by (Ritchie et al., 1975) to estimate suspended sediments (SS) using equation, $Y=A+BX$ or $Y=AB^X$ where, Y is the reflectance/radiance from the remote sensing sensor and X is the water quality parameter that needs to be determined, i.e., chlorophyll, suspended sediment, etc. A and B are empirical coefficients derived from the statistical relationship between spectral reflectance value and the corresponding *in situ* water quality parameter. Techniques like linear regression, use of single-band method, combination of more than 1 band like band ratios, artificial neural network (ANN), principal component analysis (PCA) are generally used to solve for the empirical algorithm. These methods have been widely used to estimate chlorophyll, TSS, SDD, CDOM, water clarity and turbidity. Empirical methods are easy to use as no understanding of the underlying physics is needed to model the atmospheric and underwater optical properties. The simplicity and reliability of empirical methods makes them preferable over other complex methods (Matthews, 2011). Empirical methods can also be used to model parameters which are not optically active, e.g., phosphorus (Torbick et al., 2013). The negative side of the use of empirical methods is their inability to distinguish the non-unique signals from the parameters that reflect in similar spectrum, e.g., TSS, and Chl-a (Matthews, 2011). Another disadvantage is the use of the empirical approach is that the shallow waters cause high reflectance values probably due to bed reflectance and results in overestimation of water column constituents (Mbuh, 2019).

7.3.2 SEMI-EMPIRICAL METHOD

The use of semi-empirical method is considered an improvement over empirical method because here the band selection is based on the prior knowledge of how our parameter of interest will affect reflectance in different spectral bands. Spectral characteristic identification aids in model creation which can be applied for assessment of a larger area/ reservoir. It combines the measured *in situ* spectral data along with the statistical relationships (Li, 2009). Semi-empirical methods are utilized in combination in hyperspectral remote sensing as it allows the identification of narrow spectral bands, and hence, it requires a lot of manpower for data collection.

The semi-empirical methods have been used widely for retrieval of chlorophyll-a (chl-a) concentration in high turbid waters in which the wavelengths from dissolved organic matter (DOM) and tripton overlap with that of chlorophyll (Gitelson et al., 2003; Reif, 2011) using band ratio algorithms. Though the semi-empirical methods have higher accuracy than the empirical methods (Huang et al., 2018), they require thorough *in situ* data for validation and testing and can only be applied to water bodies with similar conditions, i.e., they are scene and time-dependent. The applicability of these algorithms to different environments calls for accurate atmospheric corrections and background information which is difficult to obtain on a local basis (Feng et al., 2005).

7.3.3 ANALYTICAL OR PHYSICAL METHOD

The analytical models are complex theoretical models based on the radiation transfer theory. The absorption and scattering coefficients of the reflectance leaving the water surface are studied at multiple wavelengths. The relationships are established between the reflectance leaving the surface of water and the concentration of the constituent in water and their inherent optical properties (IOPs) (Giardino et al., 2007; Knaeps et al., 2010; Reif, 2011). An analytical model was first proposed by Schiebe et al. (1992) based on the IOPs of water and the relationship between the spectral and physical characteristics of the surface water to determine suspended sediment concentrations. The model had the form

$Rj = Bj[1 - e^{(c/Si)}]$ where Bi and Si were statistically determined coefficients, Rj is the reflectance obtained from the RS in i^{th} wave band and c is the suspended sediment concentration. The analytical methods have the ability to measure simultaneously different water quality parameters provided IOPs of the parameters are known and a huge amount of *in situ* data is available. When measured *in situ* data is not available, inversion methods are used to optimize the unknown parameters to determine the water quality parameter through inversion of radiative transfer equation (RTE) (Giardino et al., 2007).

7.3.4 SEMI-ANALYTICAL METHOD

This method is similar to the analytical method but here the analytical techniques are empirically parameterized with the *in-situ* measurements. The three-band model, utilizing absorption coefficients from three wavelengths was used to estimate chlorophyll in terrestrial vegetation (Gitelson et al., 2003) and turbid waters (Dall'Olmo et al., 2003) and retrieval of suspended sediment concentration (Kong et al., 2015). A Quasi Analytical model was used by Matin et al. (2019) to study the variability of backscattering coefficient to estimate chl-a in coastal regions of Kerala in India. The semi-analytical method is more convenient, flexible, and offers higher inversion accuracy.

Model inversion accuracy is an important component for the development of the four methods described here especially in shallow waters to separate the reflectance from the surface and the bed of the water body (Reif, 2011). Scientists, over the world have used many methods to improve the accuracy of these methods by using analysis of the measured spectra, band combinations/band-ratios, time segmentation, water space division, etc. (Wang and Yang, 2019). It is difficult to conclude which method is better because the choice of method depends on the resources available and the amount of accuracy needed to estimate these water quality parameters. In general, analytical and semi-analytical methods are a better choice over other methods as they combine both the physical and spectral aspects of the modeling, but they are complex and computationally intensive in comparison to the empirical methods (Reif, 2011).

7.4 SENSORS FOR WATER QUALITY ANALYSIS

The different types of sensors employed for water quality analysis are given in the Figure 7.2 and Table 7.1.

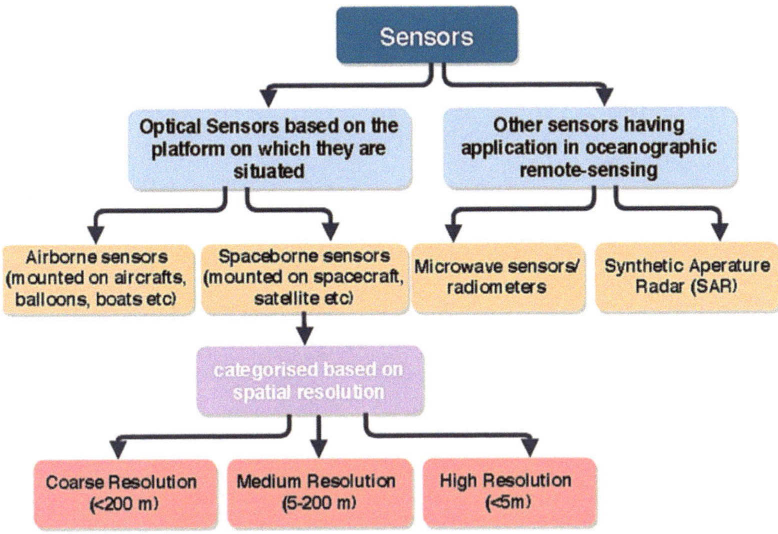

FIGURE 7.2 Classification of the different types of sensors used in water quality studies.

The optical sensors are the ones that convert light to an electric signal which is then read by an integrated measuring device. Optical sensors have been extensively adopted in water resource management because of availability of optimum spatial and spectral resolution along with higher data availability (Huang et al., 2015). They are classified based on the mounting of the platforms. Airborne sensors work within the periphery of the earth's atmosphere while the space-borne sensors work beyond the boundary of the earth's atmosphere. Airborne sensors are perfect for monitoring small waterbodies because they offer higher spectral and temporal resolution and are much more flexible than spaceborne sensors. The commonly used airborne sensors along with their spatial resolution applied for water quality assessments are listed in Table 7.2. More detailed information on these airborne sensors can be found in (Gholizadeh et al., 2016).

TABLE 7.1 Commonly Used Airborne Sensors Used in Water Quality Assessments

Sl. No.	Sensor	Type	Resolution (m)
1.	Airborne visible infrared imaging spectrometer (AVIRIS)	Hyperspectral	17
2.	Airborne prism experiment (APEX)	Hyperspectral	2–5
3.	Airborne imaging spectrometer (AISA)	Hyperspectral	1
4.	Compact airborne spectrographic imager (CASI-1500)	Hyperspectral	0.5–3
5.	Digital airborne imaging spectrometer (DAIS 7915)	Hyperspectral	3–20
6.	Daedalus multispectral scanner (Daedalus MSS)	Multispectral	25
7.	Environmental protection system (EPS-H)	Hyperspectral	Minimum 1 m
8.	HySpex hyperspectral cameras (HySpex)	Hyperspectral	0.5 m at 2,000 m
9.	Hyperspectral digital imagery collection experiment (HYDICE)	Hyperspectral	0.8–4
10.	PROBE-1 in the U.S. (HyMap)	Hyperspectral	3–10
11.	Multispectral infrared and visible imaging spectrometer (MIVIS)	Multispectral	3–8

Source: Adapted from: Gholizadeh et al. (2016).

Satellite sensors are generally employed to study larger water bodies because of high spectral resolution. Space-borne sensors are further classified based on the spatial resolution of the sensors. Spatial resolution is the area on the ground that is clearly discernable from a certain height within a pixel and how precisely these details are captured for the construction of a digital image. The three categories are high (<5 m), medium (5–200 m), and coarse (>200 m) resolution. The commonly used space-borne sensors falling in these three categories are given in the table along with their spatial and temporal resolution. More detailed information on other aspects of these sensors can be found in (Gholizadeh et al., 2016; Huang et al., 2018).

TABLE 7.2 Commonly Used Spaceborne Sensors Used in Water Quality Assessments

Sensor Group	Satellite/ Sensor	Full Form	Spatial Resolution (m)	Temporal Resolution (days)
High resolution sensor	IKONOS	Greek word for image	1.4	1.5–3
	QuickBird	–	0.61–2.24	2.7
	WorldView	Digital GlobeWorldView-2	1.85–0.46	1.1
		Digital GlobeWorldView-1	0.5	1.7
	NOAA WorldView-3	–	1.24–3.7–0.31	1–4.5
	RapidEye	–	5	1–5.5
	ZY-3	–	2.1–5.8	5
	GF-1/GF-2	–	1–16	4–5
Medium resolution sensor	Landsat	Landsat-5 MSS	80	18
		Landsat-5 TM	30–120	16
		Landsat-7 ETM^+	30–15–60	16
		Landsat-8 OLI/TRS	30–15–100	16
	ASTER	Advanced spaceborne thermal emission and reflection radiometer	15–90	16
	Sentinel-2 MSI	Sentinel-2 multi spectral instrument	10–60	5
	SPOT	Systeme probatoire d'observation dela tarre	2.5–20	26
Coarse-resolution sensor	NOAA/ AVHRR	National Oceanic and atmospheric administration/ advanced very high-resolution radiometer	1,100	0.5
	SeaWiFS	Sea-viewing wide field-of-view sensor	1,130	1
	AATSR	Advanced along track scanning radiometer	1,000	3–6

TABLE 7.2 *(Continued)*

Sensor Group	Satellite/ Sensor	Full Form	Spatial Resolution (m)	Temporal Resolution (days)
	MODIS	Moderate resolution imaging spectroradiometer	250–1,000	0.5
	Suomi NPP-VIIRS	Suomi national polar-orbiting-visible infrared imaging radiometer	375–750	0.5
	MERIS	Medium resolution imaging spectrometer	300	3
	Sentinel-3 OCLI	Sentinel-3 Ocean and land color instrument	300	2

Source: Adapted from: Gholizadeh et al. (2016); Huang et al. (2018).

The coarse resolution sensors are characterized by low accuracy because of high information generalization but higher temporal resolution and broad coverage. A popular coarse resolution sensor MODIS has been used extensively used for detecting surface water (Peng et al., 2005; Chen et al., 2013) and lake water variations (Peng et al., 2005; Feng et al., 2012; Cai et al., 2016; Lu et al., 2017) because of its high spatial resolution and short revisit time. VIIRS has also been used as an upgradation to AVHRR and MODIS (Huang et al., 2015). Medium resolution sensors are widely used in studies concerned with the dynamics of most kinds of surface water bodies because of its optimum spatial and temporal resolution. Landsat series is one of the most successfully used satellite series in history. Starting with the launch of Landsat-1 in 1972 to the latest Landsat-8, launched in 2013, the spatial resolution of these satellites has improved from 80 m to 15 m over the years. The quality of sensors has also improved from the use of multispectral scanner (MSS) in Landsat-1, thematic mapper (TM) in Landsat 4 and 5, enhanced thematic mapper (ETM+) in Landsat-7 to the latest operational land imager (OLI) in Landsat-8. The latest OLI sensor has already been applied for surface water studies (Acharya et al., 2016; Ji et al., 2015; Li et al., 2016; Singh et al., 2015). The Landsat series has been most preferred because of its universal appeal and easy accessibility. Another medium resolution satellite series, i.e., SPOT is loaded with sensors having resolutions (10 m) higher than that of Landsat imagery has limited application because of the non-availability of freely distributed data. High resolution sensors like IKONOS, QuickBird, RapidEye, and WorldView have been used for the detection of small water bodies because of its finer spatial resolution of 1 m or less. They are generally not applicable to large water bodies because of its small scene coverage, limited availability of images and shadow problem in images (Sawaya et al., 2003).

The selection of appropriate optical sensors should be done after careful consideration on parameters like spatial and temporal resolution. There is no doubt that the higher resolution sensors have higher accuracy, but they generally have low temporal resolution, that hinders the intensive real-time monitoring. Therefore, high resolution is not always a good option. On the other hand, the coarse resolution sensors have lower accuracy but have higher temporal resolution. It is therefore important to choose the data from the sensor with suitable resolutions keeping in view the scale of water bodies to be studied.

The other types of sensors include synthetic aperture radar (SAR) and microwave radiometers (MWR). Passive MVRs uses long wavelength radiation called microwaves. Because they are not dependent on radiation from sunlight, they can work day and night in any weather as microwaves can penetrate both cloud and vegetation coverage. These types of sensors are generally used by oceanographers to study salinity and surface temperature of the waterbodies. SAR creates 2D and 3D images of objects, and can be used either as a spaceborne or airborne sensor. SAR has been used for detecting oil spills in water, for studying wind speed at sea-surface, ocean topography and regional ice monitoring but there use is limited in water quality assessment studies (Gholizadeh et al., 2016).

7.5 WATER QUALITY PARAMETER ESTIMATION THROUGH REMOTE SENSING

Water quality studies determine the physical, chemical, and biological characteristics of waterbodies and identify the possible contaminants that deteriorate the water quality (Usali and Ismail, 2010). Some of the reasons behind the water quality deterioration are heavy metals, waste discharges nutrients, pesticides, sediments, and microorganisms. Most of the research conducted previously have emphasized on optically active parameters, like chlorophyll-a (chl-a), turbidity, and TSS that reflect light in a particular wavelength. In a systematic review conducted by (Huang et al., 2018) to review the water quality monitoring and evaluation studies conducted in China using remote sensing, out of the 403 studies taken into consideration, 212 studies were only focused on chl-a and another 171 studies were for SS. In contrast, only a limited amount of literature has been reported for optically inactive parameters like TP, Nitrogen, pH, Electrical conductivity, BOD, and COD, etc., because of their weak reflectance characteristics and low signal noise ratio. The assessment of these parameters is also necessary for estimating water quality indices and further efforts are needed to assess these parameters. Some of the commonly estimated water quality parameters through remote sensing are discussed in this chapter. The most commonly measured qualitative water quality parameters through remote sensing are listed in Table 7.3.

TABLE 7.3 Commonly Evaluated Water Quality Parameters via Remote Sensing

	Water Quality Parameter	**Abbreviation**	**Units**
Optically Active Water Quality Parameters	Chlorophyll-a	(Chl-a)	milligram/liter
	Turbidity	(TUR)	NTU
	Total suspended solids	(TSS)	milligram/liter
	Total organic carbon	(TOC)	milligram/liter
	Temperature	T	°C
	Electrical conductivity	(EC)	μs/cm
	Colored dissolved organic matter	(CDIM)	milligram/liter
	Secchi disk depth	(SDD)	meter
	Sea surface salinity	(SSS)	PSU
Optically Inactive Water Quality Parameters	Chemical oxygen demand	(COD)	milligram/liter
	Dissolved organic carbon	(DOC)	milligram/liter
	Total phosphorus	(TP)	milligram/liter
	Biological oxygen demand	(BOF)	milligram/liter
	Ortho-phosphate	(PO_4)	milligram/liter

Source: Gholizadeh et al. (2016).

7.5.1 OPTICALLY ACTIVE WATER QUALITY PARAMETERS

7.5.1.1 CHLOROPHYLL

As discussed earlier, Chlorophyll-a (chl-a) is one of the most intensely studied water quality parameters as it acts as an important water trophic level indicator, especially for lakes (An et al., 2013). Chl-a is present as a photosynthetically active pigment in plants, algae, phytoplankton, and cyanobacteria. The chl-a concentration is thus a representative of these algal, phytoplankton, and cyanobacterial blooms driven by eutrophication phenomenon in freshwater lakes. So, monitoring chlorophyll concentration through remote sensing helps quantify the eutrophication of lakes both spatially and temporally. In most cases, detection of chlorophyll through remote sensing is based on empirical relationships between reflectance in narrow bands and chlorophyll concentrations. The *in-situ* measurements are conducted to calibrate and validate these empirical relationships. Chl-a absorbs energy in the violet-blue and orange-red region of the electromagnetic spectrum (EM) while reflecting green causing chlorophyll to appear green.

A large number of wavelengths and algorithms have been used to monitor chlorophyll concentrations in estuaries, oceans, and freshwaters. Initially, bio-optical algorithms were designed. In ocean waters, whose IOPs are dominated by phytoplankton, i.e., Case 1 waters, spectral bands in the blue to green region were adequate for identifying chl-a concentrations. However, advanced approaches are needed for detection chl-a in Case 2 waters, whose IOPs are affected by CDOMs, mineral particles in addition to phytoplankton. The retrieval algorithms for chl-a concentrations differs depending on the type of water (Gitelson et al., 2009; Gordon and Wang, 1994). In a study conducted by Schalles et al. (1997) to estimate algal loads, he showed increased reflectance with the increase in chlorophyll concentration across most of the wavelengths but decreased reflectance of chlorophyll for the wavelength range of 675 to 680 nm. Some researchers have also reported that at short wavelengths, i.e., blue, the spectral response decreases with increase in chl-a concentration (Brivio et al., 2001; Bukata et al., 1995; Dekker and Peters, 1993).

The different types of bands used in multispectral and hyperspectral imaging are shown in Figure 7.3 (Suarez et al., 2018). The single band of red (600–700 nm), green (500–600 nm) and blue (400–500 nm) of Landsat

5-TM were used as linear models for chl-a studies by Ritchie et al. (1990); Dekker and Peters (1993); and Lathrop and Lillesand (1986), respectively. As a further improvement, band combinations and band ratios are also used widely to quantify chl-a (Dekker et al., 1991). Combination of bands and band ratios can significantly reduce the air and water surface interactions, irradiance, and atmospheric effects in the remotely sensed signal (Dekker and Peters, 1993). Different authors have reported different spectral responses of chl-a in different types of water, but it mainly reflects in visible and NIR region. Most commonly, the chl-a studies show high absorption in the blue (450–475 nm) and red (670 nm) region and the reflectance peaks in green (550 nm) and NIR (700 nm) region (Gholizadeh et al., 2016). Another sensitive indicator for chlorophyll concentration lies at the reflectance height of 690 nm above the base line from 670 nm to 850 nm (Gitelson et al., 1995). The ratio between green and red was used by Allee and Johnson (1999); Brezonik et al. (2005); Hellweger et al. (2004) using Landsat 5-TM data. Similar studies were conducted using ratios between Near Infrared Region (NIR) and red (Hunter et al., 2010; Le et al., 2013) between green and blue (Giardino et al., 2007), between blue and red (Han and Jordan, 2005) respectively. Similarly, combinations of more than three bands were also applied for chl-a studies (Ekercin, 2007).

Large uncertainties exist in chlorophyll detection during cyanobacterial blooms, also known as blue-green algae. The distribution of cyanobacterial blooms has been studied using broadband sensors like AVHRR (Hakansson and Moberg, 1994), multispectral sensors like SeaWiFS (Siegel and Gerth, 2000) and SARs (Svejkovsky and Shandley, 2001).

While several studies have claimed to achieve satisfying results for chl-a estimation using broadband sensors, a lot of other authors have claimed that the estimation of chl-a is difficult in waters dominated by SS (Dekker et al., 1991; Ritchie et al., 1994) because the spectral response is dominated by SS in highly turbid and eutrophic waters (Harma et al., 2001). It is also difficult to extract chl-a from ocean color measurements due to the independent variation in CDOM concentrations in inland waterways like lakes, rivers, and reservoirs having high chl-a concentrations. The three-band model has also been used to predict chl-*a* concentration in case 2 waters (Bhatti et al., 2010). The reflectance ratio of R700/R670 has been used in a number of studies to develop different algorithms to retrieve chl-*a* in turbid waters (Bhatti et al., 2010; Gitelson et al., 2008).

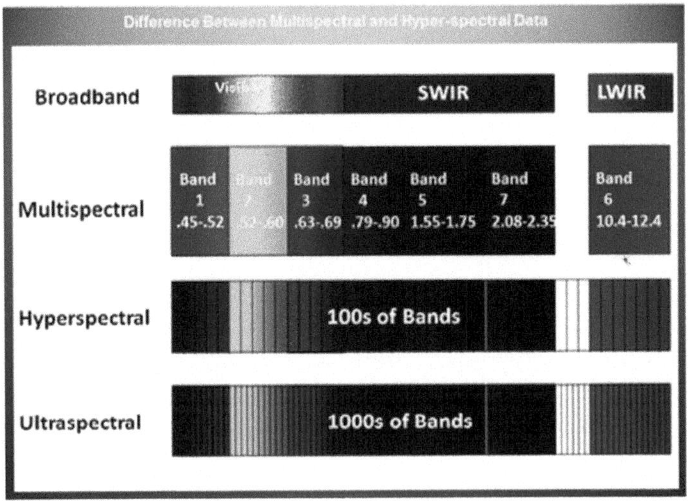

FIGURE 7.3 Image showing difference between multispectral and hyperspectral bands.
Source: Suarez et al. (2018)

These limitations suggest that new approaches need to be exploited to map chl-a in case 2 waters, i.e., application of hyperspectral data. Hyperspectral sensors are being used as the future sensors for measurement of chl-a concentration in water. The chlorophyll absorption occurs in short wavelengths and hyperspectral imaging allows us to detect small changes in the narrow bands of the EM spectrum (Harrington and Repic, 1995). The hyperspectral data from satellite sensors like SeaWiFS, IKONOS, ocean color, temperature scanner (OCTS) and modular optical scanner (MOS) can be used to differentiate different groups of algae. Field and laboratory studies with hyperspectral data have been employed to differentiate blue-green and green algae (Dekker et al., 1995). Two and three band models have been used to estimate chl-a concentration in turbid waters using satellites like MODIS, SeaWiFS, MERIS, and RapidEye (Choe et al., 2011). Combination of four bands, i.e., Band 1 to band 4 of IKONOS data was also used for predicting chl-a concentration in Turkey (Ekercin, 2007).

Many satellites and sensors have been used over the years for chl-a estimation, but the use of Landsat TM data is one of the most widely used data for chl-a determination because of its easy accessibility and free data availability in comparison to other sensors. Hyperspectral imaging is a way forward in chl-a estimation through remote sensing however it is not easily accessible. Further improvements in hyperspectral imaging will allow us to monitor chlorophyll in a more efficient way.

7.5.1.2 TOTAL SUSPENDED SOLIDS (TSS)/TURBIDITY

The particles having a size greater than 2 microns are termed as TSS. It comprises of both inorganic materials like sand, silt, sediments, and chemical precipitates; and organic matter like plankton, bacteria, and algae. The breakdown of organic material through decomposition contributes to the TSS concentration in rivers. The turbidity has often been used as a water quality indicator to estimate these suspended sediment concentrations (Wass et al., 1997). The presence of SS and dissolved colored material causes water to appear opaque, muddy, and hazy in appearance. This type of water is called turbid water. Turbidity is commonly referred to as the opposite of water clarity.

The SS is the most common pollutant in surface waters and one of the earliest and most commonly studied water quality indicators through remote sensing (Wang and Yang, 2019). It is a key water quality indicator because of the presence of heavy metals and micropollutants. *In situ*, laboratory, satellite, and aircraft measurements have shown that the reflectance due to SS in surface water is affected by color, texture, and sediment type (Novo et al., 1989), sensor view, sun angles (Ritchie et al., 1975) and water depth (Blanchard et al., 1973). Many researchers have tried to establish the relationship between suspended sediment and the sensor reflectance. The SS present in surface waters tends to increase the reflectance in the visible and NIR region of the EM spectrum (Ritchie et al., 1976). The radiance absorption by suspended sediment is much smaller as compared to the chlorophyll. The SS are mainly responsible for the scattering part whereas chl-a, colored dissolved or particulate matter are responsible for the absorption part of the EM spectrum (Myint and Walker, 2002). The scattering is more prominent if more particles are present and less otherwise.

Different techniques of Remote sensing are widely used to map and monitor the concentration of SS in large water bodies. Studies have shown that single bands and algorithms sensitive to SS concentration can estimate the TSS provided suitable band is chosen. In an *in-situ* study conducted by

Ritchie et al. (1976), he concluded that the wavelengths between 700 nm and 800 nm are most suitable to determine suspended sediment in surface waters. Some scientists also claim that the optimum wavelength depends on the concentration of SS (Curran and Novo, 1988). Ritchie et al. (1976) found that a linear relationship exists between SS and its reflectance at any wavelength provided SS concentration is between 0 mg/L and 50 mg/L. However, it exhibits a curvilinear relationship if the concentration of SS lies between 50 mg/L and 150 mg/L. Also, different SS reflect light differently causing variation in color. In such cases, a combination of different bands and band ratios can be employed for TSS retrievals (Doxaran et al., 2002; Nechad et al., 2010). The presence of dark-colored humus acids due to vegetation decay results in high turbidity and TSS readings. Therefore, single bands can be combined with band ratios to increase accuracy in such waterbodies.

Many studies have reported that the first four bands of Landsat give a good estimation of TSS. Dekker et al. (2002) showed that Landsat-4 TM gives a good estimation of suspended matter. The wavelength in the red region is found to be the best estimator of turbidity. Various satellite platforms such as Landsat, MODIS, SPOT, etc., have been used for TSS estimation. The single band of NIR and red were used by Onderka (2008); and Papoutsa et al. (2014), respectively using Landsat 7-ETM+ and green band of Landsat 5-TM was used by Khorram et al. (1991). Combination of bands such as green and red, blue, and red, NIR, and red were studied by Akbar et al. (2010); and Cox et al. (1998), respectively using different satellites. The higher resolution satellite like IKONOS has also been used to estimate turbidity in Charles River of USA (Hellweger et al., 2007). A combination of more than two bands have also been used successfully for TSS estimation in many studies (He et al., 2008).

The use of empirical relationships and algorithms to estimate TSS from other water quality parameters has also been studied, but few studies have extended the applicability of these equations in a different environment. An algorithm developed by Ritchie and Copper (1991) for a watershed was found to be applicable for some coming years until an event occurs that changes the quality of SS in the given watershed.

The literature review showed that Landsat has been used much more in comparison to other sensors. The hyperspectral images have been used for the studies involving higher spectral resolution. Over the years, the methods used to interpret and evaluate images have been improved from the use of simple linear regression models in earlier studies to the complex non-linear multiple regression models, PCAs, and the use of ANNs in modern studies.

7.5.1.3 COLORED DISSOLVED ORGANIC MATTERS (CDOM)

Organic carbon in marine waters also known as dissolved organic matter (DOM) is a major contributor to the global carbon cycle (Downing et al., 2009). CDOM, also called Gilvin, yellow matter, gelbstoff, or aquatic humus is the general term given to the naturally occurring soluble organic substances that can pass through a 0.45 μm filter. CDOM is generally brown in color but gives yellowish brown color when present in higher concentrations (Aiken, 1985). CDOM is also the largest reservoir of dissolved organic carbon (DOC). CDOM is the dominant photoactive portion of DOM that affects the color of water and is used as a tracer to map the spatial distribution of DOC in aquatic ecosystems. Additionally, CDOM helps determine the photochemical characteristics of ocean ecosystems along with chlorophyll and non-algal particles (NAPs). In general, CDOM dominate the watercolor along with chl-a and TSS.

CDOM absorbs light in the UV and visible portion of the EM spectrum. The absorption spectrum of CDOM overlaps with the wavelengths at which chl-a is measured. The absorptivity of CDOM increases quasi-exponentially with decreasing wavelengths and is increasingly subdued in green and red regions (below 500 nm). Also, at higher CDOM concentrations, it significantly absorbs red lights (Strombeck and Pierson, 2001).

Similar to the estimation of chlorophyll and TSS, CDOM concentrations can also be estimated by the single bands, combination of two or more bands and band ratios. Many such studies have been conducted for estimation of CDOM concentrations using sensors such as Landsat 5-TM, EO-1 Hyperion with MIM (matrix inversion method), SeaWiFS, MODIS, HICO, etc. Many new techniques like neural networks, fuzzy modeling, and multi-band quasi-analytical algorithm (QAA) have also been applied for CDOM estimations (Gholizadeh et al., 2016).

Remote sensing of CDOM provides a rapid, efficient, and extensive approach to monitor the spatial and temporal dynamics of CDOM. The inversion algorithms are widely used to retrieve CDOM concentrations. The CDOM absorption coefficient at 440 nm accounts for 50% of the total absorption of CDOM; hence, it is usually taken as representative of CDOM concentrations in ocean studies (Corbett, 2007). The algorithms used to measure chl-a concentrations are inverted empirically to derive its covariance with chlorophyll (Hoge et al., 1995). Waters with a low concentration of CDOM have shown relatively higher accuracy probably because of uniform phytoplankton degradation (Mannino et al., 2008).

However, since both CDOM and chlorophyll absorb light in the similar spectral range, it is difficult to differentiate between the two especially in river and coastal waters. Therefore, hyperspectral remote sensing is used for CDOM estimation due to their spectral response of water IOPs. Hyperspectral imaging has an advantage of spectral responses to water IOPs and their broad spectrum of narrow bands. It increases the accuracy of estimation by incorporating bottom effects and computational techniques. The identification of the wavelength corresponding to the CDOM concentration from the narrow bands of hyperspectral RS is a difficult task and it is accomplished by techniques such as spectral indices, derivative analysis, band selection, or hyperspectral transformation to reduce the dimensions of hyperspectral data. Ground truthing of the remotely sensed data is done using *in situ* hyperspectral measurements, spectroradiometers, and shipboard data acquisition approach. The hyperspectral remote sensing is used in combination with semi-analytical and empirical models.

The use of semi-analytical algorithms for CDOM is based on the assumption that the spectral slope is constant for CDOM (Maritorena et al., 2000), but such a hypothesis does not always hold true, especially in turbid waters. In such cases, empirical approaches are used, but their applicability is limited in space in which they are applied.

7.5.1.4 SECCHI DISK DEPTH (SDD)

A white circular disc 20 cm in diameter, later modified to a black and white disk 20 cm in diameter, is called a Secchi disk. When a Secchi disk is immersed into a water body, the depth at which the disk is no longer visible is referred to as Secchi disk depth (SDD) (Z_{SD}, m). It works on the principle of light attenuation (Mobley, 1994). SDD is used as an indicator of water clarity or transparency and is also related to water turbidity. The changes in Zsd values have been linked to variation in eutrophication and phytoplankton biomass (Fleming and Laamanen, 2012) hence, it is also referred to as an index of water quality and eutrophication. The previous studies have also shown that TSS has an inverse correlation with the Secchi depth of the waterbodies.

In situ determination of Zsd values can be an unproductive despite of its low cost and easy operation because it not does not provide its distribution in space and time. The remote sensing has been used to map the distribution of Zsd values in large areas. Empirical and semi-analytical approaches are the two basic approaches are employed to retrieve Zsd values from remote sensing data. The empirical approach uses the regression principles as discussed earlier but the semi-analytical approach is based on classical theory, also called underwater visibility theory. Lee et al. (2015) found some flaws in the classical theory which was strongly based on the bean attenuation coefficient and proposed a new theory known as Lee15 algorithm based on the diffuse attenuation coefficient (Kd) at a wavelength corresponding to the maximum transparency for Zsd retrieval. The Kd varies with IOPs. He used the QAA (version 6) to obtain the values of IOPs. Many different algorithms have been developed to find Zsd values using the remote sensing data from Landsat-TM, IKONOS, MSS, etc., but Landsat-TM remains the most frequently used sensors for Zsd estimations.

Many different relationships have been established between the SDD and other water quality variables. In a study conducted by Khattab and Merkel (2015), they derived algorithms between SDD and different water quality indicators for the upper 6 meters of Mosul Lake in Iraq. The SDD was highly correlated with turbidity (R^2 >0.9) and pH (R^2 > 0.8) respectively. Significant correlation was found between Secchi disk visibility and the TDS, temperature, DOC, EC, and phosphorus concentration at most of the depths (R^2 > 0.7) through signal regression analysis. Similar to the previously discussed water quality parameters, Zsd values can also be estimated from spectral bands, a combination of bands and band ratios (Gholizadeh et al., 2016).

7.5.1.5 WATER TEMPERATURE

Water temperature is a key factor that regulates the physical, chemical, and biological processes in all the aquatic ecosystems as well as the air-water interactions. The solubility of various chemical constituents is affected by the water temperatures. It especially influences the DO concentration in water, as the solubility of DO decreases with increase in temperature. So, during the summer season, the DO concentrations are low as compared to the winter season.

Lakes are considered as an indicator of climate change because of its sensitivity to respond to variations in energy exchanges with the atmosphere. Lake surface water temperature (LSWT) are traditionally measured by *in situ* sensors or gages; however, because of limitations of expenses, time consumed, and spatial heterogeneity of these systems, they are not generally preferred. Thermal infrared (TIR) bands were previously employed to detect thermal pollution from anthropogenic activities but with the improvement in the abilities of satellite sensors in terms of spectral, radiometric, temporal, and spatial resolutions have made it possible to measure LSWT in the top layer (upper 100 μm, also called the skin temperature) of the water bodies with determined accuracy and uncertainties. Thermal infrared (TIR) bands measure LSWT but their availability is limited to a few sensors that detect thermal radiation in the wavelengths of 3–14 μm (Atwell et al., 1971; Chen et al., 1998; Robinson et al., 1984). The choice of bands should be done after carefully considering the noise and atmospheric effects. The choice between spaceborne or airborne sensor should be based on the spatial (pixel size) and temporal (revisit interval) resolution and radiometric precision of the sensor. Airborne sensors with finer pixel size are suitable for small water bodies like rivers, whereas space-borne sensors are apt for larger waterbodies. Some of the commonly used thermal sensors for surface temperature estimation include TIR bands of Landsat sensors including TM, ETM+ and OLI/TRIS, MODIS, ASTER, and AVHRR (Gholizadeh et al., 2016). Tavares et al. (2019) in a comparative analysis of two thermal sensors, i.e., MODIS, and Landsat 7 ETM+ to estimate the surface water temperatures recommended MOD11 (RMSE 1.07C) for large lakes and AtmCorr Landsat-derived LSWT (RMSE of 1.07°C) for small lakes. He also showed that LWST derived from Landsat using the RTE is very sensitive to atmospheric parameters and emissivity. Passive microwave techniques like microwave radiometers can be used in cloudy conditions with an accuracy of 1.5–2°C. A combination of point *in situ* measurements and the remotely sensed data can provide accurate estimates of the surface temperatures at low costs.

7.5.1.6 SALINITY

The distribution of sea surface salinity (SSS) gives information about the physical and biochemical processes in marine ecosystems. It plays an important part in global water balance, evaporation studies and air-water interactions. Methods like microwave remote sensing (microwave radiometers and airborne radiometers) and optical remote sensing are used for SSS determination (Font et al., 2008; Talone et al., 2010), e.g., soil moisture and ocean salinity (SMOS) is used to measure SSS with high accuracy. Microwave imaging radiometer using aperture synthesis (MIRAS) is used as one of the main instruments on the SMOS sensing at wavelengths of 20–30 cm having a spatial resolution of 40 km and a three-day temporal resolution (Barre et al., 2008). Another sensor used in SSS studies is Aquarius with a spatial resolution of 100–150 km and a 7-day temporal resolution. Optical methods include developing algorithms from satellites like Landsat-TM, MODIS through regression analysis. Sun et al. (2019) applied a regionalized algorithm to data obtained from geostationary ocean color imager (GOCI) to estimate SSS in the yellow sea. Since salinity has no direct color signal, indirect methods are used like establishing empirical relationships between different water quality parameters (sensitive to color signal) and salinity, e.g., SSS, and CDOM have been found to have an inverse relationship (Yu et al., 2017). The literature suggests that identification of appropriate algorithms, calibration, and validation is a key to estimate SSS with higher accuracy.

7.5.2 OPTICALLY INACTIVE WATER QUALITY PARAMETERS

7.5.2.1 DISSOLVED OXYGEN (DO)

Dissolved oxygen (DO) is an important parameter as too high or too low affects the living conditions of the organisms that depend on oxygen to survive. The combined effect of water temperature, salinity, DO, turbidity, and nutrient availability directly affects the phytoplankton and plant communities (Carr et al., 2016; Pesce et al., 2018). Respiring and decaying organisms reduce the amount of DO in the system while photosynthetic plants, stream flow and aeration increase the DO concentrations in marine ecosystems. Decrease in DO concentrations over the years due to rising ocean temperatures has caused a negative effect on the marine ecosystems, especially in coastal and oceanic environments (Breitburg et al., 2018; Diaz and Rosenberg, 2008).

The relationship between water quality data like chemical oxygen demand (COD), nutrients, DO, and remote sensing data is quite complex, and methods like neural network, artificial intelligence, and machine learning have been used (Kim et al., 2014; Sharaf et al., 2017). Kim et al. (2020) used a multiple-regression approach to measure DO concentrations. He used the satellite data from MODIS and VIIRS and correlated it to the long-term monitoring datasets of 15 water quality parameters measured *in situ* in coastal waters of Korea. These results showed that the DO and water temperature were highly correlated, and the *in situ* DO can be estimated from the sea surface temperature (SST) values of present and a month prior and the information on chlorophyll-a.

7.5.2.2 BIOLOGICAL OXYGEN DEMAND (BOD) AND CHEMICAL OXYGEN DEMAND (COD)

Both BOD and COD represent the oxygen demanding strength of the wastewaters (Shah et al., 2020). BOD is defined as the oxygen consumed by the micro-organisms while decomposing organic matter under aerobic conditions at a given temperature. The microorganisms present in the wastewater released from the sewage treatment plants (STPs) use the DO from the waterbodies to decompose the organic matter and hence can reduce the amount of DO to a point that can be lethal to marine life. COD is defined as the amount of oxygen required to chemically oxidize both the biologically active, i.e., bacteria, and biologically inactive matter in water. The values of COD are greater than BOD because COD calculates the total oxygen required to decompose both the organic and inorganic matter present in water. The estimation of these optically inactive water quality parameters has been done mostly by correlating it with optically active constituents. Gholizadeh et al. (2016) used an empirical method of regression and selected bands based on spectral response of target objects to assess the COD and BOD. Several linear, exponential, and logarithmic models are studied to establish relationships between *in situ* measurements and the remote sensing data, but these relationships are not implicit. Compared to the optically active water quality parameters, very few studies have focused on the estimation of BOD and COD (Wang et al., 2004; Yang et al., 2011). Sharaf et al. (2017) used a back propagation neural network (BPNN) to map the concentration of BOD and COD using satellite data from Landsat-8. The developed algorithm was able to estimate BOD (R^2 = 0.93) and COD (R^2 = 0.937) with reasonable accuracy in the testing phase. Zhang et al. (2020) also obtained similar results for BOD and COD estimation using hybrid BPNN and hyperspectral data from unmanned aerial vehicle (UAV). He also compared the hybrid BPNN model with the other models like ANN-BP, semi-analytical and empirical regression method and found that the R^2, RMSE, and MPAE values were most favorable in case of hybrid BPNN method.

7.5.2.3 TOTAL PHOSPHORUS (TP)

Total phosphorus (TP) act as an indicator of plant nutrient availability that promotes phytoplankton growth and eutrophication of lakes. Agricultural runoffs containing fertilizers and wastewater effluents contribute largely to the TP in waterbodies. It is therefore required to reduce the TP concentration to limit the eutrophication process. TP can be derived both directly and indirectly. Many studies have been reported where the TP has been determined directly through statistical relationships between the reflectance from single bands or band ratios and *in situ* observations, but the results have not been synonymous in terms of characteristic bands and the procedure used to derive estimation algorithms (Gong et al., 2008; Isenstein and Park, 2014; Kutser et al., 1995). Since TP does not has a direct influence on the spectrum and monitoring TP through remote sensing is not very clear, so the values are found out indirectly by correlating the reflectance values with other optically active water quality parameters like chl-a, TSS, and CDOM (Li et al., 2017; Song et al., 2012; Wu et al., 2010) and then estimating the TP using empirical algorithms. TP increases with increase in chl-a concentration (Chen et al., 2003; McQueen et al., 1986; Vollenweider, 1976) whereas TP increases with the decrease in SDD concentrations (Heiskary and Wilson, 2005).

Most of the studies related to developing algorithms to estimate TP are only applicable to the study area being studied and the algorithms fail to perform in other environments (Gao et al., 2015; Hui and Yao, 2016; Wu et al., 2010). Band combinations like blue and green, red, and green available on sensors like Landsat-TM, MODIS, CASI, and SPOT are widely used to monitor TP. The hyperspectral data sets can also be used for detection of TP in smaller water bodies.

7.6 CONCLUSIONS

The remote sensing techniques combined with the conventional *in situ* measurements are a cost-effective and efficient way to monitor the Spatio-temporal variations of the water quality parameters in different waterways. The sensor technology has improved from the use of multi-spectral sensors having only 5–10 bands to hyperspectral sensor which can have as many as 224 bands that can cover narrow regions of the spectrum with reasonable accuracy. The selection on the type of sensor to be used for a particular study should be based on the temporal, radio-metric, spectral, and spatial resolution and the amount of precision and accuracy required by the study. The literature review suggests that the Landsat series has been most commonly employed for different water quality studies because of easy and free data availability and universal appeal. Optically active water quality parameters like chlorophyll, SS, CDOM, temperature, and salinity have been effectively monitored by remote sensing techniques, but the use of remote sensing for optically inactive water quality parameters is relatively limited because they do not influence the spectrum directly. Different empirical and analytical approaches are employed to relate optically active and inactive parameters. Although analytical approaches offer much more in terms of the physics involved, they are relatively complex and difficult to understand. These approaches have different limitations in terms of the environments in which they are applicable. In short, remote sensing is an indispensable tool for water quality measurements, but it cannot fully replace the conventional methods. Therefore, combining remote sensing with the *in-situ* measurements is necessary along with adequate calibration and validation of the remotely sensed data. Further research is needed in the field of remote sensing to understand the dynamics of the optically inactive water quality parameters.

KEYWORDS

- **artificial neural network**
- **colored dissolved organic matter**
- **dissolved oxygen**
- **near-infrared**
- **Secchi disk depth**
- **total phosphorus**
- **total suspended solids**

REFERENCES

Acharya, T. D., Lee, D. H., Yang, I. T., & Lee, J. K., (2016). Identification of water bodies in a Landsat 8 OLI image using a J48 decision tree. *Sensors, 16*(7), 1075.

Aiken, G. R., (1985). Humic substances in soil, sediment and water. *Geochemistry, Isolation and Characterization.*

Akbar, T., Hassan, Q., & Achari, G., (2010). A remote sensing based framework for predicting water quality of different source waters. *Proceedings of ISPRS Commission I Mid-Term Symposium, Image Data Acquisition.*

Allee, R. J., & Johnson, J. E., (1999). Use of satellite imagery to estimate surface chlorophyll a and Secchi disc depth of Bull Shoals Reservoir, Arkansas, USA. *International Journal of Remote Sensing, 20*(6), 1057–1072.

An, R., Liu, Y., Qu, C., Ja, Q. B., Liang, X., Xu, X., Wang, Z., & Jiang, D., (2013). Estimation of chlorophyll-a concentration of case - waters from hyperspectral remote sensing data in NDCI Method. *Journal of Lake Sciences, 25*(3), 437–444.

Atwell, B. H., MacDonald, R. B., & Bartolucci, L. A., (1971). Thermal mapping of streams from airborne radiometric scanning 1. *JAWRA Journal of the American Water Resources Association, 7*(2), 228–243.

Barré, H. M. J. P., Duesmann, B., & Kerr, Y. H., (2008). SMOS: The mission and the system. *IEEE Transactions on Geoscience and Remote Sensing, 46*(3), 587–593.

Bhattf, A. M., Rundquist, D., Schalles, J., & Ramirez, L. (2010). Application of hyperspectral remotely sensed data for water quality monitoring: Accuracy and limitation. In *Accuracy 2010 – Proceedings of the 9th International Symposium on Spatial Accuracy Assessment in Natural Resources and Environmental Sciences* (pp. 349–352). International Spatial Accuracy Research Association (ISARA).

Blanchard, B. J., BJ, B., & RW, L., (1973). Spectral reflectance of water containing suspended sediment. *Remote Sensing and Water Resource Management,* 1283–1289.

Bonansea, M., Ledesma, M., Rodriguez, C., & Pinotti, L., (2019). Using new remote sensing satellites for assessing water quality in a reservoir. *Hydrological Sciences Journal, 64*(1), 34–44.

Breitburg, D., Levin, L. A., Oschlies, A., Grégoire, M., Chavez, F. P., Conley, D. J., Garçon, V., et al., (2018). Declining oxygen in the global ocean and coastal waters. *Science, 359*(6371).

Brezonik, P., Menken, K. D., & Bauer, M., (2005). Landsat-based remote sensing of lake water quality characteristics, including chlorophyll and colored dissolved organic matter (CDOM). *Lake and Reservoir Management, 21*(4), 373–382.

Brivio, P. A., Giardino, C., & Zilioli, E., (2001). Determination of chlorophyll concentration changes in Lake Garda using an image-based radiative transfer code for Landsat TM images. *International Journal of Remote Sensing, 22*(2, 3), 487–502.

Bukata, R. P., Jerome, J. H., Kondratyev, A. S., & Pozdnyakov, D. V., (1995). *Optical Properties and Remote Sensing of Inland and Coastal Waters*. CRC press.

Cai, X., Feng, L., Hou, X., & Chen, X., (2016). Remote sensing of the water storage dynamics of large lakes and reservoirs in the Yangtze River Basin from 2000 to 2014. *Scientific Reports, 6*, 36405.

Carr, J. A., D'Odorico, P., McGlathery, K. J., & Wiberg, P. L., (2016). Spatially explicit feedbacks between seagrass meadow structure, sediment and light: Habitat suitability for seagrass growth. *Advances in Water Resources, 93*, 315–325.

Chen, Y. D., Carsel, R. F., McCutcheon, S. C., & Nutter, W. L., (1998). Stream temperature simulation of forested riparian areas: I. Watershed-scale model development. *Journal of Environmental Engineering, 124*(4), 304–315.

Chen, Y., Fan, C., Teubner, K., & Dokulil, M., (2003). Changes of nutrients and phytoplankton chlorophyll-a in a large shallow lake, Taihu, China: An 8-year investigation. *Hydrobiologia, 506*(1–3), 273–279.

Chen, Y., Huang, C., Ticehurst, C., Merrin, L., & Thew, P., (2013). An evaluation of MODIS daily and 8-day composite products for floodplain and wetland inundation mapping. *Wetlands, 33*(5), 823–835.

Choe, E. Y., Lee, J. W., & Lee, J. K., (2011). Estimation of chlorophyll-a concentrations in the Nakdong River using high-resolution satellite image. *Korean Journal of Remote Sensing, 27*(5), 613–623.

Corbett, C. A., (2007). *Colored Dissolved Organic Matter (CDOM) Workshop Summary*.

Cox, Jr. R. M., Forsythe, R. D., Vaughan, G. E., & Olmsted, L. L., (1998). Assessing water quality in Catawba River reservoirs using Landsat thematic mapper satellite data. *Lake and Reservoir Management, 14*(4), 405–416.

Curran, P. J., & Novo, E. M. M., (1988). The relationship between suspended sediment concentration and remotely sensed spectral radiance: A review. *Journal of Coastal Research*, 351–368.

Dai, X., Zhou, Y., Ma, W., & Zhou, L., (2017). Influence of spatial variation in land-use patterns and topography on water quality of the rivers inflowing to Fuxian Lake, a large deep lake in the plateau of southwestern China. *Ecological Engineering, 99*, 417–428.

Dall'Olmo, G., Gitelson, A. A., & Rundquist, D. C., (2003). Towards a unified approach for remote estimation of chlorophyll-a in both terrestrial vegetation and turbid productive waters. *Geophysical Research Letters, 30*(18).

Dar, M. U. D., Bhat, S. A., Ali, & Shah, A. I., (2017). Modeling climate change impact; a study on different procedures and strategies: A review. *Int. J. Pure Appl. Biosci., 5*, 183–200.

Dar, M. U. D., Shah, A. I., Ali, S., & Bhat, S. A., (2020). *Woodchip Bioreactors for Nitrate Removal in Agricultural Land Drainage: Agricultural Waste Book* (pp. 99–118). Taylor &franc.

Dekker, A G., Malthus, T. J., & Hoogenboom, H. J., (1995). In: Danson, F. M., & Plummer, S. E., (eds.), *The Remote Sensing of Inland Water Quality, Advances in Remote Sensing*. John Wiley and Sons, Chichester, United Kingdom.

Dekker, A. G., & Peters, S. W. M., (1993). The use of the thematic mapper for the analysis of eutrophic lakes: A case study in the Netherlands. *International Journal of Remote Sensing, 14*(5), 799–821.

Dekker, A. G., Malthus, T. J., & Seyhan, E., (1991). Quantitative modeling of inland water quality for high-resolution MSS systems. *IEEE Transactions on Geoscience and Remote Sensing, 29*(1), 89–95.

Dekker, A. G., Vos, R. J., & Peters, S. W. M., (2002). Analytical algorithms for lake water TSM estimation for retrospective analyses of TM and SPOT sensor data. *International Journal of Remote Sensing, 23*(1), 15–35.

Diaz, R. J., & Rosenberg, R., (2008). Spreading dead zones and consequences for marine ecosystems. *Science, 321*(5891), 926–929.

Downing, B. D., Boss, E., Bergamaschi, B. A., Fleck, J. A., Lionberger, M. A., Ganju, N. K., Schoellhamer, D. H., & Fujii, R., (2009). Quantifying fluxes and characterizing compositional changes of dissolved organic matter in aquatic systems in situ using combined acoustic and optical measurements. *Limnology and Oceanography: Methods, 7*(1), 119–131.

Doxaran, D., Froidefond, J. M., Lavender, S., & Castaing, P., (2002). Spectral signature of highly turbid waters: Application with SPOT data to quantify suspended particulate matter concentrations. *Remote Sensing of Environment, 81*(1), 149–161.

Ekercin, S., (2007). Water quality retrievals from high resolution IKONOS multispectral imagery: A case study in Istanbul, Turkey. *Water, Air, and Soil Pollution, 183*(1–4), 239–251.

Feng, H., Campbell, J. W., Dowell, M. D., & Moore, T. S., (2005). Modeling spectral reflectance of optically complex waters using bio-optical measurements from Tokyo Bay. *Remote Sensing of Environment, 99*(3), 232–243.

Feng, L., Hu, C., Chen, X., Cai, X., Tian, L., & Gan, W., (2012). Assessment of inundation changes of Poyang Lake using MODIS observations between 2000 and 2010. *Remote Sensing of Environment, 121*, 80–92.

Fleming-Lehtinen, V., & Laamanen, M., (2012). Long-term changes in Secchi depth and the role of phytoplankton in explaining light attenuation in the Baltic Sea. *Estuarine, Coastal and Shelf Science, 102*, 1–10.

Font, J., Camps, A., & Ballabrera-Poy, J., (2008). Microwave aperture synthesis radiometry: Paving the path for sea surface salinity measurement from space. In: *Remote Sensing of the European Seas* (pp. 223–238). Springer.

Gao, Y., Gao, J., Yin, H., Liu, C., Xia, T., Wang, J., & Huang, Q., (2015). Remote sensing estimation of the total phosphorus concentration in a large lake using band combinations and regional multivariate statistical modeling techniques. *Journal of Environmental Management, 151*, 33–43.

Gholizadeh, M. H., Melesse, A. M., & Reddi, L., (2016). A comprehensive review on water quality parameters estimation using remote sensing techniques. *Sensors, 16*(8), 1298.

Giardino, C., Brando, V. E., Dekker, A. G., Strömbeck, N., & Candiani, G., (2007). Assessment of water quality in Lake Garda (Italy) using Hyperion. *Remote Sensing of Environment, 109*(2), 183–195.

Giardino, C., Bresciani, M., Villa, P., & Martinelli, A., (2010). Application of remote sensing in water resource management: The case study of Lake Trasimeno, Italy. *Water Resources Management, 24*(14), 3885–3899.

Gitelson, A. A., Dall'Olmo, G., Moses, W., Rundquist, D. C., Barrow, T., Fisher, T. R., Gurlin, D., & Holz, J., (2008). A simple semi-analytical model for remote estimation of chlorophyll-a in turbid waters: Validation. *Remote Sensing of Environment, 112*(9), 3582–3593.

Gitelson, A. A., Gritz, Y., & Merzlyak, M. N., (2003). Relationships between leaf chlorophyll content and spectral reflectance and algorithms for non-destructive chlorophyll assessment in higher plant leaves. *Journal of Plant Physiology, 160*(3), 271–282.

Gitelson, A. A., Gurlin, D., Moses, W. J., & Barrow, T., (2009). A bio-optical algorithm for the remote estimation of the chlorophyll-a concentration in case 2 waters. *Environmental Research Letters, 4*(4), 45003.

Gitelson, A. A., Yacobi, Y. Z., Karnieli, A., & Kress, N., (1995). Remote estimation of chlorophyll concentration in polluted marine waters in Haifa Bay, Southeastern Mediterranean. *Air Toxics and Water Monitoring, 2503*, 44–54.

Gong, S. Q., Huang, J. Z., Li, Y. M., Lu, W. N., Wang, H. J., & Wang, G. X., (2008). Preliminary exploring of hyperspectral remote sensing experiment for nitrogen and phosphorus in water. *Guang Pu Xue Yu Guang Pu Fen Xi= Guang Pu, 28*(4), 839–842.

Gordon, H. R., & Wang, M., (1994). Retrieval of water-leaving radiance and aerosol optical thickness over the oceans with SeaWiFS: A preliminary algorithm. *Applied Optics, 33*(3), 443–452.

Hakansson, B. G., & Moberg, M., (1994). Cover. The algal bloom in the Baltic during July and August 1991, as observed from the NOAA weather satellites. *Title Remote Sensing, 15*(5), 963–965.

Hameed, M., Saqib, N. E., Bandit, B. P., & Shakeel, A. B., (2017). Assessing the impact of climate change on surface water resources of Wular Lake. *Journal of Pharmacognosy and Phytochemistry, 6*, 1967–1972.

Han, L., & Jordan, K. J., (2005). Estimating and mapping chlorophyll-a concentration in Pensacola Bay, Florida using Landsat ETM+ data. *International Journal of Remote Sensing, 26*(23), 5245–5254.

Härmä, P., Vepsäläinen, J., Hannonen, T., Pyhälahti, T., Kämäri, J., Kallio, K., Eloheimo, K., & Koponen, S., (2001). Detection of water quality using simulated satellite data and semi-empirical algorithms in Finland. *Science of the Total Environment, 268*(1–3), 107–121.

Harrington, J. A., & Repic, R. L., (1995). Hyperspectral and video remote sensing of Oklahoma lakes. *Papers and Proceedings of Applied Geography Conferences, 18*, 79–86.

He, W., Chen, S., Liu, X., & Chen, J., (2008). Water quality monitoring in a slightly-polluted inland water body through remote sensing—Case study of the guanting reservoir in Beijing, China. *Frontiers of Environmental Science & Engineering in China, 2*(2), 163–171.

Heiskary, S. A., & Wilson, C. B., (2005). *Minnesota Lake Water Quality Assessment Report: Developing Nutrient Criteria*. Water Assessment & Environmental Information Section, Environmental Analysis.

Hellweger, F. L., Miller, W., & Oshodi, K. S., (2007). Mapping turbidity in the Charles River, Boston using a high-resolution satellite. *Environmental Monitoring and Assessment, 132*(1–3), 311–320.

Hellweger, F. L., Schlosser, P., Lall, U., & Weissel, J. K., (2004). Use of satellite imagery for water quality studies in New York Harbor. *Estuarine, Coastal and Shelf Science, 61*(3), 437–448.

Hoge, F. E., Williams, M. E., Swift, R. N., Yungel, J. K., & Vodacek, A., (1995). Satellite retrieval of the absorption coefficient of chromophoric dissolved organic matter in continental margins. *Journal of Geophysical Research: Oceans, 100*(C12), 24847–24854.

Huang, C., Chen, Y., Wu, J., Li, L., & Liu, R., (2015). An evaluation of Suomi NPP-VIIRS data for surface water detection. *Remote Sensing Letters, 6*(2), 155–164.

Huang, C., Chen, Y., Zhang, S., & Wu, J., (2018). Detecting, extracting, and monitoring surface water from space using optical sensors: A review. *Reviews of Geophysics, 56*(2), 333–360.

Hui, J., & Yao, L., (2016). Analysis and inversion of the nutritional status of China's Poyang Lake using MODIS data. *Journal of the Indian Society of Remote Sensing, 44*(5), 837–842.

Hunter, P. D., Tyler, A. N., Carvalho, L., Codd, G. A., & Maberly, S. C., (2010). Hyperspectral remote sensing of cyanobacterial pigments as indicators for cell populations and toxins in eutrophic lakes. *Remote Sensing of Environment, 114*(11), 2705–2718.

Isenstein, E. M., & Park, M. H., (2014). Assessment of nutrient distributions in Lake Champlain using satellite remote sensing. *Journal of Environmental Sciences, 26*(9), 1831–1836.

Ji, L., Geng, X., Sun, K., Zhao, Y., & Gong, P., (2015). Target detection method for water mapping using Landsat 8 OLI/TIRS imagery. *Water, 7*(2), 794–817.

Khattab, M. F. O., & Merkel, B., (2015). Secchi disc visibility and its relationship with water quality parameters in the photosynthesis zone of Mosul Dam Lake, Northern Iraq. *FOG-Freiberg Online Geoscience*.

Khorram, S., Cheshire, H., Geraci, A. L., & Rosa, G. L. A., (1991). Water quality mapping of Augusta Bay, Italy from Landsat-TM data. *International Journal of Remote Sensing, 12*(4), 803–808.

Kim, Y. H., Im, J., Ha, H. K., Choi, J. K., & Ha, S., (2014). Machine learning approaches to coastal water quality monitoring using GOCI satellite data. *GIScience & Remote Sensing, 51*(2), 158–174.

Kim, Y. H., Son, S., Kim, H. C., Kim, B., Park, Y. G., Nam, J., & Ryu, J., (2020). Application of satellite remote sensing in monitoring dissolved oxygen variabilities: A case study for coastal waters in Korea. *Environment International, 134*, 105301.

Knaeps, E., Raymaekers, D., Sterckx, S., & Odermatt, D., (2010). *An Intercomparison of Analytical Inversion approaches to Retrieve Water Quality for Two Distinct Inland Waters*.

Kong, J. L., Sun, X. M., Wong, D. W., Chen, Y., Yang, J., Yan, Y., & Wang, L. X., (2015). A semi-analytical model for remote sensing retrieval of suspended sediment concentration in the Gulf of Bohai, China. *Remote Sensing, 7*(5), 5373–5397.

Kutser, T., Arst, H., Miller, T., Käärmann, L., & Milius, A., (1995). Telespectrometrical estimation of water transparency, chlorophyll-a and total phosphorus concentration of Lake Peipsi. *International Journal of Remote Sensing, 16*(16), 3069–3085.

Lathrop, R., & Lillesand, T. M., (1986). Use of thematic mapper data to assess water quality in green bay and central Lake Michigan. *Photogrammetric Engineering & Remote Sensing, 52*(5), 671–680.

Le, C., Hu, C., Cannizzaro, J., English, D., Muller-Karger, F., & Lee, Z., (2013). Evaluation of chlorophyll-a remote sensing algorithms for an optically complex estuary. *Remote Sensing of Environment, 129*, 75–89.

Lee, Z., Shang, S., Hu, C., Du, K., Weidemann, A., Hou, W., Lin, J., & Lin, G., (2015). Secchi disk depth: A new theory and mechanistic model for underwater visibility. *Remote Sensing of Environment, 169*, 139–149.

Li, W., (2009). Method of water quality remote sensing and its application. *Energy and Environment, 5*(5), 62–64.

Li, Y., Gong, X., Guo, Z., Xu, K., Hu, D., & Zhou, H., (2016). An index and approach for water extraction using Landsat--OLI data. *International Journal of Remote Sensing, 37*(16), 3611–3635.

Li, Y., Zhang, Y., Shi, K., Zhu, G., Zhou, Y., Zhang, Y., & Guo, Y., (2017). Monitoring spatiotemporal variations in nutrients in a large drinking water reservoir and their relationships with hydrological and meteorological conditions based on Landsat 8 imagery. *Science of the Total Environment, 599*, 1705–1717.

Lindell, T., Pierson, D., & Premazzi, G., (1999). *Manual for monitoring European lakes using remote sensing techniques*. Joint Research Centre, Ispra, Italy.

Lu, S., Jia, L., Zhang, L., Wei, Y., Baig, M. H. A., Zhai, Z., Meng, J., Li, X., & Zhang, G., (2017). Lake water surface mapping in the Tibetan plateau using the MODIS MOD09Q1 product. *Remote Sensing Letters, 8*(3), 224–233.

Mannino, A., Russ, M. E., & Hooker, S. B., (2008). Algorithm development and validation for satellite-derived distributions of DOC and CDOM in the US Middle Atlantic Bight. *Journal of Geophysical Research: Oceans, 113*(C7).

Maritorena, S., Morel, A., & Gentili, B., (2000). Determination of the fluorescence quantum yield by oceanic phytoplankton in their natural habitat. *Applied Optics, 39*(36), 6725–6737.

Matin, S., Dash, S. K., & Usha, T., (2019). A semi-analytical approach to understand remote sensing-based backscattering characteristics for Kerala coast using in situ observation. In: *Applications and Challenges of Geospatial Technology* (pp. 43–56). Springer.

Matthews, M. W., (2011). A current review of empirical procedures of remote sensing in inland and near-coastal transitional waters. *International Journal of Remote Sensing, 32*(21), 6855–6899.

Mbuh, M. J. (2019). Use of Hyperspectral Remote Sensing to Estimate Water Quality. In *Processing and Analysis of Hyperspectral Data*. IntechOpen.

McQueen, D. J., Post, J. R., & Mills, E. L., (1986). Trophic relationships in freshwater pelagic ecosystems. *Canadian Journal of Fisheries and Aquatic Sciences, 43*(8), 1571–1581.

Mobley, C. D., (1994). *Light and Water: Radiative Transfer in Natural Waters'* (Vol. 565). Academic Press. New York.

Myint, S. W., & Walker, N. D., (2002). Quantification of surface suspended sediments along a river-dominated coast with NOAA AVHRR and SeaWiFS measurements: Louisiana, USA. *International Journal of Remote Sensing, 23*(16), 3229–3249.

Nechad, B., Ruddick, K. G., & Park, Y., (2010). Calibration and validation of a generic multisensor algorithm for mapping of total suspended matter in turbid waters. *Remote Sensing of Environment, 114*(4), 854–866.

Novo, E. M. M., Hansom, J. D., & Curran, P. J., (1989). The effect of sediment type on the relationship between reflectance and suspended sediment concentration. *Remote Sensing, 10*(7), 1283–1289.

Onderka, M., (2008). *Remote Sensing and Identification of Places Susceptible to Sedimentation in the Danube River*. Available Online: Citeseerx. Ist. Psu. Edu/Viewdoc/Download.

Papoutsa, C., Retalis, A., Toulios, L., & Hadjimitsis, D. G., (2014). Defining the Landsat TM/ETM+ and CHRIS/PROBA spectral regions in which turbidity can be retrieved in inland waterbodies using field spectroscopy. *International Journal of Remote Sensing, 35*(5), 1674–1692.

Peng, D., Xiong, L., Guo, S., & Shu, N., (2005). Study of Dongting Lake area variation and its influence on water level using MODIS data/Etude de la variation de la surface du Lac Dongting et de son influence sur le niveau d'eau, grâce à des données MODIS. *Hydrological Sciences Journal, 50*(1).

Pesce, M., Critto, A., Torresan, S., Giubilato, E., Santini, M., Zirino, A., Ouyang, W., & Marcomini, A., (2018). Modeling climate change impacts on nutrients and primary production in coastal waters. *Science of the Total Environment, 628*, 919–937.

Qi, L., Hu, C., Duan, H., Barnes, B. B., & Ma, R., (2014). An EOF-based algorithm to estimate chlorophyll a concentrations in Taihu Lake from MODIS land-band measurements: Implications for near real-time applications and forecasting models. *Remote Sensing, 6*(11), 10694–10715.

Reif, M. K., et al., (2011). *Remote Sensing for Inland Water Quality Monitoring: A US Army Corps of Engineers Perspective*.

Ritchie, J. C., & Cooper, C. M., (1991). An algorithm for using Landsat MSS for estimating surface suspended sediments. *Water Resources Bulletin, 27*, 373–379.

Ritchie, J. C., Cooper, C. M., & Schiebe, F. R., (1990). The relationship of MSS and TM digital data with suspended sediments, chlorophyll, and temperature in Moon Lake, Mississippi. *Remote Sensing of Environment, 33*(2), 137–148.

Ritchie, J. C., McHenry, J. R., Schiebe, F. R., & Wilson, R. B., (1975). Relationship of reflected solar radiation and the concentration of sediment in the surface water of reservoirs. *Remote Sensing of Earth Resources*.

Ritchie, J. C., Schiebe, F. R., & McHenry, J. R., (1976). Remote sensing of suspended sediments in surface waters. *Photogramm. Eng. Remote Sens, 42*, 1539–1545.

Ritchie, J. C., Schiebe, F. R., Cooper, C. M., & Harrington, Jr. J. A., (1994). Chlorophyll measurements in the presence of suspended sediment using broadband spectral sensors aboard satellites. *Journal of Freshwater Ecology, 9*(3), 197–206.

Ritchie, J. C., Schiebe, F. R., Wilson, R. B., & May, J., (1975). Sun angle, reflected solar radiation and suspended sediments in north Mississippi reservoirs. *Remote Sensing of Earth Resources, 4*, 555–564.

Ritchie, J. C., Zimba, P. V., & Everitt, J. H., (2003). Remote sensing techniques to assess water quality. *Photogrammetric Engineering & Remote Sensing, 69*(6), 695–704.

Robinson, I. S., Wells, N. C., & Charnock, H., (1984). The sea surface thermal boundary layer and its relevance to the measurement of sea surface temperature by airborne and spaceborne radiometers. *International Journal of Remote Sensing, 5*(1), 19–45.

Sawaya, K. E., Olmanson, L. G., Heinert, N. J., Brezonik, P. L., & Bauer, M. E., (2003). Extending satellite remote sensing to local scales: Land and water resource monitoring using high-resolution imagery. *Remote Sensing of Environment, 88*(1, 2), 144–156.

Schalles, J. F., Schiebe, F. R., & Starks, P. J., (1997). *Estimation of Algal and Suspended Sediment Loads (Singly and Combined) Using Hyperspectral Sensors and Integrated Mesocosm Experiments*.

Schiebe, F. R., Harrington, Jr. J. A., & Ritchie, J. C., (1992). Remote sensing of suspended sediments: The Lake Chicot, Arkansas project. *International Journal of Remote Sensing, 13*(8), 1487–1509.

Shah, A. I., Din, D. M. U., Bhat, R. A., Singh, J. P., Singh, K., & Bhat, S. A., (2020). Prospectives and challenges of wastewater treatment technologies to combat contaminants of emerging concerns. *Ecol. Eng., 152*, 105882. https://doi.org/10.1016/j.ecoleng.2020.105882.

Sharaf El, D. E., Zhang, Y., & Suliman, A., (2017). Mapping concentrations of surface water quality parameters using a novel remote sensing and artificial intelligence framework. *International Journal of Remote Sensing, 38*(4), 1023–1042.

Siegel, H., & Gerth, M., (2000). Remote-sensing studies of the exceptional summer of 1997 in the Baltic Sea: The warmest august of the century, the Oder flood, and phytoplankton blooms. In: *Elsevier Oceanography Series* (Vol. 63, pp. 239–255). Elsevier.

Singh, K., Ghosh, M., & Sharma, S. R., (2015). WSB-DA: Water surface boundary detection algorithm using Landsat 8 OLI data. *IEEE Journal of Selected Topics in Applied Earth Observations and Remote Sensing, 9*(1), 363–368.

Song, K., Li, L., Wang, Z., Liu, D., Zhang, B., Xu, J., Du, J., Li, L., Li, S., & Wang, Y., (2012). Retrieval of total suspended matter (TSM) and chlorophyll-a (Chl-a) concentration from remote-sensing data for drinking water resources. *Environmental Monitoring and Assessment, 184*(3), 1449–1470.

Strömbeck, N., & Pierson, D. C., (2001). The effects of variability in the inherent optical properties on estimations of chlorophyll a by remote sensing in Swedish freshwaters. *Science of the Total Environment, 268*(1–3), 123–137.

Suárez, P. L., Sappa, A. D., & Vintimilla, B. X., (2018). Adaptive Harris corner detector evaluated with cross-spectral images. *International Conference on Information Theoretic Security*, 732–744.

Sun, D., Su, X., Qiu, Z., Wang, S., Mao, Z., & He, Y., (2019). Remote sensing estimation of sea surface salinity from GOCI measurements in the southern yellow sea. *Remote Sensing, 11*(7), 775.

Svejkovsky, J., & Shandley, J., (2001). Detection of offshore plankton blooms with AVHRR and SAR imagery. *International Journal of Remote Sensing, 22*(2, 3), 471–485.

Talone, M., Sabia, R., Camps, A., Vall-llossera, M., Gabarró, C., & Font, J., (2010). Sea surface salinity retrievals from HUT-2D L-band radiometric measurements. *Remote Sensing of Environment, 114*(8), 1756–1764.

Tavares, M. H., Cunha, A. H. F., Motta-Marques, D., Ruhoff, A. L., Cavalcanti, J. R., Fragoso, C. R., Martin, B. J., et al., (2019). Comparison of methods to estimate lake-surface-water temperature using Landsat 7 ETM+ and MODIS imagery: Case study of a large shallow subtropical lake in Southern Brazil. *Water, 11*(1), 168.

Torbick, N., Hession, S., Hagen, S., Wiangwang, N., Becker, B., & Qi, J., (2013). Mapping inland lake water quality across the Lower Peninsula of Michigan using Landsat TM imagery. *International Journal of Remote Sensing, 34*(21), 7607–7624.

Usali, N., & Ismail, M. H., (2010). Use of remote sensing and GIS in monitoring water quality. *Journal of Sustainable Development, 3*(3), 228.

Vollenweider, R. A., (1976). *Advances in Defining Critical Loading Levels for Phosphorus in Lake Eutrophication.* Memoirs of the Italian Institute of Hydrobiology, Dr. Marco De Marchi Verbania Pallanza.

Wang, H., Zhao, D., Wang, L., & Huang, F., (2012). Advance in remote sensing of water quality. *Marine Environmental Science, 31*(2), 285–288.

Wang, X., & Yang, W., (2019). Water quality monitoring and evaluation using remote sensing techniques in China: A systematic review. *Ecosystem Health and Sustainability, 5*(1), 47–56.

Wang, X., Zhang, F., Ghulam, A., Trumbo, A. L., Yang, J., Ren, Y., Jing, Y., et al., (2017). Evaluation and estimation of surface water quality in an arid region based on EEM-PARAFAC and 3D fluorescence spectral index: A case study of the Ebinur Lake Watershed, China. *Catena, 155*, 62–74.

Wang, Y., Xia, H., Fu, J., & Sheng, G., (2004). Water quality change in reservoirs of Shenzhen, China: Detection using LANDSAT/TM data. *Science of the Total Environment, 328*(1–3), 195–206.

Wass, P. D., Marks, S. D., Finch, J. W., Leeks, G. J. L., & Ingram, J. K., (1997). Monitoring and preliminary interpretation of in-river turbidity and remote sensed imagery for suspended sediment transport studies in the Humber catchment. *Science of the Total Environment, 194*, 263–283.

Wu, C., Wu, J., Qi, J., Zhang, L., Huang, H., Lou, L., & Chen, Y., (2010). Empirical estimation of total phosphorus concentration in the mainstream of the Qiantang River in China using Landsat TM data. *International Journal of Remote Sensing, 31*(9), 2309–2324.

Yang, B., Liu, Y., Ou, F., & Yuan, M., (2011). Temporal and spatial analysis of cod concentration in East Dongting Lake by using of remotely sensed data. *Procedia Environmental Sciences, 10*, 2703–2708.

Yu, X., Xiao, B., Liu, X., Wang, Y., Cui, B., & Liu, X., (2017). Retrieval of remotely sensed sea surface salinity using MODIS data in the Chinese Bohai Sea. *International Journal of Remote Sensing, 38*(23), 7357–7373.

Zhang, Y., Wu, L., Ren, H., Deng, L., & Zhang, P., (2020). Retrieval of water quality parameters from hyperspectral images using hybrid Bayesian probabilistic neural network. *Remote Sensing, 12*(10), 1567.

CHAPTER 8

Biosand Filters for Household Wastewater Treatment

PUNEET SHARMA,[1] MEHRAJ U. DIN DAR,[2] and N. L. KUSHWAHA[3]

[1]*Assistant Professor, Soil, and Water Engineering, Krishi Vigyan Kendra, Pathankot (Gho) – 145023, Punjab Agricultural University, Ludhiana – 141004, Punjab, India,*
E-mail: sharmapuneet840@gmail.com

[2]*Department of Soil and Water Engineering,*
Punjab Agricultural University, Ludhiana – 141004, Punjab, India,
E-mail: mehrajudindar24@gmail.com

[3]*Scientist, Division of Agricultural Engineering ICAR-IARI, New Delhi, India, E-mail: nand.kushwaha@icar.gov.in*

ABSTRACT

Biosand filtration technology is a water purifying technology for drinking purpose getting popularized in the third world countries of Africa and Asia, where it is an economically viable source of clean drinking water as compared to other conventional alternatives like boiling. This technology has come up as an institutional reform in the society of developing countries as boiling of water is mostly fueled by forest wood in African countries like Ethiopia. Therefore, it is also contributing to reducing carbon emission and economic stress. A lot of biosand filtration models have been developed which roots to modern history as well, but in this chapter, some of the successful filtration models have been mentioned. This technology's success in improving water quality conditions depends on mass participation as per the guidelines of

Advanced Technologies for Water Quality Treatment and Management. Mehraj U. Din Dar, PhD, Aamir Ishaq Shah, MTech, Shakeel Ahmad Bhat, PhD & Syed Rouhullah Ali, PhD (Eds.)

8.2.1.1 WORKING

The hydraulic head of the filter is kept high at the inlet reservoir, which pushes water into the diffuser chamber from where filtration occurs and decreasing a water flow through the media. The flow rate is low as there is gravitational force of water only acting force on the entire system. The wastewater includes oxygen component nutrients and other contaminants of which oxygen present in the contaminated water is absorbed by the microbes present in the biofilm. The other components, including large, suspended particles and harmful microbes are captured in the top section of media and purified water passes through the filtration media at a slow flow rate.

8.3 RETENTION TIME

Also known as 'idle time' this time is generally referred to as 80% of the daily cycle, as during this cycle major microbial count is decreased. The major removal of contaminants occurs in the section where biofilm is present. When the hydraulic head at the input is equivalent to the outlet pressure, the flow stops. Generally, the flow should be kept constant to keep the media wet and to keep the microbes in the biolayer active. This time responsible for the microbes to consume pathogens and nutrients are contaminants present in wastewater poured. If this period is generally accepted from more than two days that is 48 hours depending on the standard size of the filtration unit, then the biolayer is expected to consume all the microbes, including the desirable also and ultimately reducing its efficacy to clean water leading to decrease in efficiency of the system. Pathogens other than present in biolayer get killed due to the absence of food for nutrients by getting in trapped in the inert media.

8.4 REGULAR CLEANING

With regular use of the filtration unit to clean wastewater, the microbes present in the wastewater get concentrated in the filter media over time. As water passes through the filtration media, the biofilm presents at the diffuser chamber leading to the decrease in flow rate which is inconvenient to the user on the consumer side as the quantity of desirable quality water is required at a fast rate. It is recommended by CAWST as if the flow rate is less than 0.1 liter per the servicing of the filtration unit is suggested. The servicing includes the cleaning of media the diffusion chamber and the inside of the filtration unit. This process is done repetitively until desired flow rate is achieved. Cleaning should also be done for the outside surface of the filter. One of the major factors playing a role in servicing of biosand filters includes education and awareness level of the user as the long-term sustainability of the filter is very much dependent on these factors. For example, in Haiti several cleaning programs based on usage of biosand filters to provide safe drinking water to the people include regular servicing protocol after the prescribed time of the servicing depending on the type and scale of biosand filters under operation under these special appointees visit at regular intervals of time to the beneficiary homes to provide proper training and awareness to use the filtration unit safely. Such awareness programs led to use of long-term use of biosand filters among public beneficiaries (Duke et al., 2006).

8.5 TYPES OF CONTAMINANTS REMOVAL

8.5.1 TURBIDITY

Turbidity removal depends on the composition of media that is the composition of various ratios of sand silt and clay. Some studies suggested removal of turbidity to a significant level. In a study based on the local water supplies from 30 sources show the decrease in turbidity of 1.45 units being filtered using slow sand filters using biolayers (Makutsa et al., 2001).

8.5.2 HEAVY METALS

There is need to carry out more studies on the removal of heavy metals from wastewater using biosand filters as information on removal of many heavy metals is still limited (Sobsey et al., 2008).

8.5.3 BACTERIA REMOVAL

In studies carried out over the removal of bacterial count from household wastewater indicate that biosand filters are capable of removing *E. coli* bacteria at significant level making it safe for drinking purpose which depends on the biofilm formation and the supply of wastewater (Andersson, 2017). Research also shows that biosand filters remove a higher number of bacteria in lab conditions as compared to field conditions. Research conducted in the households of Dominican Republic the average equalizer reduction was 93% from wastewater (Souter et al., 2009).

8.5.4 VIRUS REMOVAL

Removal of viral load from the water depends on the retention time of wastewater being passed through the filtration unit. A recent study indicated that approximately 99% viral load was removed after a retention period of 150 days (Andrew et al., 2013).

8.5.5 PROTOZOA

Studies indicated the removal of 99% of protozoa after retention time of 29 days, making it at a par with slow sand filter (Hendricksn and Bellamy, 1991).

8.5.5.1 MATERIALS

Concrete is the most widespread material used for the construction of biosand filters as it is easily available around the world, and it is of low cost. CAWST has suggested several designs for construction with concrete for the filter of version 9 biosand filter has the maximum loading rate. Recent searches have biosand filters maximum loading rate decreases by significant rate to keep constant hydraulic head in contact with media (Stauber et al., 2009). Biosand filters are also available using plastic containers or barrels of food grade and medical grade with ultraviolet resistance as plastic barrel makes it light in weight and more portable in nature. Stainless steel is another material which is being employed by several engineering agencies like as Sehgal Foundation an NGO from India that has used stainless steel filter based on the principle of biosand and filtration process developed filtration unit called 'Jal Kalp' which offers increased filtration rate and better portability as compared to the concrete models (WHO, 2006; Ram et al., 2007). Concrete models are prone to breaking and are difficult to carry, which makes it unsuitable as compared to stainless steel models. It also depends on the skill of the manufacturer to build concrete filter whereas quality of material is also important; but in case of stainless-steel filter such issues do not arise as skill of manufacturer is the only factor. Therefore, stainless steel has clear advantages of strength, durability, portability, and even appearance as compared to the same capacity of concrete filtration model of biosand filter. Water quality test study carried out on 'Jal Kalp' model against microbes like *E. coli*, other color forms, turbidity, and iron contamination indicate the effectiveness of 'Jal Kalp' stainless steel model as comparison to plastic as well as concrete model in Indian subcontinent study area. This filter model like other biosand filter models can be improved depending on the financial input in the construction of the filter.

8.5.5.2 LIMITATIONS

Water quality decreases with high water intake rate. If the water supply is stopped for a few hours, biofilm microbes which are desirable are unable to survive because of unfavorable conditions created by dryness of biofilm, thus affecting the efficiency of the purification. If the turbidity is more than 30 NTU in input wastewater the clogging of filter media may occur which may require prefiltering of input water before bio-filtration. Depending on the quality of water media has limitation of cleaning wastewater input to a certain contamination level. Smooth vertical surfaces in the filtration unit may cause bypassing of wastewater through the filtration media; leading to contaminated water as the output. There could be regions where filtration media or biofilm could not be easily available, therefore alternate media such as rice husk could be used, but that could add to the cost of the technology application. Disposal of the used contaminated media is also a challenge which could itself act as pollutant (WHO, 2006).

8.5.5.3 ADAPTATION

The success of the technology is dependent on the adoption of technology by the masses which is equally true for the Bio-sand filtration technology. This technology needs to be promoted to the masses by effective communication which is based on the communication process (Roma et al., 2014; Wells and Harris, 2007; WHO, 2006). To promote this technology there are mainly four components required to communicate on the importance and necessity of clean potable drinking water issue concerned with health and sanitation, i.e., communicator, message, channel, and receiver as illustrated as Figure 8.2. The process is initiated by the personal appointed to deliver the details of bio-sand filter technology.

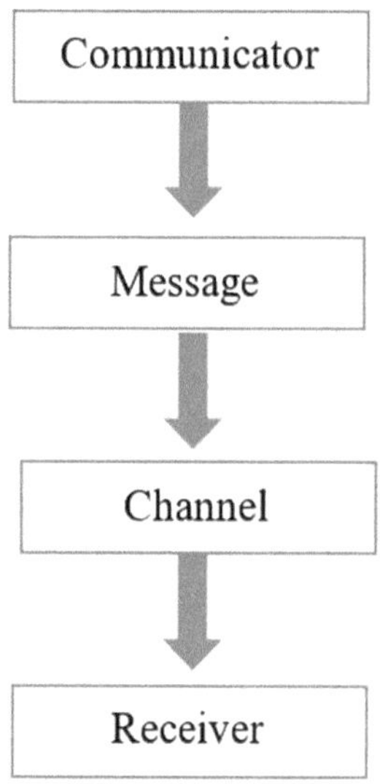

FIGURE 8.2 Stages involved in the process of communication.

Communicator, the personnel appointed should be well aware of the topics, i.e., hygiene, importance of hygiene, role bio-sand filter technology can contribute in improving drinking water condition at the household level. The communication skills of the communicator play a pivotal role in convincing the masses to adopt this technology in their personal spaces. To promote bio-sand filtration technique, the communicator should deliver the information in the form message addressing target masses by displaying them model bio-sand filters unit in working condition, for the receiver to get better clarity regarding the working of technologies (Roma et al., 2014; Wells and Harris, 2007; WHO, 2006). The communicator should be clear about the objective of delivering the message on the bio-sand filter unit as all the receivers of the message may not be interested in listening to the technical details of the unit as they could be more interested in utility and cost of the technology. Also, the message being delivered to the receiver should be based on strong scientific research being validated for region specific sites. The communicator is supposed to develop good functional relationship with local authorities with good rapport responsible for health and sanitation related to cleanliness of water so that effective communication of higher degree could take place. The communicator should be able to analyze the extent of awareness regarding bio-sand filtration of the receiver of information so that planning and execution of the information delivery process to be completed. The following flow chart should be kept in mind while preparing a message for effective communication on bio-sand filters so that general queries in the mind of the receiver should be answered as illustrated in Figure 8.3. The information being delivered in the form of the message must communicate the basic details of the bio-sand filter which the receiver could understand. The communicator must check the receivers' information being enhanced by asking for frequent feedback regarding technology being displayed.

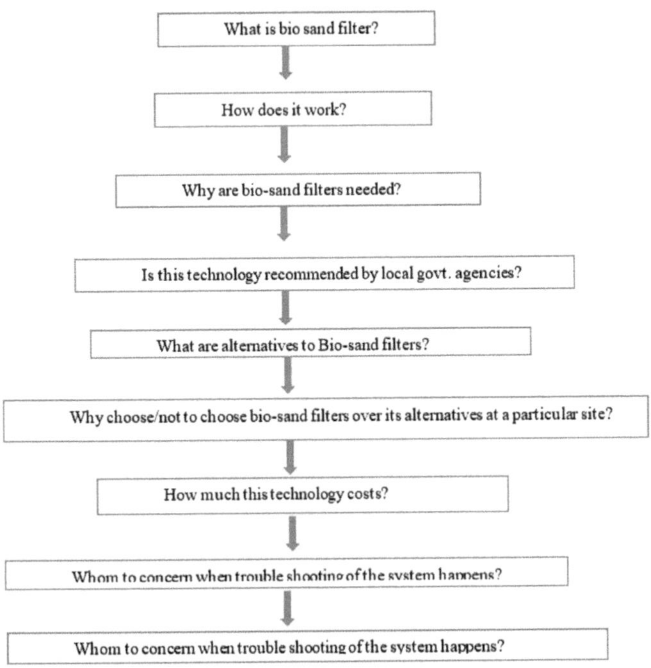

FIGURE 8.3 Stagewise points to be considered in the planning by the communicator.

The message containing basic information regarding the technology should be lucid and must be focused on the advantage of bio-sand filter technology to be adopted over the alternative ways of purifying water under the given conditions. The message must highlight repetitively the importance of adopting water purification technique of bio-sand filtration but keeping in content of comparison to other water purifying technologies and advantages or limitations of the bio-sand filter in the target region. Masses should be appealed based on facts based on research along with emotional appeal as well as such kind of technology is meant for the overall health development of the society which is fundamental in nation building. The key characteristics which a message based on the topic of bio-sand filtration should hold are represented as flowchart illustrated in Figure 8.4.

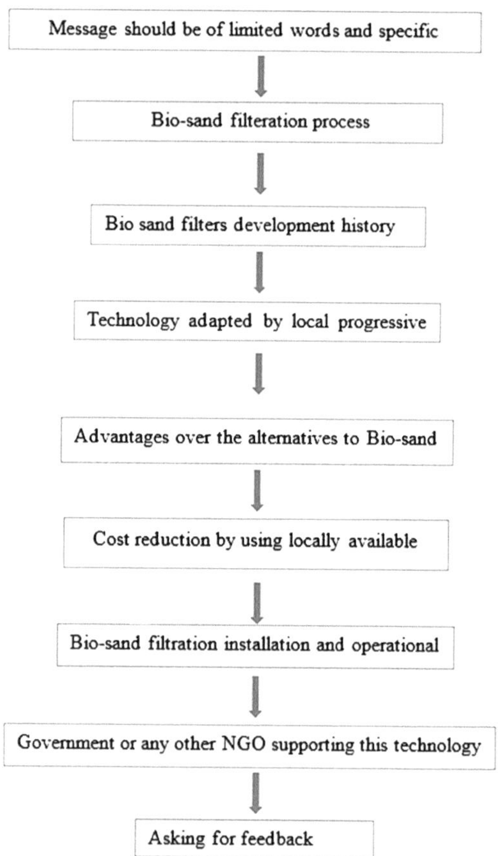

FIGURE 8.4 Characteristics of message to be communicated.

Channel of communicating this specific technology involves language of message delivery communicator, language understood by translator if any or language understood by receiver which is a cyclic process is illustrated as Figure 8.5. It is important that messages containing information to be delivered in regional languages or the lucid manner which may create significant awareness among the masses (Roma et al., 2014; Wells and Harris, 2007; WHO, 2006). Another mode of communication which could be used to deliver information are as follows: Newspaper articles, magazine articles, leaflets, descriptive folders, television programs, live working demonstrations, booklets, radio programs, community level plays, skits, videos uploading on social media platforms, organizing pilot project demonstrations to create awareness, interviewing successful operators of bio-sand filters.

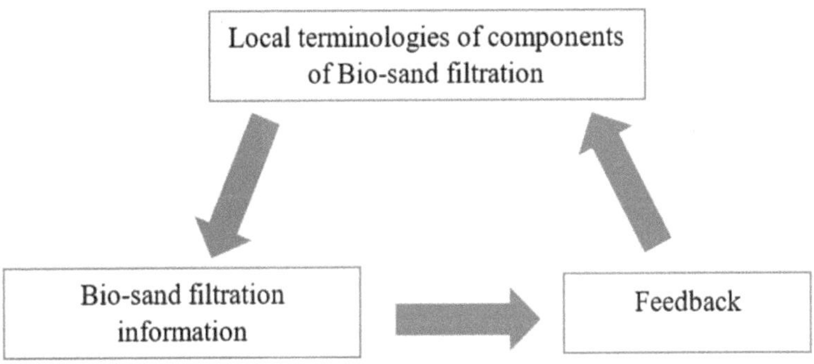

FIGURE 8.5 Cycle to be followed to deliver message using suitable channel.

The receiver of the information is the main center of communication as the key idea behind the entire idea of communication to create awareness. Before information is transmitted to the receiver following steps illustrated as Figure 8.6 in the form of a flow chart should be followed to meet the baseline information already acquired by the receiver (Roma et al., 2014; Wells and Harris, 2007; WHO, 2006).

8.6 FUTURE CONSIDERATION FOR THE TECHNOLOGY

Removal of harmful pathogens have been proved to be controlled by bio-sand filter technology based on the supportive results of several researchers

(Stauber et al., 2009). It is still recommended to carry out research on the removal of heavy metals from the wastewater using bio-sand filters.

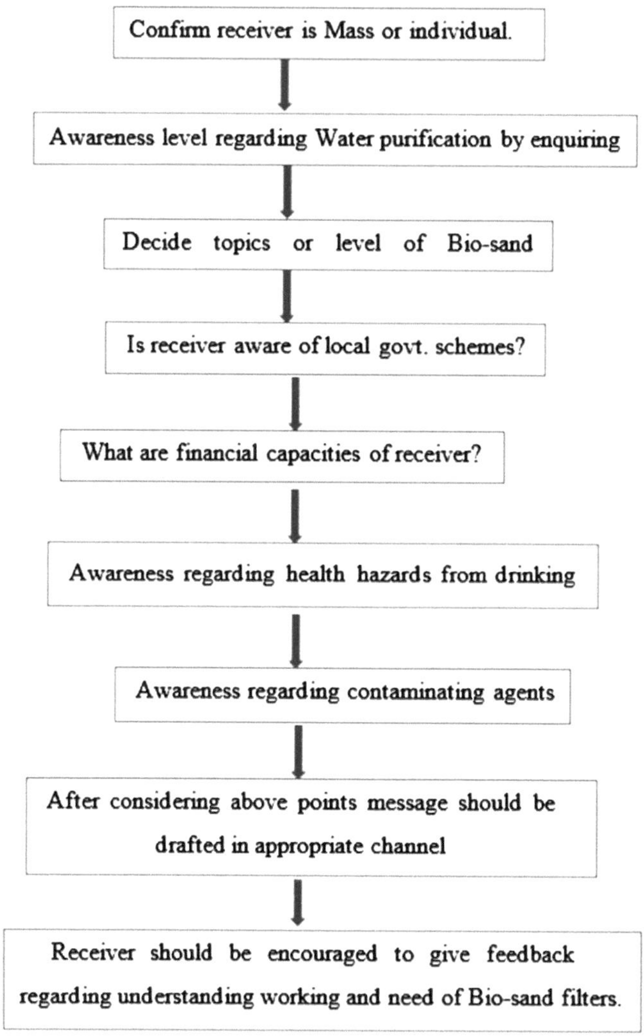

FIGURE 8.6 Steps to be followed while communicating message to receiver.

Also, there is need for standardization of the designs as well as process of filtration so that there is no limitation of using information regarding experiences of adaptation of technologies under various situation which will help to improve water quality conditions in the developing countries. By getting this information, it could be recommended that how many cycles of water passing does contaminate media or biofilm to make filter unsuitable for use.

Region specific socio-economic research studies are carried out to identify the niche target section of the receiver of this technology. As already this technology has indicated promising results in the developing countries; but still significant part of the needy population is unaware of this technology which need to be addressed for every possible platform of media. A major challenge in this technology is its adoption which could only be executed by organizing institutional reforms targeting hygiene and clean drinking water condition as its goal.

Governments of the world especially need to make policies and allocate funds towards the research and extension of the technologies. For instance, financial support could be provided to promote the adaptation of this technology.

KEYWORDS

- **biosand**
- **communicator**
- **filter**
- **microbes**
- **wastewater**
- **water quality**

REFERENCES

Andersson, L., (2017). *Evaluation of Biosand Filter as a Water Treatment Method in Ghana: An Experimental Study Under Local Conditions in Ghana* (Dissertation). Retrieved from http://urn.kb.se/resolve?urn=urn:nbn:se:kau:diva-62836 (accessed on 7 July 2022).

Andrew, J. S., Wampler, P. J., Rediske, R. R., & Molla, A. R., (2013). An assessment of long-term biosand filter use and sustainability in the Artibonite valley near Deschapelles, Haiti. *J. Water. Sanit. Hyg. Dev., 3*(1), 51–60.

Bradley, I., Straub, A., Maraccini, P., Markazi, S., & Nguyen, T. H., (2011). Iron oxide amended biosand filters for virus removal. *Water Research, 45*, 4501–4510.

Chan, N., Young-Rojanschi, C., & Li, S., (2018). Effect of water-to-cement ratio and curing method on the strength, shrinkage and slump of the biosand filter concrete body. *Water Sci. Technol. 77*(6), 1744–1750.

Clasen, T. F., Brown, J., &Collin, S. M., (2006). Preventing diarrhoea with household ceramic water filters: Assessment of a pilot project in Bolivia. *Int. J. Environ. Health Res., 16*(3), 231–239.

Clasen, T., Brown, J., Suntura, O., & Collin, S., (2004). Safe household water treatment and storage using ceramic drip filters: A randomized controlled trial in Bolivia. *Water Sci. Technol., 50*(1), 111–115.

Clasen, T., Garcia, P. G., Boisson, S., & Collin, S., (2005). Household based ceramic water filters for the prevention of diarrhea: A randomized, controlled trial of a pilot program in Colombia. *Am. J. Trop. Med. Hyg., 73*(4), 790–795.

Duke, W. F., Nordin, R. N., Baker, D., & Mazumder, A., (2006). The use and performance of BioSand filters in the Artibonite Valley of Haiti: A field study of 107 households. *Rural Remote Health, 6*(3), 570.

Earwaker, P., (2006). *Evaluation of BioSand Filters in Ethiopia.* Master's thesis, Cranfield University Institute of Water and Environment, Silsoe, Bedfordshire, England.

Elliott, M., Stauber, C., Koksal, F., DiGiano, F., & Sobsey, M., (2008). Reduction of *E. coli*, echovirus type 12 and bacteriophages in an intermittently operated 2 household-scale slow sand filter. *Water Research, 42*, 10, 11.

Hendricks, D. W., & Bellamy, W. D., (1991). Microorganism removals by slow sand filtration. In: Logsdon, G. S., (ed.), *Slow Sand Filtration.* American Society of Civil Engineers, New York.

Makutsa, P., Nzaku, K., Ogutu, P., Barasa, P., Ombeki, S., Mwaki, A., & Quick, R. E., (2001). Challenges in implementing a point-of-use water quality intervention in rural Kenya. *Am. J. Public Health, 91*(10), 1571–1573.

Ram, P. K., Kelsey, E., Rasoatiana, Miarintsoa, R. R., Rakoto, M. O., Dunston, C., & Quick, R. E., (2007). Bringing safe water to remote populations: An evaluation of a portable point-of-use intervention in rural Madagascar. *Am. J. Public Health, 97*(3), 398–400.

Roma, E., Bond, T., & Jeffrey, P., (2014). Factors involved in sustained use of point-of-use water disinfection methods: A field study from Flores Island, Indonesia. *J Water & Health, 12*(3), 573–583.

Sobsey, M., Stauber, C., Casanova, L., Brown, J., & Elliott, M., (2008). Point of use household drinking water filtration: A practical, effective solution for providing sustained access to safe drinking water in the developing world. *Environ. Sci. Tech., 43*(3), 970, 971.

Souter, P. F., Cruickshank, G. D., Tankerville, M. Z., Keswick, B. H., Ellis, B. D., Langworthy, D. E., Metz, K. A., et al., (2003). Evaluation of a new water treatment for point-of-use household applications to remove microorganisms and arsenic from drinking water. *J. Water Health, 1*(2), 73–84.

Stauber, C., Ortiz, G. M., Loomis, D. P., & Sobsey, M., (2009). A randomized controlled trial of the concrete biosand filter and its impact on diarrheal disease in Bonai, Dominican republic. *The American Journal of Tropical Medicine and Hygiene, 80*(2), 286–293.

Wells, N. M., & Harris, J. D., (2007). Housing quality, psychological distress, and the mediating role of social withdrawal: A longitudinal study of low-income women. *J Environmental Psychology, 27*(1), 69–78.

WHO, (2006). *Linking Technology Choice with Operation and Maintenance*. (pp. 77–79). Water treatment. WHO Press, World Health Organization, Geneva, Switzerland.

CHAPTER 9

Earthworm-Assisted Bio-Remediation of Wastewater Treatment

MUDASIR SHAFI and SYED ROUHULLAH ALI

College of Agricultural Engineering and Technology, SKUAST–K, Jammu and Kashmir – 190025, India,
E-mail: agasyedrouhullah@gmail.com (S. R. Ali)

ABSTRACT

Due to fast population expansion and wastewater production as a result of that expansion, freshwater supply is diminishing annually due to anthropogenic activities. Urban and industrial wastewater that has not been cleaned is released into the surrounding environment. It pollutes and reduces the quality surface waters. Villages account for around 70% of India's population, and they need better sanitation. Rural regions are usually un-sewered, owing to a lack of water supply required for the efficient functioning of the water transport network as well as the dispersed population. Traditional technologies for wastewater treatment having a considerable carbon impact on the environment require mechanical processes and should be rendered economical while also being simple to operate. To get beyond the roadblocks that come with it, earthworm-assisted bio-remediation, which uses earthworms in a filter-bed to treat effluent waste, has shown to be a viable option in contrast to standard "treatment and recycling" techniques. Because of its simplicity, the application of the earthworm bio-remediation method in sewage-water treatment is simple to implement in underdeveloped nations and processes water to appropriate levels. In this

Advanced Technologies for Water Quality Treatment and Management. Mehraj U. Din Dar, PhD, Aamir Ishaq Shah, MTech, Shakeel Ahmad Bhat, PhD & Syed Rouhullah Ali, PhD (Eds.)

chapter, the application of earthworm assisted bio-filtration technology for the treatment and reuse of local and industrial wastes utilizing different filter designs and methods are presented. Furthermore, new possibilities for improving the efficacy of the earthworm assisted bio-remediation technology for "wastewater treatment and recycling" are discussed.

9.1 INTRODUCTION

Increased worldwide population, urban settlements, and industry have resulted in overall deterioration of the environment, particularly decay of water quality (Bhat et al., 2017, 2020). Wastewater from rural regions is disposed off in its natural state, with no treatment. It is usually dumped onto highways, local watercourses, cultivable lands, and public places around settlements in most underdeveloped/developing nations. Water contamination results from the discharge of unfiltered pollutants and waste from industries into local waterways (Goel, 2006; Shah et al., 2020). Degradation of "dissolved oxygen (DO)" as well as eutrophication are issues caused by wastewater containing organics such as "biological oxygen demand (BOD), chemical oxygen demand (COD), and minerals" (Zheng et al., 2013; Bhat et al., 2020). Also, water-borne infections are transmitted by coming into contact with water infected by pathogens discharged from sewage into surface waterways. This results in the creation of harmful surroundings around living beings due to loss of freshwater supplies and the degradation of river ecosystems (Reddy and Smith, 1987; Wang et al., 2012). Also, due to scarcity of freshwater resources, the amount of accessible water per capita is decreasing as the population grows (Bhat et al., 2017; Dar et al., 2017; Hameed et al., 2017). As a result, at a given level of treatment, wastewater created in homes as well as in other locations needs to be recycled and used again. Due to water shortages and pollution caused by anthropogenic activities, wastewater treatment and then using it again for commercial, farming, and non-drinking applications is essential (Dar et al., 2020; Pimentel et al., 2004).

Anaerobic and aerobic methods are utilized to clean wastewater all over the world (Speece, 1983). Bacteria transform organic matter to produce methane gas and CO_2 in the anaerobic respiration, whereas aerobic microbes transform organics to form biomass and CO_2. As compared to the aerobic process, the anaerobic one has more effectiveness for wastewater with large COD, uses lesser energy, and generates minimal sludge. However, when it comes to acclimatizing the changes in pH, heat, and organic loading rates (OLR), the anaerobic method falls short of aerobic method (Bhat et al., 2020; Metcalf et al., 1991). Furthermore, restarting of the aerobic method requires lesser duration and functions at temperatures ranging from 25 to 35°C, whereas 30°C optimal temperature is required for the anaerobic method (Singh et al., 2019b). "Oxidation ponds, lagoons, stabilization ditches, activated sludge, up-flow anaerobic sludge beds, sequencing biological reactors, and land treatment" are some of the conventional waste cleaning and recycling technologies utilized in rural regions. Both traditional wastewater treatment procedures, on the other hand, need a large initial investment, recurrent expenses, specialized labor, more restarting time, and automated and power consuming processes. Furthermore, waste produced by traditional methods requires further treatment until it is discharged to the surroundings. Physical and chemical treatment techniques are also employed in around the globe in addition to biological therapy (Adin and Asano, 1998; Noumsi et al., 2005). On the other hand, these procedures are ineffective in removing organic and nutritional contaminants from wastewater (Ra et al., 2000). Many poor countries cannot afford to build and operate big, expensive sewage treatment facilities (STPs), and even wealthy countries must focus on decentralized solutions in the future to ensure sustainable wastewater management. Furthermore, most nations choose sewage treatment technologies that can deliver a high-quality effluent at a low cost. As a result, in the current environment, an affordable and long-term wastewater treatment method is required whist also having simplicity of operation and functioning.

On-site wastewater treatment seems to be a cost-effective and energy-efficient option, as well as easy and dependable enough for even the most inexperienced operators to use (Schudell and Boller, 1989). The use of earthworms to filter effluents and waste came out to emerge as an environmentally benign and cost-effective option as compared to traditional wastewater treatment, and is known as earthworm bio-remediation (Carballeira et al., 2017). The microbial transformation of liquid/wastewater forming a value-added final produce is known as "earthworm bio-remediation," while as "vermicomposting" is the earthworm assisted breakdown of solidified sewage. In other words, earthworm bio-remediation is the entry of worms into a filter system containing appropriate bedding materials for decomposing organics (Tomar and Suthar, 2011). The combined activity of worms and microbes in earthworm bio-remediation results in the breakdown of organic contaminants in wastewater. The worms decompose the organics to humus across the first layer, and then the waste is filtered via the filter medium, which promotes microorganism development, and then secondary treatment takes place. Throughout the procedure hardly any sludge is generated, rather economically and environmentally beneficial "vermicompost" is produced. Furthermore, the procedure is odorless, with the earthworm bioremediated water produced being excellent for irrigating agricultural fields as well as for use in parks and gardens. Recent research has demonstrated that the earthworm bio-remediation technology can be a viable and long-term wastewater treatment and recycling option (Bhat et al., 2020).

9.2 OVERVIEW OF EARTHWORM BIO-REMEDIATION TECHNOLOGY

Earthworms and microorganisms work together to cleanse wastewater in the "earthworm assisted bio-remediation" process. In addition to microbiological deterioration, earthworms provide a variety of useful functions such as grinding ingested dirt with ingested organics, adding gut bacteria, and excreting soil and organic matter in the form of vermicast (Xing et al., 2014). It functions as an "aerator, grinder, crusher, chemical decomposer, and bio-accelerator" in wastewater, encouraging the growth of "beneficial decomposer bacteria." Earthworms have millions of bio-degrader bacteria in their guts, which in addition to minerals like "nitrogen and phosphorus" are excreted into the soil (Sinha et al., 2002; Singleton et al., 2003). For the household and industrial wastewater treatment, an earthworm bio-remediation system consists of an "active worm zone" and a "filter medium bed" that sustains a microbiological population. *Eisenia fetida, Lumbricus rubellus*, *Eudrilus eugeniae*, and *Eisenia andrei* are among the earthworm species used in earthworm bio-remediation technology, with a filter bed made up of soil, compost, and cow dung. Different types of components like "sand, gravel, cobblestone, and quartz sand" are widely utilized in the design of filter medium, through which the wastewater flows (Singh et al., 2019b; Xing et al., 2011). In an earthworm bio-remediation system, wastewater passes through the earthworm active zone first, then through the filter medium bed. The earthworms digest the suspended particles caught on top of the earthworm bio-filter and are then given to the immobilized soil microorganisms. Wastewater is cleaned in this method as it percolates through the earthworm bed, where it is degraded both physically and microbially, while the organic materials pass via the gut of the vermicompost.

The "earthworm assisted bio-remediation systems" may be classified into two categories based on the path in which the effluent travels: "horizontal flow systems (HFS) and vertical flow systems (VFS)." As illustrated in Figures 9.1 and 9.2, wastewater in "HFS" travels horizontally via the medium, whereas wastewater in "VFS" travels vertically via the medium. For wastewater treatment, a mechanism that uses fusion of "HFS" and "VFS" in the same order is also utilized. In this hybrid system, wastewater flows from a horizontal system to a vertical system or the other way around, as depicted in Figures 9.3(a) and (b).

To enhance wastewater treatment efficiency, researchers are now concentrating on a combined macrophyte earthworm bio-remediation technology. It is based on the idea of swamps and marshes utilizing diverse types of flora such as "*Canna indica*, *Phragmites australis*, *Typha angustifolia*, *Saccharum spontaneum*, and *Cyperus rotundus*," and others in combination to earthworm bio-remediation method for treating effluent waste. When a macrophyte absorbs a considerable quantity of nutrients for its development, it is removed from wastewater (Bhat et al., 2020; Chen et al., 2016; Samal et al., 2017a; Wang et al., 2010b). In Figure 9.4, a "macrophyte-assisted" earthworm bio-remediation mechanism is depicted schematically.

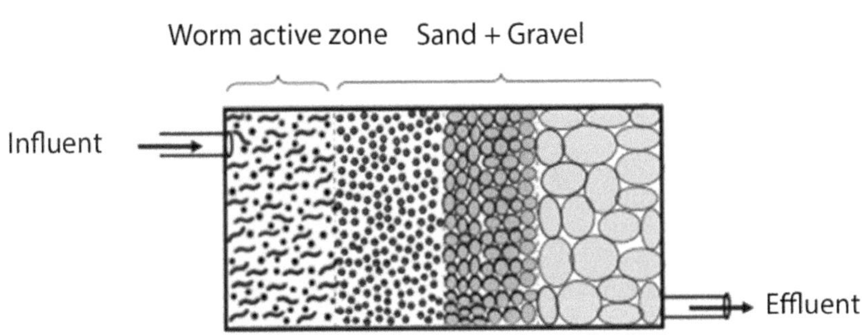

FIGURE 9.1 Layout of a "horizontal-flow" earthworm bio remediation mechanism.

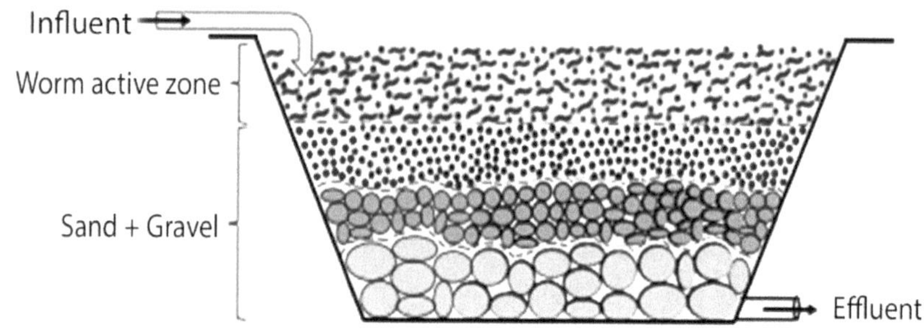

FIGURE 9.2 Layout of a "vertical-flow" earthworm bio remediation mechanism.

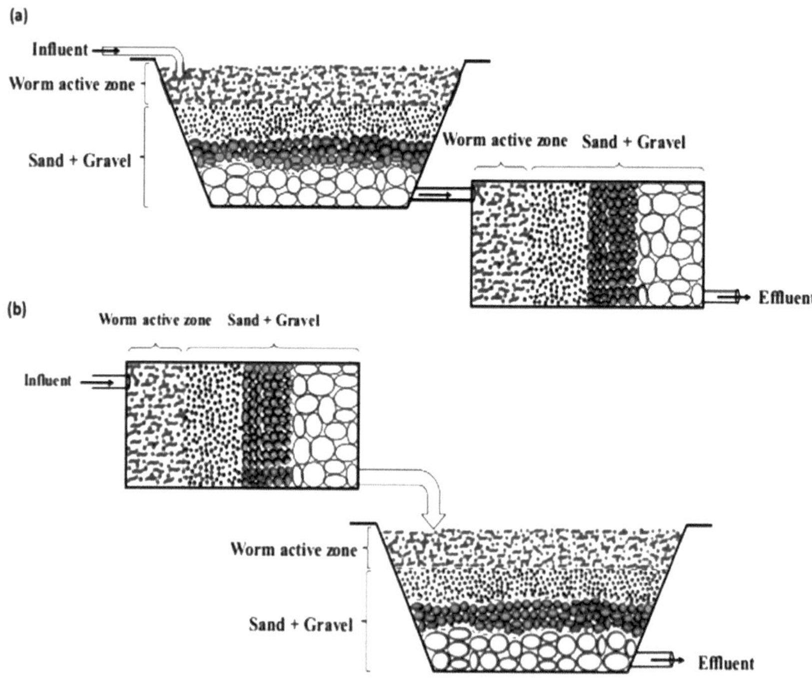

FIGURE 9.3 Layout of a "hybrid earthworm bioremediation system" on the basis of flow path of the effluent: (a) VFS followed by HFS; (b) HFS followed by VFS.

The plants' root or "rhizospheric" zone contributes to the establishment of a varied microbial population that degrades organic pollutants by providing a favorable environment for their growth. The macrophytes transport oxygen from air to the rhizospheric zone, where it is used by the micro-biological population as discovered by the scientists (Bezbaruah and Zhang, 2005; Bhat et al., 2020). The excess oxygen level keeps the microorganisms and earthworms in an aerobic state, which helps to speed the breakdown of organic pollutants.

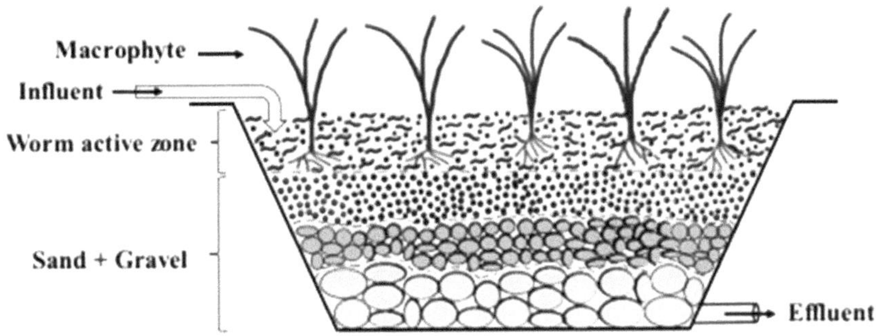

FIGURE 9.4 Layout of a "macrophyte-assisted" earthworm bio remediation mechanism.

9.2.1 EARTHWORMS

In India, there are around 500 kinds of earthworms as compared to 3,000 worldwide. Earthworms known as "soil-eaters" are excellent decomposers, with more than "600 million years" of trash and nature maintenance experience (Julka, 1986). Thus, there is no surprise in the fact that "Charles Darwin" dubbed the earthworms as "the unheralded warriors of mankind," and "Aristotle" described them "the intestine of the earth," implying that they digested a diverse range of organics, including waste organic materials of the soil (Martin, 1976). The "earthworms" are segmented creatures with no bones that are long, cylindrical, thin, bilaterally symmetrical, and have "100–200" roughly cylindrical rings or segments, which are lined by small hair-like spikes and weigh about "1,400–1,500 mg" after 2–3 months (Bhat et al., 2020). The earthworms are expected to live around "3–7 years," based upon the breed along with the environmental condition. According to research, they have a rapid reproductive rate and may double, their population in 60–70 days (Sinha et al., 2008). Millions of beneficial micro-organisms, such as nitrogen-fixing and decomposer bacteria, live in the guts of earthworms. The availability of "organic matter, soil moisture, and soil pH" determines the presence of the worms in soil. They breathe through their skin, and they like a dark, wet environment. They can withstand temperatures ranging from 5 to 29°C. Earthworms may thrive in environments having temperatures of 20–25°C and a water content of 60 to 75%. The earthworm activeness is decreased by lower temperatures, while higher temperatures may quickly burn and wipe them out. These worms are sexually dimorphic creatures with a fast rate of reproduction. If given the ideal circumstances of temperature, moisture, and food sources, earthworms may grow by "256 worms every 6 months" from one single worm. They have the potential to generate 300–400 young during their life cycle (Hand, 1988). Earthworms use their muscular gizzard to grind dirt, detritus, food, and other materials to smaller sizes of 2–4 micrometers. The worms perform a critical function in the earthworm bio-remediation process as they rapidly eat organic matter in the effluent water. As a result, the earthworm bio-remediation bed's population density, maturity, and health play a vital part in the treatment process. Maintaining an optimal worm density is critical for the earthworm bio-remediation system's efficient operation. The quantity of earthworms per unit area in the earthworm bio-filter bedding unquestionably influences the treatment effectiveness of the earthworm bio-remediation process (Li et al., 2008). As a result, a sufficient number of worms should be introduced at first to decompose the entering effluent and generate an adequate "humus filter." According to research, at least "15,000–20,000 worms" per cubic meter are necessary to get an earthworm bio-remediation system up and running (Sinha et al., 2008).

The worms may be categorized into three groups depending upon their eating and excretion habits, as well as the spatial structure in the soil layers., as given below:

1. **EPIGEIC:** These earthworm species have a rapid reproductive rate and are voracious eaters of organic materials, despite their tiny body size. Because they live on the surface, they do not alter the soil structure, therefore they are commonly employed in vermicomposting. "*Eisenia fetida* (brandling, red wiggler, or dung worm), *Eisenia andrei* (red tiger), *Lumbricus rubellus* (red worms)," etc., are some examples of epigeic earthworms.
2. **ENDOGEIC:** Endogenic species are darkly colored and have a wide range of body sizes. They eat dirt rather than organic materials and dwell in large horizontal burrows. Despite their ability to digest dead plant roots, these species are rarely utilized in vermicomposting. They actively participate in soil formation processes

such as mixing and aeration due to their burrowing habit. Some examples include *Allolobophora chlorotica* (green worms), *Aporrectodea caliginosa* (gray-worm), etc.

3. **ANECIC:** Burrowing earthworms known as anecics develop a complex tunnel system beneath the land surface that aids in the thorough blending of top layer materials with that of lower layers. These worms are large in structure, live for longer periods, but grow and reproduce more slowly than "epigeic" worms. Anecic worms are sometimes employed in decomposing of effluents, although they are generally utilized along with epigeic worms. For example, "*Aporrectodea longa* (black-headed worms), *Lumbricus terrestris* (lob worms)," etc.

The compatibility of "*Eisenia fetida, Eudrilus eugeniae,* and *Perionyx excavates*" is given by Reinecke et al. (1992) in Table 9.1.

TABLE 9.1 Description of Worms Used for Wastewater Treatment

Parameters	*Eisenia fetida*	*Eudrilus eugeniae*	*Perionyx excavatus*
Duration of life cycle (days)	70	60	46
Growth rate (mg/worm/day)	7	12	3.5
Maximum body mass (mg/worm)	1,500	4,294	600
Maturation attained at age (days)	50	40	21
Start cocoon production (days)	55	46	24
Cocoon production (worm/day)	0.35	1.3	1.1
Incubation period (days)	23	16.6	18.7
Hatching success in water (%)	73	50	63.4
Mean number of hatching (cocoon)	2.7	2.7	1.1
Number of hatchings from one cocoon	1–9	1–5	1–3

Source: Reinecke et al. (1992).

9.3 EARTHWORM BIO-REMEDIATION TECHNIQUE MECHANISM

Earthworms and microorganisms are used in the earthworm bio-remediation process. Macrophyte-assisted earthworm bio-remediation, which evolved from the fundamental technology, is emerging to be an environmentally acceptable option for "wastewater treatment and recycling." The functioning of different strata and elements of a "macrophyte-assisted earthworm bio-remediation system" has been depicted in Figure 9.5 so as to figure out how the system works. Earthworms eat the sediments that remain on the filter bed and convert them to "humus" (Singh et al., 2017). The formation of a microbiological coating on the filter-bed aids in the breakdown of pollutants. In general, an earthworm bio-remediation system is made up of two parts: earthworms and a filter bed. The filter bed which provides food via sorption from the effluent, aids in the development of worms while a microbiological layer like coating is produced due to the limited porous nature of the bed. In earthworm bio-remediation, the worm active zone is called the "aerobic zone," whereas the filter bed is known as the anoxic zone (Samal et al., 2018a; Singh et al., 2017; Wang et al., 2010a, b).

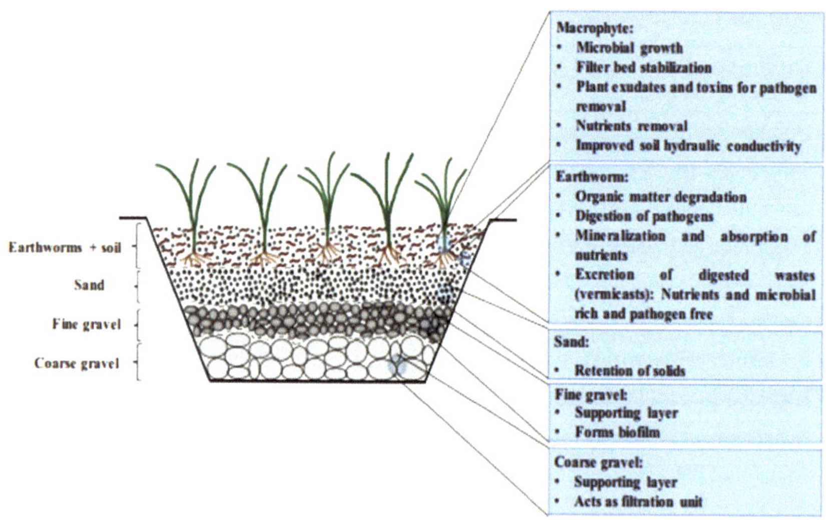

FIGURE 9.5 Schematic representation of the role of different layers and components in macrophyte assisted vermifiltration system.

The borrowing activity of earthworms raises the oxygen level in the filter bed. Furthermore, the increased volume of "soil" particulates and the increased porous nature of the filter-bed allow earthworms to hold extra biological contaminants, facilitating in more degradation (Singh et al.,

2018; Bhat et al., 2020). Ingestion, grinding, digesting, and excretion are all processes that earthworms use to process waste, and these processes have a variety of "physical, chemical, and biological" impacts upon the interior ecology of the "earthworm active zone" (Singh et al., 2017). Apart from this, the intake and crushing operations of earthworms convert input wastes into tiny particulates of 2–4 micrometers in size, which are then digested in the gut thanks to the activities of bacteria along with intestinal secretions (Sinha et al., 2010; Wang et al., 2011). In the gizzard and gut of earthworms, enzymes such as "protease, lipase, amylase, cellulase, and chitinase" are released, resulting in the biological breakdown of cellulose and protein components in the effluent. The waste intake items are expelled to the surroundings in the form of nutrient-rich "vermicast," since the earthworm stomach contains a complex microbial community. Microbes in the biofilm breakdown nutrients kept on it, as well as nutrients in the vermicast, for their population development (Sinha et al., 2008, 2010). To keep their body surface, moist, earthworms produce mucus (a sticky fluid made up of different metabolites), that aids in the absorption of "oxygen," too. These worms may transform huge amounts of organic materials into complex amorphous solids including phenolic chemicals, a process known as humification. These humic chemicals found in vermi-bed aid in gathering of elements and comprise organics with complicated molecular structures such as "aromatic rings, carbonyl groups, phenolic, and alcoholic hydroxyl" groups. The same chemical structure bonds to various metal ions, thus assisting in the elimination of metals (Bhat et al., 2020). The clay particles are granulated by earthworms, boosting the system's "hydraulic conductivity," which increases the "organic" presence in the earthworm biofilter. The crushing actions of the worms results in the increase of filter medium volume, allowing "organic and inorganic" contaminants to be absorbed from the effluent. This method has been proven to be suitable for diluted wastes such as effluents and wastewater. Apart from this, the worms swallow and eat the excess dangerous as well as inefficient microorganisms present in "wastewater," to form earthworm biofiltered effluent free of germs. Due to their burrowing nature, earthworms also prevent the medium from choking and retain a culture of efficient bio-degrader bacteria to work (Bhawalkar, 1995; Bhat et al., 2020).

Furthermore, the earthworm bio-remediation method has been shown to reduce pathogens significantly. These worms possess the ability to eliminate infections found in the things they eat. The decrease in pathogen number in "earthworm assisted bio-remediation" is mostly due to enzymatic and microbiological activity (Hartenstein, 1978; Samal et al., 2017a; Sinha et al., 2010; Swati and Hait, 2018).

Treatment of different kinds of effluent/wastewater using earthworm bio-remediation mechanism is shown in Table 9.2.

TABLE 9.2 Treatment of Wastewaters Using Earthworm Bio-Remediation

Sr. no.	Types of wastewater	Earthworm species	Organics removal (%)	Nutrient removal (%)	Bed material and size	HLR (m^3/m^2d)	HRT (h)
1	Sewage	*Eisenia fetida*	BOD 98, COD 45, TSS 90	-	Pure soil, sand (10–12 mm), gravel (7.5, of 3.5–4.5)	-	1–2
2	Dairy industry effluent	*Earthworm*	BOD 98, COD 80–90, TSS 90–95, TDS 90–92	–	Pure soil, sand (10–12 mm), gravel (7.5, of 3.5–4.5)	–	6–10
3	Synthetic sewage	*Eisenia fetida*	COD 83.6	TN 63, TP 86.7, NH3-N 70.5	Cobblestones (6–10 cm), soil, sawdust	0.2	48, 72, 96
4	Urban wastewater	*Perionyx sansibaricus*	COD 80–90, TSS 88.6, TDS 99.8	No3 92.7, Po3-4 98.3	Surface vegetation, soil, dried leaves, sawdust, small stones (5–7 cm), large stones (10–15 cm)	–	1
5	Cheese whey	*Eisenia fetida*	BOD 76, COD 82, TSS 77	TN 60, TP 77	Ligneous mature compost, stones	0.04	-
6	Sewage	*Eisenia fetida*	BOD 98, COD 70, TDS 95	–	Garden soil, sand, aggregates (3–5, 7–8 cm)	–	2
7	Rural domestic sewage	*Eisenia fetida*	BOD 78, COD 67.6, TSS 89.8	NH+ 4 92.1	Ceramsite (3–5 mm)	4.2	1
8	Synthetic wastewater	*Eisenia fetida*	BOD 96, COD 90, TDS 82	–	Vermicompost, riverbed material (6–8 mm), sand (1– 2 mm), gravels (10–12.5 mm)	1.5, 2, 2.5, 3	

Source: Patel et al. (2018).

The worms consume residual particulates in the filter, considerably decreasing amount of "BOD by more than 90% and COD by 80–90%," as well as a considerable drop in nutrient content. The worm types "*Eisenia fetida* and *Eudrilus eugeniae*" used to treat wastewater generated from homes resulted in reductions of BOD by around 88% and 70%, TSS by 78% and 67%, and TDS by 75% and 66%, respectively, while as scientific research found that pollutants including BOD5, COD, and TSS were removed about "55–66%, 47–65%, and 57–78%," respectively, from household wastewater (Li et al., 2009; Kumar et al., 2016; Xing et al., 2010). "*Eisenia fetida*" ranks among the frequently earthworms used to clean "household wastewater." Separate research on "treatment of household wastewater" depicted that using "*Eisenia fetida*" reduced BOD5, COD, and TSS by 78%, 68%, and 90%, respectively (Gunadi et al., 2002; Liu et al., 2013; Sinha et al., 2008). The treatment of "synthetic wastewater" utilizing a variety of "vertical sub-surface flow created wetlands" seeded with the "macrophyte Acorus calamus and the earthworm *Eisenia fetida*." In addition to earthworm bio-remediation, revealed the removal of up to "87%–COD, 86%–total nitrogen (TN) and 83%–total phosphorus (TP)" (Zhao et al., 2014).

The nitrifying and de-nitrifying microorganisms present in intestinal tracts of the worms are mostly responsible for nitrogen removal from wastewater (Ihssen et al., 2003). The removal of about 99% "*Escherichia coli* (*E. coli*), total coliform (TC), fecal coliform (FC) and fecal streptococci (FS)" from constructed effluent in an earthworm assisted bio-remediation system was reported. The filtration of domestic effluent water with earthworm bioremediation showed a reduction of FC by 99% (Arora et al., 2014; Kumar et al., 2016). Using household wastewater, during research period of one year of earthworm bio-remediation revealed a decrease in "COD by around 87% and in thermotolerant coliforms by 99%" (Furlong et al., 2014).

The earthworm bio-remediation technology, which was first restricted to the treatment of household wastewater, slowly progressed to the point that it is being investigated for treating "industrial" effluents. The earthworm assisted bio-remediation technology, when used on wastes coming out of the "food and beverage" industry, has demonstrated promising waste cleaning effectiveness and might open a new road for its use on a variety of other less toxic industrial effluents. In addition, the earthworm bio-remediation system has been used to reduce the toxicity of various industrial effluents, such as those from the petroleum and pharmaceutical industries (Dhadse et al., 2010; Singh et al., 2019a; Sinha et al., 2012). The earthworm bio-remediation process is successfully utilized in the treatment of dairy effluent, which is mostly made up of organics such as proteins, carbohydrates, and lipids. The earthworm "*Eisenia fetida*" removes about "99% BOD5 and COD in the range of 80%–90% and the removal of TDS and TSS in the range of 90–92% and 90–95%, respectively" (Sinha et al., 2007). The earthworm biofiltration of "petroleum industry" effluent depicts the elimination of "C10–C14, C15–C28 and C26–C36" by around 99% (Sinha et al., 2012). Furthermore, the use of earthworm bio-remediation to treat cheese whey waste removed around 76%-BOD, 82%-COD, and 77%-TSS. The worm "*Lumbricus rubellus*" removed COD by 89% and BOD by 90% from gelatin industrial effluent. The usage of the worm "*Eudrilus eugeniae*" to treat "herbal pharmaceutical" wastes at varied "OLR" showed removal effectiveness of around 90%-COD and 93%-BOD. Using the macrophyte "Canna indica," earthworm bio-remediation to treat synthetic dairy effluent, depicted removal rates of 81%–BOD, 76%–COD, 85%–TSS, 23%–TDS, and 43%–TN (Dhadse et al., 2010; Ghatnekar et al., 2010; Merlin and Cottin, 2009; Samal et al., 2017b).

9.4 TYPES OF WASTE WATER TREATED USING THE EARTHWORM ASSISTED BIO-REMEDIATION TECHNOLOGY

"Earthworms, soil, sand, and gravel particles" all work together in the earthworm bio-remediation process. In poor nations with a variety of wastewater treatment difficulties, the earthworm bio-remediation technique has been recommended as an alternate wastewater treatment approach.

9.4.1 URBAN AND DOMESTIC WASTEWATER

The earthworm biofilter might be a good way to stabilize surplus sludge from the household and urban "wastewater treatment" facilities. Earthworm bio-remediation is considered as environmentally benign and cost-effective method for treating household sewage in rural regions as well as herbal medicinal effluent (Dhadse et al., 2010; Li et al., 2009; Wei et al., 2003). The earthworm "*Eudrilus eugeniae*" was able to function in the filter at temperatures over 40°C, which improved the earthworm bio-filter conditions. It has been observed that the earthworm bio-filter can eliminate BOD5, COD, TSS, and coliform organisms more effectively than the filters which do not use worms. The combined earthworm bio-remediation reactor depicts to remove major chemical contaminants from wastewater more effectively than a conventional biofiltration system. The proportion of earthworms having weight above 0.3 grams is substantially related to "COD, BOD5, SS, and NH_4-N" elimination rates, suggesting that bigger worms like "*Eisenia fetida*" may contribute in a much greater way in the treating the sewage than smaller earthworms. However, the "hydraulic loading" had minimal effect on earthworm reproduction, and the rise in juveniles, as well as the reduction in elder and bigger worms, shows the reduction in "treatment efficiency" earthworm bio-filter (Adugna et al., 2014; Xing et al., 2010). Application of sawdust into soil, which may increase porosity, substantially enhanced the characteristics of sewage water. Earthworm development, growth, reproduction, and survival in a moist environment are all said to be excellent (Garkal et al., 2015).

9.4.2 DAIRY EFFLUENT TREATMENT

"Dairy wastewater" when utilized as a feed for worms shows to be excellent at eliminating 'COD, NH_4, and total-nitrogen.' The use of "*Canna indica*" and "*Eisenia fetida*" in treatment of dairy-effluent, reduces "BOD and COD by 75–81%," as well as "total-nitrogen by 24–42%," and is suitable to be utilized for a variety of sewage filtration systems (Rodgers et al., 2006; Samal et al., 2017a, b). Earthworm bio-remediation of wastewaters from breweries and milk dairies with high BOD5 and TSS levels showed that earthworms decreased elevated BOD5 loadings by 99% and TSS by 98%. "Hydraulic retention periods (HRTs)" were observed to be 3–4 hours for "brewery wastewater" and 6–10 hours for dairy wastewater. The worms must respond quickly to the movement of effluent inside their gut, including the decomposition of organics and swallowing the sediments. This is why the wastewater must be maintained in the earthworm bio-filter bed for a fair amount of time (HRT must be in terms of hours) (Sinha et al., 2007). Earthworm bio-remediation contains a hybrid of conventional filtration and "vermi-composting." The earthworm's body functions as a "biofilter," with BOD5 removal rates of 97.95%, COD removal rates of 91.64%, TSS removal rates of 76.39%, and TDS removal rates of 84.27%. The oil and grease content dropped by 84.13% (Telang and Patel, 2015). The earthworm bio-filter has micro-environments which are useful for nourishment of bacteria required in the conversion of "organic nitrogen into N_2." Also, the earthworm bio-filter's dual nature (aerobic/anaerobic) creates a perfect environment for degradation of microbes along with removing nitrogen, lowering lower strata nitrogen loads while reducing greenhouse gas emissions Earthworm bio-remediation can enable efficient treatment of dairy effluent depending on the parameters of the wastewater to be treated. Furthermore, in treating higher quantities of effluents via earthworm bio-remediation, these procedures must be optimized economically (Lai et al., 2018; Natarajan et al., 2015). In the process of earthworm bio-remediation, vermi-bed height is a crucial variable. The use of a horizontal subsurface flow earthworm bio-filter system, in conjunction with a vertical down flow earthworm bio-filter system seeded with different "macrophytes" along with worm, "*Eisenia fetida*" was able to considerably lower organic matter, minerals, and particles in "dairy" effluent (Bhat et al., 2020; Samal et al., 2018a, b).

9.4.3 SWINE WASTE TREATMENT

The earthworm assisted bio-remediation is also used to treat diluted swine manure. In comparison to raising on a slatted floor having a buildup of slurry, the wastewater may be utilized to throw out the manure, resulting in lower NH_4 levels, thus influencing concentrations of gases emitted (Li et al., 2008). In freshwater recycled manure systems, the number of earthworms may be utilized as indicators for verifying lower energy intakes, lower green-house emissions as well as low NH_4 generation (Luth et al., 2011). Pollution levels in the important hybrid-built wetland were assessed at various levels. This might be due to variations in "pig-waste" concentration, farming techniques, along with preservation factors, besides it can arise from inefficient processing of "raw piggery" sewage (Borin et al., 2013).

9.5 ADVANTAGES AND DRAWBACKS OF THE EARTHWORM ASSISTED BIO-REMEDIATION TECHNOLOGY

9.5.1 *ADVANTAGES*

The earthworm bio-remediation treatment offers several benefits over all other standard biological wastewater treatment systems while considering energy use, economics, and operational easiness.

- The earthworm assisted bio-remediation technology may be carried out with very little capital and operational expenditures, and the nutrient-rich vermicompost and earthworm biomass may be used to produce money (Xing et al., 2005).
- The earthworm bio-remediation method produces no sludge, which is impossible to achieve with any other wastewater treatment technology.
- Earthworms eat all of the germs such as "bacteria, fungi, protozoa, and nematodes" contained in effluent waste, therefore earthworm biofiltered sewage is devoid of pathogens.
- The earthworm bio-remediation process is odorless, and earthworms play a significant role in this because their burrowing efforts prevent the operation of anaerobic bacteria in the filter-bed material, by promoting aerobic environments.
- Because earthworms can bioaccumulate large amounts of hazardous compounds from sewage, such as "heavy metals and endocrine disrupting compounds (EDCs)," the earthworm bio-remediated waste is devoid of harmful substances (Bhat et al., 2020).
- Various types of effluent and wastewaters may be filtered and recycled for relatively little money in terms of operation and maintenance.

9.5.2 *DRAWBACKS*

The lab-scale functioning of the earthworm assisted bio-remediation process has been more effective thus far. During field-scale deployment, certain limitations were discovered, which have been described below:

- Because earthworms cannot live in submerged conditions, the majority of the earthworm assisted bio-remediation process takes place in subsurface vertical/horizontal reactors. Most reactors are designed such that influent enters one end and effluent exits the other in horizontal reactors, whereas influent enters from the top and effluent exits from the bottom of vertical reactors.
- Because earthworms dislike daylight and respire via the skin, their activity is practically non-existent throughout the day. As a result, caution should be exercised while situating the reactor by shielding it from direct sunlight, either using jute sacks or trying to keep it shady.
- Because the majority of earthworms are surface feeders, they tend to congregate at the top portion of the system, with limited activity in the lower layers and it might result in the filter's efficiency being reduced.
- The presence of too much moisture in the filter may cause anaerobic conditions, resulting in the death of earthworms. As a result, caution must be exercised while determining the filter's hydraulic loading rate.
- Wastewater containing a high concentration of sodium chloride (NaCl) cannot be treated in an earthworm biofilter because it is poisonous to earthworms. The deaths of earthworms might reduce the overall ability of treating the sewage, lowering the effectiveness of the filter (Hughes et al., 2007, 2008).
- Heavy metals can be accumulated in the bodies of earthworms. However, the earthworm's ability to absorb heavy metal is restricted. As a result, the efficiency of treatment of wastewater containing heavy metals is reduced (Bhat et al., 2020).
- Worm possesses a high reproduction rate, and after some time period, a scarcity of food and space can arise, reducing the effectiveness of the treatment process.
- After the reactor's working period has ended, the earthworm biofilter must be cleaned, which is a time-consuming procedure. Handling earthworms should be done with caution since they are delicate and any harm might result in their death.
- When using wastewater with high organic contents, earthworm biofilters may become clogged. Another significant restriction is the earthworm biofilters' depth, as earthworms cannot survive at greater depths.

9.5.3 FUTURE PERSPECTIVES

Earthworm bio-remediation's ability to treat both household and industrial wastewater has been widely established. An understanding of earthworm bio-remediation has been presented on the basis of scientific research, construction layouts, along with treatment process. Earthworm bio-remediation with macrophyte is a relatively new wastewater treatment method. The majority of studies have only shown vertical earthworm biofilters at a laboratory size for synthetic wastewater treatment. Earthworm bio-remediation experiments with genuine wastes from industries can be helpful in determining the effectiveness of 'organic, nutrient, and pathogen removal.' Moreover, high quality scientific-research into the various earthworm bio-remediation system design configurations for wastewater treatment is required. For scaling-up the process, a number of control factors like "earthworm stocking density, flow rate, hydraulic retention time (HRT), OLR, and filter bed design" must be adjusted. Furthermore, the vast bulk of researches have exclusively used the epigeic worm, "*Eisenia fetida.*" As numerous worms are present in the soil, it is important to investigate and study these different worm types in isolated and symbiotic trials in order to effectively remove pollutants from effluent and wastewaters.

9.6 CONCLUSIONS

The usefulness of the earthworm bio-remediation technology for sewage and industrial effluent treatment, as well as the treatment processes involved, has been thoroughly explored. The earthworm bio-remediation technique may be used to cleanse wastewater from a variety of sources or tailored to a specific wastewater. As a result, earthworm bio-remediation has been discovered to be an effective method. This is also a viable option for decentralized wastewater treatment. The presence of earthworms provides a feasible aerobic environment for the earthworm bio-remediation process to take place in. This creates a favorable environment for aerobic decomposer bacteria. The earthworm bio-remediated water included higher levels of nitrate and phosphate, making it suitable for farm and horticultural irrigation. Earthworms feed on the organic content and particulates in the effluent, converting them into useful vermicompost, and so no sludge is generated, as in previous treatment systems. As a result, the need to dispose the sludge does not arise; and to the benefit, vermicompost created is utilized as a fertilizer due to its high nitrogen and phosphate content. Furthermore, we obtain earthworm biomass, which may be utilized as fish and poultry feed. Additionally, macrophytes' potential in an integrated earthworm bio-remediation mechanism for treating effluents is also addressed. In addition, the impact of various filter-design as well as operating factors on the efficacy of the process is also discussed. The combined impact of the earthworm active zone and filter medium has been documented in efficiently removing pollutants from sewage and effluent in the "earthworm bio-remediation" technique. The studies on earthworm bio-remediation of effluent wastes document that "maximum organic and nutrient removal" efficiency of "99%-BOD, 96%-COD, 86%-nitrogen, and 83%-phosphorus" is observed. The earthworm bio-remediation approach has been shown to remove 90–99% of FC and 99% of the "TC, FS, and *E. coli*" pathogens. The components of the earthworm biofilter, such as the filter medium construction materials, worm type, as well as macrophytes used during the procedure, are extremely selective in terms of pollution removal. Furthermore, because different earthworm species coexist in nature, an earthworm bio-remediation technique for treating effluent and waste-water using symbiotic cultures must be evaluated. During earthworm bio-remediation, the impact of different procedure factors such as HLR, OLR, and number of worms is unclear. Earthworm assisted bio-remediation technique can be highly suggested as a feasible option to traditional procedures in treating organic effluents generated from village areas, smaller towns, and enterprises as long as the circumstances are suitable for earthworm proliferation. For successful wastewater treatment and recycling, extensive study is required to improve different process parameters as well as an optimal earthworm bio-filter design.

KEYWORDS

- **anthropogenic activities**
- **bio-filtration**
- **biological oxygen demand**
- **chemical oxygen demand**
- **earthworm bio-remediation**

- **earthworms**
- **recycling**
- **wastewater**
- **wastewater treatment**

REFERENCES

Adin, A., & Asano, T., (1998). The role of physical-chemical treatment in wastewater reclamation and reuse. *Water Sci. Technol., 37*(10), 79–90.

Adugna, A. T., Andrianisa, H. A., Konate, Y., Ndiaye, A., & Maiga, A. H., (2014). Greywater treatment by vermifiltration for sub-Saharan urban poor. *J. Water Sanit. Hyg. De, 4*(4), 625–632.

Arora, S., Rajpal, A., Kumar, T., Bhargava, R., & Kazmi, A. A., (2014). Pathogen removal during wastewater treatment by vermifiltration. *Environ. Technol., 35*(19), 2493–2499.

Bezbaruah, A. N., & Zhang, T. C., (2005). Quantification of oxygen release by bulrush (*Scirpus validus*) roots in a constructed treatment wetland. *Biotechnol. Bioeng., 89*(3), 308–318.

Bhat, S. A., Pandit, B. A., Khan, J. N., Kumar, R., & Jan, R., (2017). Water requirements and irrigation scheduling of maize crop using CROPWAT model. *Int. J. Curr. Microbiol. Appl. Sci, 6*, 1662–1670.

Bhat, S. A., Pandit, B., Dar, M. U. D., Jan, R., Khan, S., & Mehraj, K., (2017). Statistical comparison of reference evapotranspiration methods: A case study from Srinagar in J&K, India. *Int. J. Curr. Microbiol. App. Sci, 6*(9), 3731–3737.

Bhat, S. A., Pandit, B., Dar, M. U. D., Rouhullah, S., Ali, R. J., & Khan, S., (2017). *Comparative Study of Different Methods of Evapotranspiration Estimation in Kashmir Valley* (p. 384). Dr. PD Mistry.

Bhat, S. A., Vig, A. P., Li, F., & Ravindran, B., (2020). *Earthworm Assisted Remediation of Effluents and Wastes.* Springer Singapore.

Bhawalkar, U., (1995). *Vermiculture Eco-technology*. Bhawalkar Earthworm Research Institute (BERI), Pune.

Borin, M., Politeo, M., & De Stefani, G., (2013). Performance of a hybrid constructed wetland treating piggery wastewater. *Ecol. Eng., 51*, 229–236.

Carballeira, T., Ruiz, I., & Soto, M., (2017). Aerobic and anaerobic biodegradability of accumulated solids in horizontal subsurface flow constructed wetlands. *Int. Biodeterior. Biodegrad., 119*, 396–404.

Chen, Z., Hu, S., Hu, C., Huang, L., Liu, H., & Vymazal, J., (2016). Preliminary investigation on the effect of earthworm and vegetation for sludge treatment in sludge treatment reed beds system. *Environ. Sci. Pollut. Res., 23*(12), 11957–11963.

Dar, M. U. D., Bhat, S. A., Ali, & Shah, A. I., (2017). Modeling climate change impact; a study on different procedures and strategies: A review. *Int. J. Pure Appl. Biosci., 5*, 183–200.

Dar, M. U. D., Shah, A. I., Ali, S., & Bhat, S. A., (2020). *Woodchip Bioreactors for Nitrate Removal in Agricultural Land Drainage: Agricultural Waste Book* (pp. 99–11). Taylor &franc.

Degremont, G., (1991). *Water Technical Handbook*. Lavoisier Publishing, Paris.

Dhadse, S., Satyanarayan, S., Chaudhari, P. R., & Wate, S. R., (2010). Vermifilters: A tool for aerobic biological treatment of herbal pharmaceutical wastewater. *Water Sci. Technol., 61*(9), 2375–2380

Furlong, C., Templeton, M. R., & Gibson, W. T., (2014). Processing of human faeces by wet vermifiltration for improved on-site sanitation. *J Water Sanit. Hyg. De., 4*(2), 231–239.

Garkal, D. J., Mapara, J. V., & Prabhune, M., (2015). Domestic wastewater treatment by bio-filtration: A case study. *Journal Geogr. Geol., 4*(1), 799–811.

Ghatnekar, S. D., Kavian, M. F., Sliharma, S. M., Ghatnekar, S. S., Ghatnekar, G. S., & Ghatnekar, A. V., (2010). Application of vermi-filter-based effluent treatment plant (pilot scale) for biomanagement of liquid effluents from the gelatine industry. *Dynam. Soil Dynam. Plant, 4*(1), 83–88.

Goel, P. K., (2006). *Water Pollution: Causes, Effects and Control*. New Age International.

Gunadi, B., Blount, C., & Edwards, C. A., (2002). The growth and fecundity of *Eisenia fetida* (Savigny) in cattle solids pre-composted for different periods. *Pedobiologia, 46*(1), 15–23.

Hameed, M., Saqib, N. E., Bandit, B. P., & Shakeel, A. B., (2017). Assessing the impact of climate change on surface water resources of Wular Lake. *Journal of Pharmacognosy and Phytochemistry, 6*(5), 1067–1072.

Hand, P., (1988). Earthworm biotechnology. In: Greenshields, R., (ed.), *Resources and Application of Biotechnology: The New Wave*. MacMillan Press Ltd, New York.

Hartenstein, R., (1978). Utilization of soil organisms in sludge management. In: *Conference Proceedings* (p 171). US Department of commerce, national technical information services, PB286932. Springfield, Virginia.

Hughes, R., Nair, J., & Ho, G., (2008). The toxicity of ammonia/ammonium to the vermifiltration wastewater treatment process. *Water Sci. Technol., 58,* 1215–1220.

Hughes, R., Nair, J., Mathew, K., & Ho, G., (2007). Toxicity of domestic wastewater pH to key species within an innovative decentralised vermifiltration system. *Water Sci. Technol., 55,* 211–218.

Ihssen, J., Horn, M. A., Matthies, C., Gößner, A., Schramm, A., & Drake, H. L., (2003). N_2O-producing microorganisms in the gut of the earthworm *Aporrectodea caliginosa* are indicative of ingested soil bacteria. *Appl. Environ. Microbiol., 69*(3), 1655–1661.

Julka, J. M., (1986). Earthworm resources in India. In: *National Seminar on Organic Waste Utilization* (pp. 1–7).

Kumar, T., Rajpal, A., Arora, S., Bhargava, R., Hari, P. K. S., & Kazmi, A. A., (2016). A comparative study on vermifiltration using epigeic earthworm *Eisenia fetida* and *Eudrilus eugeniae*. *Desalin. Water Treat., 57*(14), 6347–6354.

Lai, E., Hess, M., & Mitloehner, F., (2018). Profiling of the microbiome associated with nitrogen removal during vermifiltration of wastewater from a commercial dairy. *Front. Microbiol., 9,* 1964.

Li, Y. S., Robin, P., Cluzeau, D., Bouché, M., Qiu, J. P., Laplanche, A., Hassouna, M., Morand, P., Dappelo, C., & Callarec, J., (2008). Vermifiltration as a stage in reuse of swine wastewater: Monitoring methodology on an experimental farm. *Ecol. Eng., 32*(4), 301–309.

Li, Y. S., Xiao, Y. Q., Qiu, J. P., Dai, Y. Q., & Robin, P., (2009). Continuous village sewage treatment by vermifiltration and activated sludge process. *Water Sci. Technol., 60*(11), 3001–3010.

Liu, J., Lu, Z., Zhang, J., Xing, M., & Yang, J., (2013). Phylogenetic characterization of microbial communities in a full-scale vermifilter treating rural domestic sewage. *Ecol. Eng., 61*, 100–109.

Luth, P. R., Germain, P., Lecomte, M., Landrain, B., Li, Y., & Cluzeau, D., (2011). Earthworm effects on gaseous emissions during vermifiltration of pig fresh slurry. *Bioresour. Technol., 102(*4), 3679–3686.

Martin, J. P., (1976). *Darwin on Earthworms: The Formation of Vegetable Moulds*. Bookworm Publishing. ISBN 0-916302-06-7.

Merlin, G., & Cottin, N., (2009). Performance of a compost biofilter containing earthworms to treat cheese whey. *Environ. Technol., 30*(10), 995–1002.

Metcalf, L., Eddy, H. P., & Tchobanoglous, G., (1991). *Wastewater Engineering: Treatment, Disposal, and Reuse* (3rd edn.). McGraw-Hill, New York.

Natarajan, N., Kannadasan, N., & Krishnamoorthy, R., (2015). Effect of earthworms on dairy wastewater treatment through vermifiltration. *Int. J. Nano Corr. Sci. Eng., 2*(5), 91–98.

Noumsi, I. M., Akoa, A., Eteme, R. A., Nya, J., Ngniado, P., Fonkou, T., & Brissaud, F., (2005). Mosquito development and biological control in a macrophyte-based wastewater treatment plant. *Water Sci. Technol., 51*(12), 201–204.

Patel, J. B., (2018). Wastewater treatment by vermifiltration: A review. *Int. J. Latest Technol. Eng. Manag. Appl. Sci., 6(*1), 2278–2540.

Pimentel, D., Berger, B., Filiberto, D., Newton, M., Wolfe, B., Karabinakis, E., Clark, Set al., (2004). Water resources: Agricultural and environmental issues. *Bioscience, 54*(10), 909–918.

Ra, C. S., Lo, K. V., Shin, J. S., Oh, J. S., & Hong, B. J., (2000). Biological nutrient removal with an internal organic carbon source in piggery wastewater treatment. *Water Res., 34*(3), 965–973.

Reddy, K. R., & Smith, W. H., (1987). Aquatic plants for water treatment and resource recovery. In: *Conference on Research and Applications of Aquatic Plants for Water Treatment and Resource Recovery*. Magnolia Publishing Inc, Orlando, Fla. (USA).

Reinecke, A. J., & Viljoen, S. A., (1993). Effects of worm density on growth and cocoon production of the African nightcrawler (*Eudrilus eugeniae*) oligochaeta. *Eur. J. Soil Biol., 29*(1), 29–34.

Rodgers, M., Xiao, L. W., & Mulqueen, J., (2006). Synthetic dairy wastewater treatment using a new horizontal-flow biofilm reactor. *J. Environ. Sci. Health A, 41*(5), 751–761.

Samal, K., Dash, R. R., & Bhunia, P., (2017a). Treatment of wastewater by vermifiltration integrated with macrophyte filter: A review. *J. Environ. Chem. Eng., 5*(3), 2274–2289.

Samal, K., Dash, R. R., & Bhunia, P., (2017b). Performance assessment of a *Canna indica* assisted vermifilter for synthetic dairy wastewater treatment. *Process Saf. Environ. Prot., 111*, 363–374.

Samal, K., Dash, R. R., & Bhunia, P., (2018a). Effect of hydraulic loading rate and pollutants degradation kinetics in two stage hybrid macrophyte assisted vermifiltration system. *Biochem Eng. J., 132*, 47–59.
Samal, K., Dash, R. R., & Bhunia, P., (2018b). A comparative study of macrophytes influence on performance of hybrid vermifilter for dairy wastewater treatment. *J. Environ. Chem. Eng., 6*(4), 4714–4726.
Schudell, P., & Boller, M., (1989). On-site wastewater treatment with intermittent buried filters. *International Specialized Conference on Design and Operation of Small Scale Wastewater Treatments Plants*, 95–104.
Shah, A. I., Din, D. M. U., Bhat, R. A., Singh, J. P., Singh, K., & Bhat, S. A., (2020). Prospectives and challenges of wastewater treatment technologies to combat contaminants of emerging concerns. *Ecol. Eng., 152*, 105882. https://doi.org/10.1016/j.ecoleng.2020.105882.
Shokouhi, R., Ghobadi, N., Godini, K., Hadi, M., & Atashzaban, Z., (2020). Antibiotic detection in a hospital wastewater and comparison of their removal rate by activated sludge and earthworm-based vermifilteration: Environmental risk assessment. *Process Saf. Environ. Prot., 134*, 169–177
Singh, R., Bhunia, P., & Dash, R. R., (2017). A mechanistic review on vermifiltration of wastewater: Design, operation and performance. *J. Environ. Manag., 197*, 656–672.
Singh, R., Bhunia, P., & Dash, R. R., (2018). Understanding intricacies of clogging and its alleviation by introducing earthworms in soil biofilters. *Sci. Total Environ., 633*, 145–156.
Singh, R., Bhunia, P., & Dash, R. R., (2019a). Optimization of organics removal and understanding the impact of HRT on vermifiltration of brewery wastewater. *Sci. Total Environ., 651*, 1283–1293.
Singh, R., Samal, K., Dash, R. R., & Bhunia, P., (2019b). Vermifiltration as a sustainable natural treatment technology for the treatment and reuse of wastewater: A review. *J. Environ. Manag., 247*, 140–151.
Singleton, D. R., Hendrix, P. F., Coleman, D. C., & Whitman, W. B., (2003). Identification of uncultured bacteria tightly associated with the intestine of the earthworm *Lumbricus rubellus* (*Lumbricidae*; *Oligochaeta*). *Soil Biology and Biochemistry, 35*(12), 1547–1555.
Sinha, R. K., Agarwal, S., Chauhan, K., Chandran, V., & Soni, B. K., (2010). Vermiculture technology: Reviving the dreams of sir Charles Darwin for scientific use of earthworms in sustainable development programs. *Technol. Invest., 1*(03), 155–172.
Sinha, R. K., Bharambe, G., & Bapat, P., (2007). Removal of high BOD and COD loadings of primary liquid waste products from the dairy industry by vermifiltration technology using earthworms. *Indian J. Environ. Prot., 27*(6), 486–501.
Sinha, R. K., Bharambe, G., & Chaudhari, U., (2008). Sewage treatment by vermifiltration with synchronous treatment of sludge by earthworms: A low-cost sustainable technology over conventional systems with potential for decentralization. *Environmentalist, 28*(4), 409–420.
Sinha, R. K., Chandran, V., Soni, B. K., Patel, U., & Ghosh, A., (2012). Earthworms: Nature's chemical managers and detoxifying agents in the environment: An innovative study on treatment of toxic wastewaters from the petroleum industry by vermifiltration technology. *Environmentalist, 32*(4), 445–452.

Sinha, R. K., Herat, S., Agarwal, S., Asadi, R., & Carretero, E., (2002). Vermiculture and waste management: Study of action of earthworms *Eisenia foetida, Eudrilus eugeniae* and *Perionyx excavatus* on biodegradation of some community wastes in India and Australia. *Environmentalist, 22*, 261–268.

Speece, R. E., (1983). Anaerobic biotechnology for industrial wastewater treatment. *Environ. Sci. Technol., 17(*9), 416A–427A.

Swati, A., & Hait, S., (2018). A comprehensive review of the fate of pathogens during vermicomposting of organic wastes. *J. Environ. Qual., 47*(1), 16–29.

Telang, S., & Patel, H., (2015). Vermi-biofiltration-a low cost treatment for dairy wastewater. *Int. J. Sci. Res., 4*(7), 595–599.

Tomar, P., & Suthar, S., (2011). Urban wastewater treatment using vermi-biofiltration system. *Desalination, 282*, 95–103.

Verma, A. K., Dash, R. R., & Bhunia, P., (2012). A review on chemical coagulation/ flocculation technologies for removal of color from textile wastewaters. *J. Environ. Manag., 93*(1), 154–168.

Wang, D. B., Zhang, Z. Y., Li, X. M., Zheng, W., Yang, Q., Ding, Y., & Zeng, G. M., (2010b). A full-scale treatment of freeway toll-gate domestic sewage using ecology filter integrated constructed rapid infiltration. *Ecol. Eng., 36*(6), 827–831.

Wang, J., Liu, X. D., & Lu, J., (2012). Urban river pollution control and remediation. *Procedia Environ. Sci., 13*, 1856–1862.

Wang, L., Guo, F., Zheng, Z., Luo, X., & Zhang, J., (2011). Enhancement of rural domestic sewage treatment performance, and assessment of microbial community diversity and structure using tower vermifiltration. *Bioresour. Technol., 102*(20), 9462–9470.

Wang, S., Yang, J., & Lou, S. J., (2010a). Wastewater treatment performance of a vermifilter enhancement by a converter slag–coal cinder filter. *Ecol. Eng., 36*(4), 489–494.

Wang, Y., Xing, M. Y., Yang, J., & Lu, B., (2016). Addressing the role of earthworms in treating domestic wastewater by analyzing biofilm modification through chemical and spectroscopic methods. *Environ. Sci. Pollut. Res., 23*, 4768–4777.

Wei, Y., Van, H. R. T., Borger, A. R., Eikelboom, D. H., & Fan, Y., (2003). Minimization of excess sludge production for biological wastewater treatment. *Water Res., 37*(18), 4453–4467.

Xing, M., Li, X., & Yang, J., (2010). Treatment performance of small-scale vermifilter for domestic wastewater and its relationship to earthworm growth, reproduction and enzymatic activity. *Afr. J. Biotechnol., 9*(44), 7513–7520.

Xing, M., Wang, Y., Liu, J., & Yu, F., (2011). A comparative study of synchronous treatment of sewage and sludge by two vermifiltrations using an epigeic earthworm *Eisenia fetida. J. Hazard Mater., 185*(2, 3), 881–888.

Xing, M., Yang, J., & Lu, Z., (2005). Microorganism-earthworm integrated biological treatment process— A sewage treatment option for rural settlements. *Five Days ICID 21st EUROPEAN Regional Conference.*

Xing, M., Zhao, C., Yang, J., & Lv, B., (2014). Feeding behavior and trophic relationship of earthworms and other predators in vermifiltration system for liquid-state sludge stabilization using fatty acid profiles. *Bioresour. Technol., 169C*, 149–154.

Zhao, Y., Zhang, Y., Ge, Z., Hu, C., & Zhang, H., (2014). Effects of influent C/N ratios on wastewater nutrient removal and simultaneous greenhouse gas emission from the combinations of vertical subsurface flow constructed wetlands and earthworm eco-filters for treating synthetic wastewater. *Environ. Sci. Processes Impacts, 16*(3), 567–575.

Zheng, C., Zhao, L., Zhou, X., Fu, Z., & Li, A., (2013). Treatment technologies for organic wastewater. W*ater Treat*, 249–286.

CHAPTER 10

Phytoremediation Technique for Agricultural Pollutants

MAHRUKH and SYED ROUHULLAH ALI

College of Agricultural Engineering and Technology, SKUAST–K, Jammu and Kashmir – 190025, India,
E-mail: agasyedrouhullah@gmail.com (S. R. Ali)

ABSTRACT

Arable land degradation has arisen as a severe concern, posing a danger to agricultural productivity. Chemical fertilizers, farm manure, pesticides, waste sludge, plastic mulch, irrigation, and other farming activities and agro-inputs are all substantial sources of toxins on agricultural land. These hazardous pollutants include a wide range of organic chemicals and heavy metals, many of which are detrimental to human health. Heavy metals, which are the most frequent types of contaminants in agricultural soil, have a significant impact on crop yields by reducing microbial activity and soil fertility. Decontamination of heavy metal contaminated soils is therefore crucial. Over the past few years, microbial pest control has acquired increasing interest as a feasible *in situ* remediation option for degraded soil restoration. Phytoremediation can help to reduce pollution and the negative effects of agricultural pollutants upon on ecosystem. Plants can support pollution clearance through a variety of mechanisms, including absorption and concentration, pollutant conversion, stability, and degradation of rhizosphere, which involves plants induced production of root bacteria that breakdown the

Advanced Technologies for Water Quality Treatment and Management. Mehraj U. Din Dar, PhD, Aamir Ishaq Shah, MTech, Shakeel Ahmad Bhat, PhD & Syed Rouhullah Ali, PhD (Eds.)

pollutants. The adoption of plants in order to lower contaminant levels in the soil is an affordable way to lessen the risk on human and ecological health of polluted soil locations. The purpose of this chapter is to outline numerous agricultural pollutants, phytoremediation methods, and their potential as remediation procedures that rely on growing plants' inherent capacity to remove toxins from the environment.

10.1 INTRODUCTION

Since the dawn of civilization, human actions have posed a danger to the planet's sustainability and stability. Human actions have potential and severe environmental repercussions (Varallyay, 1994). Human activities such as urbanization, industrialization, and shifting agricultural techniques have severely altered the nature of the environment across the world. Each of such practices damage the land, air, and water, causing a major and significant danger to ecologies and the health of humans and animals as well (Kang, 2014). Arable land degradation has arisen as a severe concern, posing a danger to agricultural productivity (Bhat et al., 2017, 2019; Zaman et al., 2018; Dar et al., 2020) Agricultural contaminants could occur as natural chemicals or energy generated in exceeding natural limits, as well as xenobiotic chemicals created by humans. A pollutant is a toxin (organophosphorus compounds, polychlorinated biphenyls (PCBs), polycyclic aromatic hydrocarbons (PAH) radionuclides, HM, etc.), or geochemical materials (sediments, dust), physical matter (radiation, heat, noises) or biological organism or produce, that are purposely or unintentionally discharged into the atmosphere with potential or actual detrimental, damaging, undesirable, or toxic implications. The extensive deterioration of the soils and water with organic substances and HM, as well as municipal effluents and consumer products, led to a severe effect on the environment and health impacts (Pandey and Singh, 2019). The decline in the health of soil and water ecosystems is largely caused by organic and inorganic contaminants. Inorganic pollutants include heavy metals (Ni, Cd, Pb, Co, Cr, Zn, Se, Cu, and, Cr), metalloids (Hg, Se, and As) (Ye et al., 2011; Pandey et al., 2011), and radionuclides (Ra, Ra, Sr, Cs and U) (Dushenkov, 2003; Cerne et al., 2011), whereas organic pollutants encompass mainly hydrocarbons (Hong et al., 2001; Davis et al., 2002). Each of these xenobiotics is non-biodegradable and can persist in soil for longer durations in soil ecosystems, representing a significant environmental risk. Since inorganic pollutants are toxic, they affect ecosystems adversely, and as a result of its mutagenic properties, DNA damage eventually happens in humans and animals (Baudouin et al., 2002).

Agriculture is a vital source of income and employment. Despite its numerous advantages, farming operations cause pollution, which has negative consequences both towards the human health and the environment (Abbasi et al., 2014). Heavy metals, polymers, toxic substances, greenhouse gases, particulate matter, microorganisms, and other toxins are all emitted by its activities (Vejan et al., 2016). Agriculture has employed chemical fertilizers, herbicides, and pesticides to increase crop production and productivity while adding toxic levels of nitrogen and phosphorus to soil and terrestrial ecosystems. Furthermore, harmful pollutants released by diverse man-made causes contaminate natural resources, resulting in a shortage of potable water as well as a degradation of land soil and water contamination due to metals, metalloids, and radio nuclei poses a hazard for plants,' humans,' and animals' health. Heavy metal pollution is a serious danger to agricultural land worldwide (Mico et al., 2006; Lone et al., 2008; Lin et al., 2012; Efremova and Izosimova, 2012a, b), limiting the amount of land appropriate for generating the world 's nutrition supplies The buildup of heavy metals, that do not disintegrate over time, many of those are harmful in almost any amount, and other toxic metals that are not accepted as important micronutrients, negatively impact crop productivity by reducing microbial growth and soil quality. For instance, Cadmium (Cd) has contaminated 13,330 hectares of agricultural land in China (Singh et al., 2011). Around 1,400,000 heavy metal polluted sites were found in Western Europe (Wei et al., 2005). Excess heavy metals contained by the soil also represent a considerable risk to the well-being of plants and animals and to human health also, when they enter the food chain. These can infiltrate the food chain by means of crops and build up within the body of humans via biomagnification, which poses a major health risk (Sarwar et al., 2010; Rehman et al., 2017). Figure 10.1 shows the impact of several heavy metals affecting humans. Inappropriate farming methods that escalate soil contamination by heavy metals makes it more difficult to generate safe food for human and animal use. It also increases land pollution.

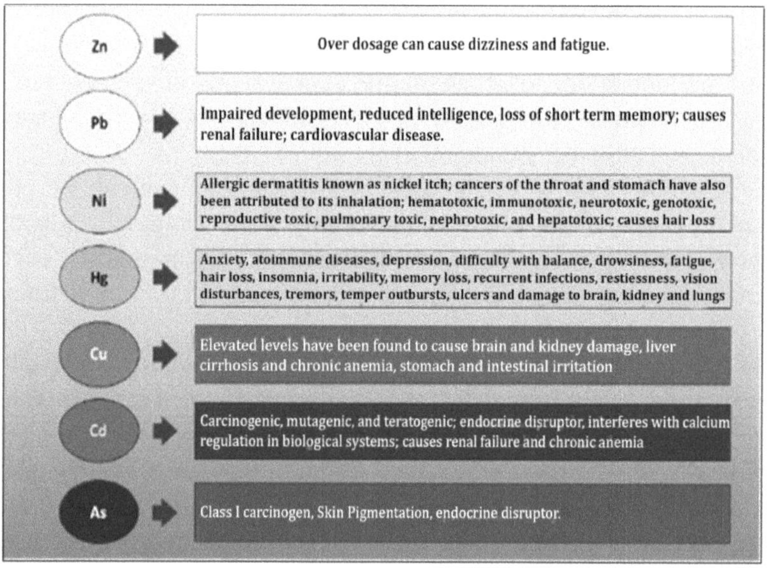

FIGURE 10.1 Impact of different heavy metals on humans.
Source: Reprinted from Qadir et al. (2021).

The extremely fast growth and extensive use of chemical products to protect plants and address certain human parasites like malaria and typhus have also caused ongoing farm and natural agrochemical contamination (McKone and Ryan, 1989). 1,1,1-trichloro-2,2-bis (p-chlorophenyl) ethane (DDT) was, for example, the main insecticide used for agriculture, forestry, and controlling malaria and typhus mosquito vectors. Its low biodegradation makes it persistent and is anticipated for a half-life of quite a few decades (Crowe and Smith, 2007). Moreover, in animals affected by it, have an accumulation of toxins in their adipose tissue. This has caused a concern that species in particular birds are at significant risk, and that they would be banned in the industrialized countries, even to battle malaria. However, scores of years after it had been prohibited in many nations, substantial quantities are still observed. Another example is Atrazine, an herbicide that has been detected in water ecosystems afterwards getting leached out from the soil and has been used widely in maize farming. It has been banned in several countries it is claimed to be responsible for endocrine disruptions, particularly in batrachians (Hayes et al., 2002). Topsoil contamination is a major problem in both developed and developing nations, posing serious implication on environmental and health of humans (FAO and ITPS, 2015). Furthermore, land contaminants might be transported to aquatic habitats over time due to leaching or wind dispersal. Contaminated areas must be remedied, and pollutant penetration into the food chain must be reduced for long-term environmental and agricultural growth. It is thus essential to consider what actions may be taken to make sure food is safe to consume, including on arable soil with a variety of heavy metal pollutants (Nicholson et al., 2003; Adefemi et al., 2012). The existing physicochemical remediation procedures are already yielding promising results; nevertheless, these procedures frequently result in secondary contamination and additional obliteration of soil productivity (Kokyo et al., 2013). These traditional soil remediation procedures are costly, time-consuming, and occasionally ecologically harmful. They are extremely tough and impractical. As a result, there is a pressing need to promote efficient soil and water treatment technologies that aim to eliminate pollutants. Plants and other creatures can be utilized to treat soil as a replacement or supplement to engineering-based remediation approaches for pollutant stabilization, removal, deterioration, or dissolution.

Phytotechnology is a collection of techniques used for soil, surface water, groundwater, or sediment contamination utilizing or including vegetation. The comparatively modest capital requirements and the esthetical quality of the planted areas have made these technologies enticing alternatives compared to traditional cleaning technology. Phytoremediation technique harnesses plants' capacity to repair polluted sites from toxins. The phytoremediation subtype of bioremediation, which is mostly used to limit the levels or deleterious implication of contaminants present in the environment, corresponds to the employment of plants and related soil organisms. The word phytoremediation originates from a Greek term "Phyto," which means plant and "Remedium," the Latin term which means a restoration, elimination, and repair. Therefore, the need arises for competent plant species for holding, breaking, or removing metal, salt, fungicide, herbicides, insecticides, organic solvents, harmful explosive materials, crude oil and extracts or various other environmental pollutants, that mitigates or removes pollutants from polluted land, water, or air by the phytoremediation process. A number of possible pollutants, e.g., petroleum hydrocarbon, organic pollutants, chlorinated substances, heavy metals, radioactive waste, agricultural waste, pentachlorophenol (PCP), polycyclic aromatic hydrocarbons (PAHs), etc., may be treated with the method for phytoreconciliation (Pivetz, 2001). Soil remediation methods based on plants can be considered as biological restoration technologies with a large self-extended root system that enriches the underground ecosystem for potential productive usage. Plants may absorb ionic substances through the root system even at low quantities in the soil. The root systems of plants are distributed throughout the soil matrix, creating a rhizo environment for heavy metals and bioavailability, allowing polluted soil to be restored and soil quality to be stabilized (Ali et al., 2013; Jacob et al., 2018; DalCorso et al., 2019). Likewise, utilizing salt tolerating flora (halophytes) to restore saline soils has been recommended as a cost-effective approach. Such halophytes can reduce salt levels in the soil, particularly at the rhizospheric level, allowing for the establishment of higher-yielding crop plants (Zuccarini, 2008). Different plant species, including snow-pine tree (*Salix viminalis*), poplar (Populus Deltoids), water hyacinth (Eichhornia crassipes), basket willows, etc., have the capability to extract cyanides from contaminated habitats (Larsen et al., 2004; Taebi et al., 2008). In phytoremediation investigations, certain food crops are renowned, like, *Zea mays* L. Czern, *Brassica napus* L., *Brassica juncea* and Czeren, *Helianthe annuus* L. (Pandey and Bajpai, 2019). The adoption of a certain species of plants in polluted regions for recommendation and growth will rely on the kind of contamination, method utilized to remove the contamination from that species, plant species' pollutant resistance capability, and many more ecological restrictions (Huang and Cunningham, 1996; Meagher, 2000; Memon et al., 2001).

10.2 MECHANISM OF PHYTOREMEDIATION

Phyto-remediation refers to any physical and biological processes influenced by plant, or soil microbial and plant associations that contribute to pollutant absorption, sequestration, degradation, and metabolization. Phytoremediation makes use of the root system's particular and exact absorption capacities, as well as translocalization, bioaccumulation, and the overall plant's capacity to store and breakdown pollutants. Plants possess a number of ways for contributing in phytoremediation. The kind of pollutant, bioavailability, and soil conditions all influence the processes and efficacy of phytoremediation (Cunningham and Ow, 1996). Because inorganic pollutants could not be chemically destroyed, various plant species can trap them in their rhizosphere or aggregate them in roots and thereafter plant biomass, where they can be removed. During their development, plants absorb metals from the contaminated soil. Plant-metal absorption could remain passive when metal ions flow through the root cell wall. It might possibly play a role in metal ion transport through the root cell membrane (Pilon-Smits, 2005). The chemical characteristics of the major metals are recognized by plant membrane proteins and bind to them for the absorption and transportation of metals (Axelsen and Palmgrem, 2001). Some of these metal ions have similar chemical compositions, so proteins mistake them for one another. The root system offers a massive surface area which, together with other non-essential pollutants, collects, and stores water and nutrients necessary for development. The rhizosphere of these plants emits a variety of biochemical compounds such as bicarbonate anions, protons, organic acids, and additional cations, which have an impact on the mobility, availability, and degradation of different soil contaminants, as well as its chemical properties (Stephenson and Black, 2014; Gerhardt et al., 2015; Ullah et al., 2015). By absorbing water from the ground, plants greatly limit pollution discharge to groundwater (Vangronsveld et al., 2009; Mench et al., 2010). There are several types of phytoremediation strategies based on distinct mechanisms (Figure 10.2).

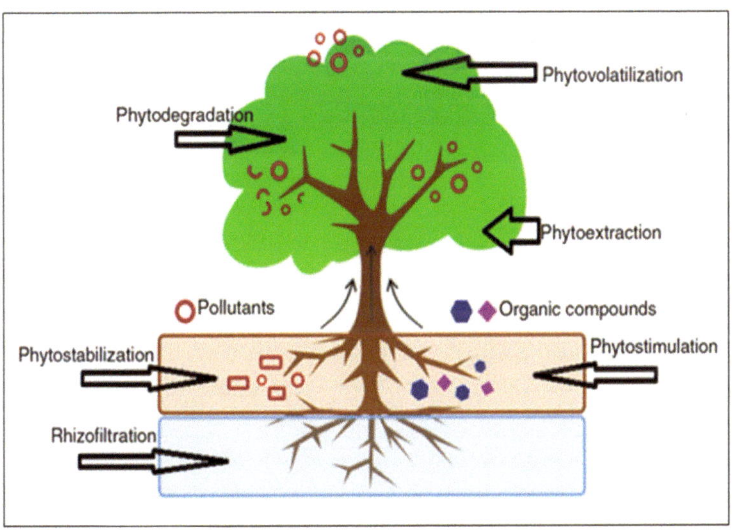

FIGURE 10.2 Phytoremediation techniques shown schematically.
Source: Reprinted with permission from Abdel-Shafy and Mansour (2018). © Springer Nature.

10.3 TYPES OF PHYTOREMEDIATION TECHNIQUES

The different types of phytoremediation techniques are listed in Table 10.1.

TABLE 10.1 Phytoremediation Technologies, Their Mechanism and Examples of Contaminants

Technique	Mechanism	Contaminants
Phytoextraction	Hyperaccumulation of contaminants into the plant	Pb, Cd, Zn, Ni, Cu
Phytostabilization	Immobilization of contaminants through sorption by roots	Cd, Zn, As, Pb, Cu, Cr, Se, U, furans, pentachlorophenol, DDT, dieldrin
Rhizofiltration	Accumulation of contaminants in rhizosphere	Metals like Zn, Ni, Cu, Pb, Cd, radionuclides – Sr, Cs, U, hydrophobic organics, and radionuclides
Phytovolatilization	Volatilization of contaminants in the transpiration stream through leaves	Se, Hg, chlorinated solvents like methylene chloride (MC), carbon tetrachloride, 1,1,1-trichloroethane (TCA), trichloroethylene (TCE), carbon tetrachloride (CT), tetrachloroethylene (PCE)
Phytodegradation	Degradation of contaminants in plant through internal enzymatic activity and photosynthetic oxidation/ reduction	Nutrients like NO_3^-, NH_4^+, PO_4^{3-}, Herbicides such as atrazine, alachlor, mixtures benzene toluene ethylbenzene and xylene (BTEX), ammunition wastes, 2,4,6-trinitrotoluene (TNT), rapid detonating explosive (RDX), chlorinated aliphatic compounds-trichloroethylene (TCE)
Rhizodegradation	Contaminant biodegradation by microbes in the rhizosphere	Polycyclic aromatic hydrocarbons (PAHs), Petroleum hydrocarbons, chlorinated solvents, pesticides, polychlorinated biphenyls (PCBs), toluene, ethylbenzene, benzene, and xylenes

10.4 PHYTOEXTRACTION

The absorption and transloading of contaminants from plant roots into the biomass of plants, including the stems, leaves, and tissues of trees, is known as phytoextraction. Phytoextraction, sometimes called phyto-accumulation, is utilized mostly for polluted soils (USEPA, 2000). The plants' procedure of extracting contaminants like heavy metals involves those some steps: (i) rhizosphere mobilizing heavy metals; (ii) plant root heavy metal ion absorption; (iii) heavy metal ion transport to aerial parts; (iv) heavy metal ions sequestered and divided in plant tissue by the plant (Ali et al., 2013). The aqueous phase through the root is one of the most common methods of chemical absorption. Plant transpiration moves organic ions and molecules from the soil and sediments to the roots (Figure 10.3). Ion transport from the soil to the root happens concurrently with water flow (National Research Council, 2003). Chemicals must pass through the plasma membrane and then into the cytoplasm of the cells to be absorbed; the plasma membrane functions as a barrier to absorption.

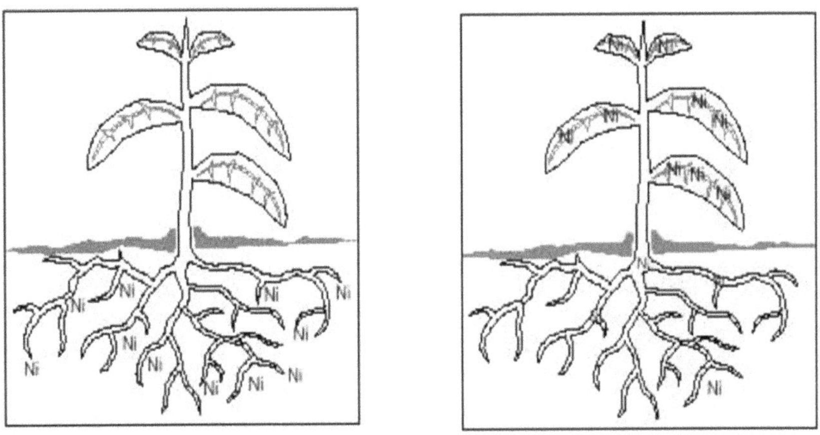

FIGURE 10.3 Image showing phytoextraction from soil.
Source: Reprinted from Etim (2012). © Modern Scientific Press Company. Open access.

For optimal phytoextraction, the plant species must be chosen carefully. Hyperaccumulator species are the best species for phytoextraction because they possess a high capacity for accumulating pollutants (Cristaldi et al., 2017). Natural heavy metal hyperaccumulators may accumulate 100 times more metals than non-hyperaccumulating species under the same conditions (Rascio and Navari-Izzo, 2011). Metal hyperaccumulators have been discovered in approximately 450 plant species from at least 45 angiosperm families (Suman et al., 2018), ranging from annual herbs to perennial shrubs and trees, including Brassicaceae, Fabaceae, Solanaceae, Lamiaceae, Asteraceae, Scrophulariaceae, and Euphorbiaceae (Salt et al., 1998; Dushenkov, 2003). Plant species that generate a large quantity of biomass in a short period of time can also be used for phytoextraction (Li et al., 2018; Nakajima et al., 2019). *Zea mays*, *Nicotiana tabacum*, *Helianthus annuus*, and *Cannabis sativa* have all been shown to remove heavy metals from contaminated soil via phytoextraction by Kayser et al. (2000); Vangronsveld et al. (2009); and Herzig et al. (2014). Phytoextraction eliminates pollutants from the soil permanently and is far less costly than other approaches. The plants must: (i) absorb into their roots enormous concentrations of heavy metals; (ii) translate heavy metal into the biomass; and (iii) create huge amounts of plant biomass, in order to be viable. In addition, remedies should have detoxification and/or tolerance mechanisms that amass large amounts of toxic metals in their shoot systems (Brennan and Shelley, 1999).

10.5 PHYTOSTABILIZATION

Heavy metal absorption by roots or heavy metal precipitation in the rhizosphere are examples of phytostabilization, both of which reduce pollutant mobility in the soil. Root absorption, precipitation, complex formation, or metal valence change or decrease in the rhizosphere zone are all ways that heavy metals might be immobilized in soils utilizing plants (Ghosh and Singh, 2005; Ali et al., 2013). Phytostabilization has been shown to be effective in managing Cd, Cr, Cu, Pb, Zn, and As polluted soils, as well as the removal of metals and other inorganic pollutants from sediments. *Festuca* spp. and *Agrostis* spp. are used to phytostabilize Cu, Zn, and Pb in the most heavily polluted soils in European countries (Mahar et al., 2016). Metal-tolerant species, such as *Brassica juncea*, *Epilobium dodonaei*, (Banuelos et al., 2005; Shiyab et al., 2009), *Vicia vilosa*, *Hordeum vulgare* (Katoh et al., 2017), *Phragmites australis* and *Typha domingensis* (Bonanno, 2013), help to stabilize metals and protect soils. *Arundo donax*, *Typha domingensis*, *Phragmites australis*, *Vossia cuspidate*, *Nasturtium officinale*, *Apium nodiflorum*, and *Zannichellia peltate* are some macrophytes that have shown to be effective in phytostabilizing heavy metal-contaminated aquatic environments (Bonanno and Vymazal, 2017).

In locations with fine-textured soils and abundant organic material, phytostabilization may be very efficient in reducing the toxicity of a diverse variety of surface pollutants (Cunningham et al., 1995). This method could potentially be employed to restore plant covers in locations wherein natural vegetation has died owing to excessive concentrations of metals in soil profile or disruptions on the material's surface. The abundance of the vegetative cover causes impacts on soil chemistry and ecology, that might lead to pollutant adsorption by plant roots in soil. Another benefit of this method is that it does not need the disposing of potentially harmful materials/biomass, and so it is highly successful when fast immobilization is necessary to safeguard surface and groundwater (Zhang et al., 2009). Nonetheless, there are many key drawbacks associated with this methodology, including pollutant retention in the soil, intensive fertilization or soil supplements, obligatory monitoring, and the stability of pollutants may be largely attributable to the soil supplements.

10.6 RHIZOFILTRATION

Rhizo filtration is a method of plant-based remediation that includes filtration of water via roots to eliminate hazardous contaminants of undesirable nutrients. Rhizofiltration is mostly used to restore slightly contaminated groundwater, surface water, and wastewater (Raskin and Ensley, 2000). It is a kind of phytoremediation, which is the technique of eliminating pollutants from contaminated water by the use of hydroponically grown (Figure 10.4) plant roots to absorb, concentrate, and precipitate them (Ramanjaneyulu et al., 2017). These plants are placed in a polluted location, within which their roots absorb the water as well as the pollutants. The plants are harvested when their roots get saturated by the pollutants. The similar fundamental approach to remediation is followed by phytoextraction and rhizofiltration. Rhizofiltration is utilized for the treatment of waters, whereas phytoextraction is employed for soil amendment. Cu^{2+}, Cd^{2+}, Cr^{6+}, Ni^{2+}, Pb^{2+}, and Zn^{2+}, that are predominantly held inside the roots, can be filtered via rhizofiltration (USEPA, 2000). The capacity of Indian mustard, sunflower, corn, spinach, tobacco, and rye, to detoxify lead from water has been confirmed, with sunflower showing the best potential (Sharma and Pandey, 2014). Indian mustard has been shown to remove lead from a broad range of concentrations (4–500 mg

L^{-1}) of lead (Raskin and Ensley, 2000). Panda (1996) used aquatic plant spp. for rhizofiltration of Zn and Ni in liquid culture medium. Water lettuce and water hyacinth were discovered to be efficient as Zn (18,041 and 14,423 ppm) and Ni hyperaccumulators (7,850 and 5,315 ppm). Low-level radioactive pollutants can also be managed to removed off the liquid streams, according to these findings. In a test in Chernobyl, Ukraine, various varieties of sunflowers were effectively employed to eliminate radioactive pollutants from pond water (Ramanjaneyulu et al., 2017). Terrestrial plants are recommended as their root systems are fibrous and considerably deeper, expanding the quantity of root area (Raskin and Ensley, 2000).

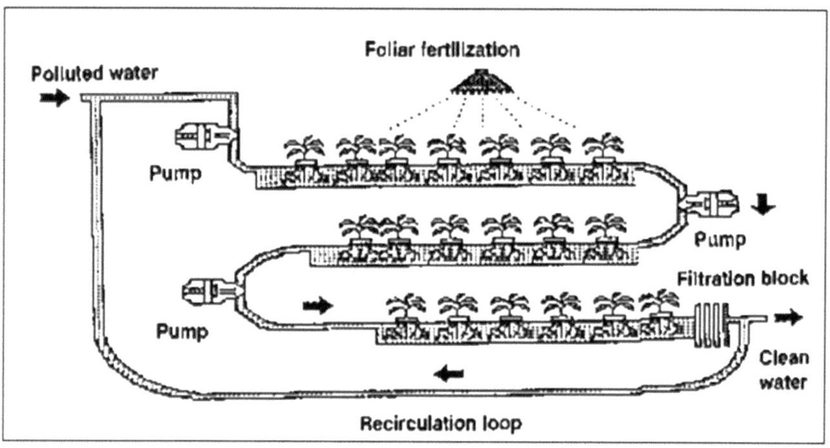

FIGURE 10.4 Engineered rhizo filtration system.
Source: Reprinted from Etim (2012). © Modern Scientific Press Company. Open access.

The flexibility to employ both terrestrial and aquatic plants for *in situ* and *ex situ* applications is one of the benefits of rhizofiltration. Another advantage is that there is no need to transfer pollutants to the shoots. As a result, non-hyperaccumulator species can be utilized. Constant pH adjustment, early plant development in a greenhouse or nursery, ongoing harvesting and plant disposal, and a thorough understanding of chemical speciation are all disadvantages. Rhizo filtration has estimated the cost of remediation to be between \$2–\$6 per 1,000 liters of water (USEPA, 2000).

10.7 PHYTOVOLATILIZATION

The use of plants to absorb soil contaminants in volatile forms and subsequently to transport it into the environment is involved in the process of phytovolatilization (USEPA, 2000). Arsenic, mercury (Hg, more toxic to Hg^{2+}, less harmful) and selenium (Se, more poisonous to $(CH_3)2se$, 600 times lower harmful), etc., are the volatilized metals through perspiration. Arabidopsis thaliana and Musk Grass is plant species that adopt the method of phytovolatilization for the removal of pollutants. Indian moutar was shown to decrease Selenium (Se) to non-toxic concentration (Banuelos and Meek, 1990; Banuelos et al., 1997). The distributions and volatilizations of selected organic pollutants through the use of hybrid poplar trees were studied by Burken and Schnoor (1999). Built wetlands are very successful in the removal of Se from selenite polluted wastewater as reported by Hansen et al. (1998). They have reported Se volatilization maximum rates at five locations of vegetated wasteland. Rabbit foot grass and cattail were Se's most effective phyto volatilizers, as a result of a Se m^2 day^{-1} volatilization rate of 180 ± 100 ug. Popular trees have been shown to volatilize 90% trichloroethylene (TCE) in use. In comparison to other methods of cleanup, after the pollutants have been eliminated by volatilization, their migration into other regions is not under control.

10.8 PHYTODEGRADATION

The utilization of plants and microorganisms for organic pollutant absorption, metabolization, and breakdown is called phytodegradation. This is called Phyto transformation as well. It involves the decomposition or integration of complex organic compounds into plant-based tissues by simple molecules (Trapp et al., 2005). Plant roots are utilized to eliminate soil pollution with organic chemicals in conjunction with microbes in this technique (Garbisu and Alkorta, 2001). The plants include enzymes that catalyze and accelerate chemical processes. They are complex chemical proteins. In certain enzymes, ammunition waste is broken up and converted while in other cases the chlorinated solvents like trichloroethylene (TCE) decompose. Different bacterial and fungal micro-organisms may help convert hazardous metals into less hazardous conditions. Phytodegradation has been found in soil, sediment, or groundwater pollutants, in order to remedy some organic pollutants, including chlorinated solvents, herbicides, and ammunition (EPA, 2000). Some plant species may effectively degrade the pollutant into less hazardous chemicals inside the plant system such as atrazine and TNT poplar tree. Plants such as *Avena sativa* and *Beta vulgaris* have been shown to be appropriate for the decontamination of TNT and GTN (Ramanjaneyulu et al., 2017). In addition, transgenic tobacco seeds that emit a Penta Ertrythritol tetranitrate reductase (PETR) enzyme can be utilized for phyto remediation and germinating and growing in the presence of GTN and TND (Gong et al., 1999).

10.9 RHIZODEGRADATION

The disintegration of the contaminants into the rhizosphere or root region of the plant, also known as phyto stimulation, is called Rhizo degradation Because of the abundance of microorganisms in the rhizosphere, the plant releases enzymes, amino acids, sugars, and other chemicals that may promote bacterial growth. The rhizodegradation process is thought to be carried out by these bacteria or other microorganisms. Treatment of a broad array of mainly organic contaminants, including pesticides, PAH, toluene, benzene, xylene, polychlorinated bipherylsethyl benzene, petroleum hydrocarbons, and chlorinated solvents, and are particularly useful in case of contaminated soils. It may be also seen as a facilitated bioremediation of plants, which stimulates the release of microbial and fungal breakdown to the root region of exudates and enzymes. Phytostimulation may also be implicated in aquatic plants that allow the active microbial degrading communities as in the example of speeding of the atrazine break-down with hornwort.

10.10 PHYTOREMEDIATION OF AGRICULTURAL POLLUTANTS

Humans are increasing the production of environmental pollutants such as pesticides, salts, oil products, acids, and heavy metals, among others. While some pollutants, such as salts and heavy metals, are found naturally in soils, they are the most significant sources of anthropogenic, induced pollution in the environment caused by the industrial and agricultural sectors (Delibacak et al., 2002; Suci et al., 2008). According to Dubois (2011), one-third to one-half of the world's agricultural fields have deteriorated, while a quarter has been severely damaged. Polluted agricultural environments significantly damage plants, micro-organisms, aquatic creatures, and essential processes such as soil productivity mineralization, immobilization, and nitrification (Batayneh, 2012). Pesticides, chlorinated solvents, heavy metals, and a vast range of other pollutants are released into the soil on a continuous basis, with the increasing international industry, urbanization, and intense agricultural land usage (FAO, 2015). The overuse of agrochemicals (fungicides, herbicides, chemical fertilizers, etc.), sewage sludge, animal feeds and plastic are mostly mediated through agricultural pollution. A substantial amount of soil pollution has resulted with the use of fertilizers and pesticides in agriculture. Although the use of these chemical compounds is considered to be an efficient means of control of pesticides and diseases, their use can have detrimental consequences on flora and fauna such as invertebrates and vertebrates (Schluz, 2004). Insecticides are by far the most significant for threatening environments and poisonous for living creatures among many groups of such chemical substances. Herbicides, bactericides, and fungicides follow (Goel and Aggarwal, 2007). Heavy metal pollution is also increased by chemical substances, such as pesticides. The plants may readily absorb and bioaccumulate in various organs, mostly heavy metals (Wang et al., 2003). Excessive heavy metal buildup affects the microbiological activity, soil fertility, soil quality and crop productivity. With food intake, the usage of water polluted with metals or air with hazardous metals, these metals may eventually come into the human systems (Jarup, 2003). Cyanide is another type of pollutant introduced to plant environment. In certain bacteria, fungus, algae, and higher plants, cyanogenic chemicals exist naturally. They are therefore present in a range of foodstuffs and plants. When eaten inadvertently or hydrolyzed by plants, cyanogenic chemicals are very poisonous (Schnepp, 2006; Barillo, 2009). They inhibit ATP production which leads to rapid death of plants or animals since no energy for regular tasks is available. Massive amounts of salt in various soil profiles are also a source of contaminants that cause salinity issues. Almost half of the irrigated land is highly salinated, making it unfit for crop production (Zhu, 2001; Bhat et al., 2107). NaCl and $MgSO_4$ are the most prevalent salts that cause soil salinity.

Increased accumulation of environmental pollutants causes biodiversity loss, greenhouse effect land degradation, pollution, deforestation, desertification, acid rain, and other environmental issues. In contrast to organic pollutants that are eventually converted to CO_2 or H_2O, inorganic pollutants such as metals and salts are likely to accumulate in a variety of ecological constituents, including lake, estuary, and marine sediments (MacDonald et al., 1996). Several pollutants may flow readily and eventually build in soil and water bodies from one environmental component to another. Most poisons may flow readily and eventually build in soil and water bodies through one of the ecological units to another. Water contamination is largely driven by wastewater, food manufacturing unit residue, agricultural usage of fertilizer and pesticides, etc. (Shah et al., 2020). Plants and aquatic animals and flora may take up these contaminants readily and transmit them to the human body wherein they lead to severe disease and disorders (Albering et al., 1999; Korte et al., 2000). However, the crops produced should be provided with drought-tolerant, disease-resistance, stress resistance to heavy metal, and improved nutritional benefits for sustainable farm output and must not just be remedied (Vejan et al., 2016). Consequently, their removal compared with the extraction of organic contaminants is significantly more challenging and needs a separate method. Phytoremediation is a potential technique for removing and plummeting the bioavailability of agricultural pollutants present in the soil. Farming and phytoremediation are inseparable bodies. Phytoremediation is in reality a farming technique that can only succeed when only a suitable approach is used.

10.11 HEAVY METALS

For their normal development and metabolic activities, plants require certain nutrients. Based on their requirements, these important nutrients are classified as macronutrients and micronutrients. Apart from these basic heavy metals of nutrients, non-nutrient heavy metals, like Pb and Cd are quite common. These metals turn into hazardous components to plants if they are present in soils above the necessary threshold levels. Plants frequently collect heavy metals at levels above their soil levels in large amounts from whence they reach the food chain. The use of excess heavy metals by plants in solar solutions leads towards various interactions at the cell level, which have toxic impacts on enzyme activities, protein structures, mineral nutrition, plant breathing and ATP content, photosynthesis, growth, and morphogenesis and reagent oxygen plants formation (ROS).

In heavy metal stress circumstances, transpiration rate and water content in treated plants have been observed to decrease (Ashraf et al., 2010). The enormous territory of China, Japan, and Indonesia, with metals such as Cd and Cu, and Zn, and North Greece, Albania, and Australia, Cr, Cu, Pb, and Ni, has been contaminated by anthropogenic agriculture operations (Song et al., 2019). The main sources of agricultural soil metals are sewage sludge, liming, untreated irrigation water and pesticides. Also, potential sources of heavy metals include fertilizers (organic and inorganic). In particular, phosphate fertilizers include Cd, Cr, Ni, Pb, and Zn at a varying amount (Nagajyoti et al., 2010). The most commonly used insecticides are organic arsenic compounds, particularly on cotton plants, whereas the major usage of inorganic arsenic is to protect wood. Wastewater for irrigation has led to the buildup in soils and plants of metals exceeding the maximum allowable animal feed in Gaza, Egypt (FAO, 2015). The Kapungwe (2013) discovered that in Zambia, soils, water, and crops pollution from metals (Co, Cr, Cu, Pb, and Ni) was reported due to the presence of heavy metal on the levels more than acceptable in soils and crops from both locations. Heavy metal pollution penetrates the human and cattle food chain and build up in them, known as bio-accumulation. One more process is also known as biomagnification. The percentage of a certain heavy metal within the food chain is increasing. When heavy metals reach an individual's system, several biological processes are disrupted. Other molecules, like the lethal reactive oxygen, can be broken down into more reactive species by these metals. Fe^{2+} and Fe^{3+} are also capable of reacting with common molecules like O_2 and H_2O_2. Some chronic consequences are carcinogenesis, mental lapse, renal, hepatic, GI, central nervous system issues, and many more (Ramanjaneyulu et al., 2017).

Phyto-remediation is the most ecologically acceptable and economically practical remediation method utilized for the mitigation of agricultural land pollution to successfully remediate vast regions of heavy metal pollution Plants collect metals from different media, including water, wastewater, and soil. The media function as pools for trace metal and plant nutrient. Some plants have a range of possible techniques that can detox heavy metals and tolerate mental stress. Mechanisms of this kind include: binding to the cells' walls, reduced absorbing, or flux pumping of the plasma membrane by metals, the repairing of stress-damaged proteins, and the vacuole metal sharing by tono-plastic transporter of different ligands such as plant cell walls, metal bindings and metallic binding proteins.

Figure 10.5 illustrates heavy metal transport and detoxification in plants. In *Thlaspi caerulescens* (Sachs, 1865), the large amounts of heavy metals accumulated in plants that initially appeared are reported and the first person to coin the name "hyperaccumulator" was Brooks et al. (1977). Chaney (1983) originally addressed the notion of hyperaccumulator plants for heavy metals from polluted soils. Environment Canada has published a Phytorem database compiling a global inventory of over 750 terrestrial and aquatic species with potential phytoremediation value (Ramanjaneyulu et al., 2017). Only around 0.2% were found as metal hyperaccumulators of all angiosperms (Ashraf et al., 2010).

Lead (Pb), Aluminum (Al), Copper (Cu), Molybdenum (Mo), Manganese (Mn), Chromium (Cr), Platinum (Pt), Zinc (Zn), Selenium (Se), Mercury (Hg) and Naphthalene are tolerated by these accumulators or species (Zn). There have been reports on a list of 26 species of cobalt hyperaccumulators from Lamiaceae, Scrophulariaceae, Asteraceae or Fabaceae families (Baker and Smith, 2000). Phytoremediation of lead (Pb) by the use of corn in polluted agricultural areas was reported by Cheng et al. (2015). The results revealed, as demonstrated by minimum impacts in the development and production of biomass, that Bright Jean 7 maize has a high tolerance to lead. Every year, up to 93.37 tons of dry matter could be produced per ha of corn and a maximum amount of 7.2 kg of lead was removable. Cadmium-polluted agricultural land has been examined by Hamzah et al. (2016) on the use of five indigenous plant species. The average Cd reduction capacity in this selected facility was up to 71.2%. Each plant, *Vetiveria zizanioides*, *Eleusine indica*, L., *Ageratum conyzoides* L., *Euphorbia hirta,* and *Chromolaena odorata,* decreased by percentage of 71.2%, 58.9%, 52.2%, 51.8% and 22.1%, respectively. They have numerous desirable properties, such as high heavy metal tolerances, large surface/volume ratios, and phototaxy expressions, genetic modification potential, and auto- and heterotrophic growing capacity (Chekroun and Baghour, 2013). Because of their efficiency in absorbing heavy metals and concentrating them in tissue, macroalgae are increasingly being used as metal biomonitors in marine and aquatic ecosystems (Gosavi et al., 2004). Cyanophyta and Chlorophyta were reported to be boron and arsenic hyperaccumulators and hyper absorber (Chekroun and Baghour, 2013). Plants used for heavy metal phytoremediation should tolerate heavy metals collected in their tissues, be fast-growing and should be able to thrive in nutrient-deficient soils with spreading branches and large root systems, easily harvestable and animal-unconsumable. But the technology of phytoremediation is constrained by reasons such as time-consuming considering chemical treatment and demands more labor and costly capital. But it has the benefit of being dependable, environmentally friendly, and sustainable.

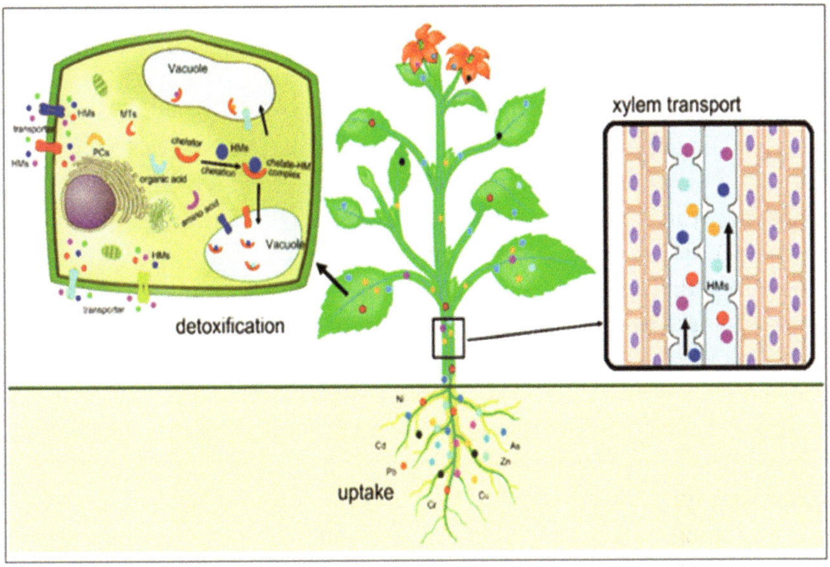

FIGURE 10.5 An illustration of the way heavy metals are collected, translocated, and sequestered in plants.
Source: Reprinted from Yan et al. (2020). https://creativecommons.org/licenses/by/4.0/

10.12 AGROCHEMICALS (PESTICIDE AND HERBICIDE)

In agriculture, huge amounts of inorganic and organic pollutants are deposited in farmlands by extensively used agrochemicals (synthetic fertilizers, insecticides, herbicides, pesticides, etc.). In addition to their biocidal and fertilizing effects, fertilizers, and pesticides supply the environment with significant amounts of heavy metals, notably in land and plants. Such compounds may have serious implication on humans and the environment, due to their presence in soil, plants, and water and also on their potential pharmacodynamic qualities. Human disorders, such as neurological disorders and many forms of malignancies are accountable of pesticides.

Moreover, as its selectivity and efficiency increase, pesticides are increasingly costly for farmers' selectivity, misuse, and over-exploitation which can lead to a risk of polluting soils, food, and drinking water for living creatures and humans. Similarly, herbicide accumulation endangers crop growth and development. DDT was typically the first chemical identified as the primary agent of bird population declines after 1945, owing to environmental impacts. Atrazine, a photosynthesis inhibitor herbicide, is mostly employed in maize and sorghum, and also in sugar cane, vines, citrus, and bananas as well as many more crops, Atrazine (2-chloro-4-(Aminoethyl)-6-(amino isopropyl), s-1,3,5-triazine) has been proven to be a major contaminant of water, which contaminates both surface (Garmouna et al., 1998) and underground water (Davoli et al., 1987), leading to its prohibition in the European Union in 2004. For male frogs particularly, Atrazine (Hayes et al., 2003) is reportedly as an endocrine disruptor and promoting the amphibian susceptibility to viruses, causing the decrease of amphibians across the world (Forson and Storfer, 2006). Food, water, soil, and sediments residues of organochlorine pesticide might lead to cancer on the whole food web (Taiwo, 2019). Dimethoate and benomyl also cause symbiotic activity of mycorrhizal fungus (Chiocchio et al., 2000). Pollution with pesticides therefore is one of scientists and environmentalists' most significant issues.

Soil microflora is used in plant remediation to improve the abilities of the soil to break down pesticides by means of bacterial or fungal growth. Numerous investigations have shown that plants are able to quickly restore solutions that include significant pesticide concentration, specifically under hydroponic circumstances, or at times when axenic plants are used (Gao et al., 2000). The pesticide is immobilized in the roots after ingested by the plant, either translocated or metabolized into the aerial portions by means of a translocation process. The medium level hydrophobic pesticide can be transmitted to the xylem vessels after roots absorb it and transport it into the shoots via the evapotranspiration flux, therefore resulting in a build-up of the chemical. In a research by Wilson et al. (2000) for example, a pesticide of the triazine family absorbed and translated the sweet flag (*Acorus gramenius*) and pickerel weed (Pontederia cordata). Erdei et al. (2005) performed research on pesticide phytoremediation from polluted soil and groundwater using *Kochia* sp. According to the findings plants have a significant innate capacity to detoxify some xenobiotic chemicals. By generating laccase and secreting it into the rhizosphere, tobacco, a transgenic plant, can remove pentachlorophenol (PCP). Tobacco plants have been genetically modified with a bacterial organophosphorus hydrolase (OPH) gene in order to clean and eliminate organic phosphorus pesticides. Transgenic plants were able to absorb more than 99% of methyl parathion after 14 days of growth, according to the study. Glutathione, cytochrome P450 sterase, O- and N-malonyltransferases, O-, and N-glucosyl transferases carboxylesterases, peroxygenases, peroxidase and transferases are among the plant enzymes involved in the phyto transformation of xenobiotics in the plant cells (Karavangeli et al., 2005). Phytoremediation has been shown to be the best approach to detoxify polluted herbicide soils with effective use of certain fungus and poplar tree (Ramanjaneyulu et al., 2017). The percent of metribuzin and metobromuron depleted in non-contaminated and polluted sites of liquid culture media was calculated by Bordjiba et al. (2001) and found the most effective depletors to be *Byssochleamys* sp., *Sordaria* sp. (Ascoycentes), *Botrytis* sp. (Dematiaceae), *Abscidia* sp. and (Zygomycents) fungi. Komives et al. (1994) observed that the safener Benoxacor protected maize against chloracent anilides by generating enhanced conjugative metabolism. Hence pesticide and herbicide polluted land and water seem to be a useful method in phytoremediation of the soil or water.

10.13 SALTS

The excessive quantities of salt are the world's biggest source of pollutants creating the salinity problem. It is believed that around 7% of the entire earth is impacted by excessive salt concentration and 20% of the total agricultural land. Though high soil salt content can have many biochemical, molecular, and physiological effects on crops, the more frequent impacts are photosynthesis inhibition, unbalanced nutrients, alterations in metabolism, accumulation disturbance of the solutes, enzyme activity and hormonal imbalances (Ashraf, 1994; Munns, 2005). Salt-affected soils generally include a range of inorganic salts, including cations like Mg^{2+}, Na+, K+, and Ca^{2+}, as well as anions such as NO^{3-}, HCO^{3-}, Cl-, SO_4^{2-}, and CO_3^{2-}, which have a detrimental influence on plant development and production owing to ion toxicity or osmotic effect on plants (Tanji, 2002; Parida and Das, 2005).

Many studies have shown that phytoremediation may effectively remove salts from polluted soils. Thus, according to scientific studies (Ashraf et al., 2010) farmers may now use plant-based techniques to eliminate salt. Kallar grass is beneficial for soil conservation and recovery in saline-sodic areas on a long-term basis (Akhter et al., 2004). Shekhawat et al. (2006) studied salt-tolerant plants *Suaeda nudiflora, Haloxylon recurvum* and *Salsola baryosma* and discovered that after three months of growth, *Haloxylon recurvum* eliminated the most Na^+ (17 g/plant) and retained the maximum biomass preceded by *Suaeda nudiflora* (15.6 g), and *Salsola baryosma* (9.6 g). Halophytes are plants that can desalinate and recover salt affected soils, as well as endure irrigation with extremely salty water (Lokhande and Suprasanna, 2012). In areas where rainfall is insufficient, halophytes can be utilized to drain salt from rhizospheres (Rabhi et al., 2009). When halophytes are cultivated in salinized soil, sodium ions are quickly absorbed by the roots and stored in the plant's above-ground parts (Rabhi et al., 2010). Cadmium-polluted saline soils may be readily phyto remediated by planting and growing Salicornia europaea (Ozawa et al., 2009). After phytoremediation, total nitrogen (TN), accessible potassium ions, available phosphorus and the organic carbon content in saline soils increase. Increases in exchangeable magnesium and calcium levels have also been seen, resulting in a rise in soil bulk density. Phytoremediation is connected to increased field capacity, soil permeability, soil porosity, infiltration rate and water holding capacity. Salt-tolerant and accumulator species have a significant impact on soil fertility and sustainability (Mishra et al., 2004).

10.14 ADVANTAGES AND LIMITATIONS OF PHYTOREMEDIATION

Phytoremediation is a relatively new method in which contaminated areas can be remediated using trees, grasses, and plants. Phytoremediation techniques, as contrasted with physical or chemical methods, are simple, cost-effective, broadly accepted publicly and adaptable in broad regions. Furthermore, phytoremediation also provides a variety of environmental and socio-economic benefits. The phytoremediation process has been discovered as a possible method for inorganic, organic, or combined pollutant contaminated remediation sites. Phytoremediation is, however, generally slower than other typical treatment procedures and relies on the climate. Plants may also absorb harmful heavy metals, which may potentially endanger wildlife and pollute the food supply chain. The potential for these plants to infiltrate food systems is worth contemplating and should be taken into account (Burken et al., 2011). In addition, this technology is still evolving and seems to be unknown to authorities. Some advantages and disadvantages of phytoremediation are discussed in subsections.

10.14.1 ADVANTAGES

- Depending on the situation, phytoremediation could be used *in situ* or *ex situ*. *In situ* treatments are commonly chosen because they decrease pollution transmission by air and waterborne wastes while minimizing disturbance of the soil and surrounding ecosystem.
- It is a green technology that, when correctly applied, adds to the landscape aspect and is eco-friendly.
- It can permanently remove a wide range of contaminants in a variety of ecological situations.
- It aids in decreasing surface runoff as well as pollutant mobilization and/or leaching into the soil.
- In comparison to traditional clean-up technologies, phytoremediation has the most cost-effective benefit. It is quite simple to execute and does not need expensive equipment or highly trained staff.
- Plant monitoring to track the phytoremediation process is simple to do.

10.14.2 LIMITATIONS

- The effectiveness of phytoremediation is limited only by the depth of the roots.
- Plant-based remediation is a considerably slower process, hence cleaning up a hazardous waste site might take several years or more, and the pollution may not be completely removed.
- The method for removing the strong sorbet is ineffective (such as PCBs).

- The products may be bioaccumulated by animals or mobilized into groundwater.
- The use of invasive, non-native species has the potential to reduce biodiversity, and plants might not be able to adapt to changing environmental and climatic circumstances.
- Wildlife feeding on polluted plants is also a significant problem.
- Pollutants created by the phytoextraction process bioaccumulate in harvested plants, which are classed as hazardous waste and must be handled and disposed of properly.
- Plant growth and phyto mass output might be limited by an un favorable environment, reducing process efficiency.

10.15 CONCLUSION

Phytoremediation has proven to be a potential approach for revegetating polluted soil that has a high level of public acceptability and offers a number of benefits over other physicochemical methods. Phytoremediation is a process for cleaning soils that have been contaminated by both inorganic and organic agricultural pollutants. Because it is a biological method, it has a bright future ahead of it. It has the ability to provide a low-cost and long-term means of boosting the economy of poor nations. The selection of the appropriate plant species is critical to the technology's success. Phytoextraction, phytodegradation, rhizodegradation, phytostabilization, rhizofiltration, and phytovolatilization are some of the possible pollutant removal methods. A key advantage of phytoremediation is, by making the leaves unpleasant and evading herbivores from using the cumulative metals in the food chain. Plants may be used for various positive uses, such as biofuels, following the phytoremediation. Suitable combinations and agronomic methods of soil, plant species, plants or varieties might prevent the transmission of trace metals to the food chain and/or industrial extract energy and metals, and provide safe farming activities. However, phytoremediation of heavy metal-affected soil has certain limits seeing as it is a time-consuming technique that takes a long time to clear up, especially in moderately and severely polluted areas. As a result, improving plant performance is an important step in developing highly effective plant remediation. Fortunately, genetic engineering is a powerful tool for adapting to various climatological and geological environments, with essential characteristics such as quick development, high biomass output, and high heavy metal toleration and accumulation. More research and skill are required before this technology can be commercialized on a large scale and food safety can be assured in a long-term way.

KEYWORDS

- **agricultural contaminants**
- **phytodegradation**
- **phytoextraction**
- **phytoremediation**
- **phytostabilization**
- **phytovolatilization**
- **rhizodegradation**
- **rhizofiltration**

REFERENCES

Abbasi, A., Sajid, A., Haq, N., Rahman, S., Misbah, Z. T., Sanober, G., Ashraf, M., & Kazi, A. G., (2014). Agricultural pollution: An emerging issue. In *Improvement of crops in the Era of Climatic Changes* (pp. 347–387). Springer, New York, NY.

Abdel-Shafy, H. I., & Mansour, M. S., (2018). Phytoremediation for the elimination of metals, pesticides, PAHs, and other pollutants from wastewater and soil. In: *Phytobiont and Ecosystem Restitution* (pp. 101–136). Springer, Singapore.

Adefemi, O. S., Ibigbami, O. A., & Awokunmi, E. E., (2012). Level of heavy metals in some edible plants collected from selected dumpsites in Ekiti State, Nigeria. *Global Advanced Research Journal of Environmental Science and Toxicology, 1*(5), 132–136.

Akhter, J., Murray, R., Mahmood, K., Malik, K. A., & Ahmed, S., (2004). Improvement of degraded physical properties of a saline-sodic soil by reclamation with kallar grass (*Leptochloa fusca*). *Plant and Soil, 258*(1), 207–216.

Albering, H. J., Van, L. S. M., Moonen, E. J., Hoogewerff, J. A., & Kleinjans, J. C., (1999). Human health risk assessment: A case study involving heavy metal soil contamination after the flooding of the river Meuse during the winter of 1993–1994. *Environmental Health Perspectives, 107*(1), 37–43.

Ali, H., Khan, E., & Sajad, M. A., (2013). Phytoremediation of heavy metals—concepts and applications. *Chemosphere, 91*(7), 869–881.

Ashraf, M. Y., & Wu, L., (1994). Breeding for salinity tolerance in plants. *Critical Reviews in Plant Sciences, 13*(1), 17–42.

Ashraf, M. Y., Ashraf, M., Mahmood, K., Akhter, J., Hussain, F., & Arshad, M., (2010). Phytoremediation of saline soils for sustainable agricultural productivity. In: *Plant Adaptation and Phytoremediation* (pp. 335–355). Springer, Dordrecht.

Axelsen, K. B., & Palmgren, M. G., (2001). Inventory of the superfamily of P-type ion pumps in *Arabidopsis*. *Plant Physiology, 126*(2), 696–706.

Baker, A., & Smith, A., (2000). Metal hyperaccumulator plants: Biological resources for exploitation in the phytoextraction of metal contaminated soils. In: *Proc. Inter Cost Workshop for Bioremediation* (pp. 3–5).

Banuelos, G. S., & Meek, D. W., (1990). *Accumulation of Selenium in Plants Grown on Selenium-Treated Soil* (Vol. 19, No. 4, pp. 772–777). American Society of Agronomy, Crop Science Society of America, and Soil Science Society of America.

Banuelos, G. S., Ajwa, H. A., Mackey, B., Wu, L., Cook, C., Akohoue, S., & Zambruzuski, S., (1997). *Evaluation of Different Plant Species Used for Phytoremediation of High Soil Selenium* (Vol. 26, No. 3, pp. 639–646). American Society of Agronomy, Crop Science Society of America, and Soil Science Society of America.

Bañuelos, G., Terry, N., LeDuc, D. L., Pilon-Smits, E. A., & Mackey, B., (2005). Field trial of transgenic Indian mustard plants shows enhanced phytoremediation of selenium-contaminated sediment. *Environmental Science & Technology, 39*(6), 1771–1777.

Barillo, D. J., (2009). Diagnosis and treatment of cyanide toxicity. *Journal of Burn Care & Research, 30*(1), 148–152.

Batayneh, A. T., (2012). Toxic (aluminum, beryllium, boron, chromium and zinc) in groundwater: Health risk assessment. *International Journal of Environmental Science and Technology, 9*(1), 153–162.

Baudouin, C., Charveron, M., Tarroux, R., & Gall, Y., (2002). Environmental pollutants and skin cancer. *Cell Biology and Toxicology, 18*(5), 341–348.

Bhat, S. A., Dar, M. U. D., & Meena, R. S., (2019). Soil erosion and management strategies. In: *Sustainable Management of Soil and Environment* (pp. 73–122). Springer, Singapore.

Bhat, S. A., Hamid, I., Dar, M. U. D., Rasool, D., Pandit, B. A., & Khan, S., (2017). Soil erosion modeling using RUSLE & GIS on micro watershed of J&K. *Journal of Pharmacognosy and Phytochemistry, 6*(5), 838–842.

Bhat, S. A., Pandit, B. A., Khan, J. N., Kumar, R., & Jan, R., (2017). Water requirements and irrigation scheduling of maize crop using CROPWAT model. *Int. J. Curr. Microbiol. Appl. Sci, 6*, 1662–1670.

Bhat, S. A., Pandit, B., Dar, M. U. D., Jan, R., Khan, S., & Mehraj, K., (2017). Statistical comparison of reference evapotranspiration methods: A case study from Srinagar in J&K, India. *Int. J. Curr. Microbiol. App. Sci, 6*(9), 3731–3737.

Bhat, S. A., Pandit, B., Dar, M. U. D., ROUHULLAH, S., ALI, R. J., & Khan, S., (2017). *Comparative Study of Different Methods of Evapotranspiration Estimation in Kashmir Valley* (p. 384). Dr. PD Mistry.

Bonanno, G., & Vymazal, J., (2017). Compartmentalization of potentially hazardous elements in macrophytes: Insights into capacity and efficiency of accumulation. *Journal of Geochemical Exploration, 181*, 22–30.

Bonanno, G., (2013). Comparative performance of trace element bioaccumulation and biomonitoring in the plant species *Typha domingensis*, *Phragmites australis* and *Arundo donax*. *Ecotoxicology and Environmental Safety, 97*, 124–130.

Bordjiba, O., Steiman, R., Kadri, M., Semadi, A., & Guiraud, P., (2001). Removal of herbicides from liquid media by fungi isolated from a contaminated soil. *Journal of Environmental Quality, 30*(2), 418–426.

Brennan, M. A., & Shelley, M. L., (1999). A model of the uptake, translocation, and accumulation of lead (Pb) by maize for the purpose of phytoextraction. *Ecological Engineering, 12*(3, 4), 271–297.

Brooks, R. R., & Wither, E. D., (1977). Nickel accumulation by *Rinorea bengalensis* (Wall.) OK. *Journal of Geochemical Exploration, 7*, 295–300.

Burken, J. G., & Schnoor, J. L., (1999). Distribution and volatilization of organic compounds following uptake by hybrid poplar trees. *International Journal of Phytoremediation, 1*(2), 139–151.

Burken, J. G., Vroblesky, D. A., & Balouet, J. C., (2011). Phytoforensics, dendrochemistry, and phytoscreening: New green tools for delineating contaminants from past and present. *Environmental Science & Technology, 45*(15), 6218–6226.

Cerne, M., Smodis, B., & Strok, M., (2011). Uptake of radionuclides by a common reed (Phragmites australis (Cav.) Trin. ex Steud.) grown in the vicinity of the former uranium mine at Zirovski vrh. *Nuclear Engineering and Design, 241*(4), 1282–1286.

Chaney, R. L., (1983). Plant uptake of inorganic waste. *Land Treatment of Hazardous Wastes.*

Chekroun, K. B., & Baghour, M., (2013). The role of algae in phytoremediation of heavy metals: A review. *J. Mater. Environ. Sci., 4*(6), 873–880.

Cheng, S. F., Huang, C. Y., Lin, Y. C., Lin, S. C., & Chen, K. L., (2015). Phytoremediation of lead using corn in contaminated agricultural land—An in situ study and benefit assessment. *Ecotoxicology and Environmental Safety, 111*, 72–77.

Chiocchio, V., Venedikian, N., Martinez, A. E., Menendez, A., Ocampo, J. A., & Godeas, A., (2000). Effect of the fungicide benomyl on spore germination and hyphal length of the arbuscular mycorrhizal fungus *Glomus mosseae*. *International Microbiology: The Official Journal of the Spanish Society for Microbiology, 3*(3), 173–175.

Cristaldi, A., Conti, G. O., Jho, E. H., Zuccarello, P., Grasso, A., Copat, C., & Ferrante, M., (2017). Phytoremediation of contaminated soils by heavy metals and PAHs. A brief review. *Environmental Technology & Innovation, 8*, 309–326.

Crowe, A. S., & Smith, J. E., (2007). Distribution and persistence of DDT in soil at a sand dune-marsh environment: Point Pelee, Ontario, Canada. *Canadian Journal of Soil Science, 87*(3), 315–327.

Cunningham, S. D., & Lee, C. R., (1995). Phytoremediation: Plant-based remediation of contaminated soils and sediments. *Bioremediation: Science and Applications, 43*, 145–156.

Cunningham, S. D., & Ow, D. W., (1996). Promises and prospects of phytoremediation. *Plant Physiology, 110*(3), 715.

DalCorso, G., Fasani, E., Manara, A., Visioli, G., & Furini, A., (2019). Heavy metal pollutions: State of the art and innovation in phytoremediation. *International Journal of Molecular Sciences, 20*(14), 3412.

Dar, M. U. D., Bhat, S. A., Meena, R. S., & Shah, A. I., (2020). Carbon footprint in eroded soils and its impact on soil health. In: *Soil Health Restoration and Management* (pp. 1–30). Springer, Singapore.

Davis, L. C., Castro-Diaz, S., Zhang, Q., & Erickson, L. E., (2002). Benefits of vegetation for soils with organic contaminants. *Critical Reviews in Plant Sciences, 21*(5), 457–491.

Davoli, E., Benfenati, E., Bagnati, R., & Fanelli, R., (1987). Analysis of atrazine in underground waters at part per trillion levels as an early warning method for contamination and for soil distribution studies. *Chemosphere, 16*(7), 1425–1430.

Delibacak, S., Elmaci, O. L., Secer, M., & Bodur, A., (2002). Fertility status, trace elements and heavy metal pollution of agricultural land irrigated from the Gediz River. *International Journal of Water, 2*(2, 3), 184–195.

Dubois, O., (2011). *The State of the World's Land and Water Resources for Food and Agriculture: Managing Systems at Risk*. Earthscan.

Dushenkov, S., (2003). Trends in phytoremediation of radionuclides. *Plant and Soil, 249*(1), 167–175.

Efremova, M., & Izosimova, A., (2012a). Contamination of agricultural soils with heavy metals. In: Jakobsson, C., (ed.), *Ecosystem Health and Agriculture. Sustainable Agriculture* (pp. 250–252). The Baltic University Program. Uppsala University, Uppsala, Sweden.

Efremova, M., & Izosimova, A., (2012b). Contamination of agricultural soils with radionuclides. In: Jakobsson, C., (ed.), *Ecosystem Health and Agriculture: Sustainable Agriculture* (pp. 253–255). The Baltic University Program. Uppsala University, Uppsala, Sweden.

EPA, (2000). *A Citizen's Guide to Phytoremediation* (p. 6). EPA 542-F-98-011. United States Environmental Protection Agency.

Erdei, L., (2005). Phytoremediation as a program for decontamination of heavy-metal polluted environment. *Acta Biologica Szegediensis, 49*(1, 2), 75–76.

Etim, E. E., (2012). Phytoremediation and its mechanisms: A review. *International Journal of Environment and Bioenergy, 2*(3), 120–136.

FAO and ITPS, (2015). *Status of the World's Soil Resources (SWSR)–Main Report* (Vol. 650). Food and agriculture organization of the United Nations and intergovernmental technical panel on soils, Rome, Italy.

Forson, D. D., & Storfer, A., (2006). Atrazine increases ranavirus susceptibility in the tiger salamander, *Ambystoma tigrinum. Ecological Applications, 16*(6), 2325–2332.

Gao, J., Garrison, A. W., Hoehamer, C., Mazur, C. S., & Wolfe, N. L., (2000). Uptake and phytotransformation of organophosphorus pesticides by axenically cultivated aquatic plants. *Journal of Agricultural and Food Chemistry, 48*(12), 6114–6120.

Garbisu, C., & Alkorta, I., (2001). Phytoextraction: A cost-effective plant-based technology for the removal of metals from the environment. *Bioresource Technology, 77*(3), 229–236.

Garmouma, M., Teil, M. J., Blanchard, M., & Chevreuil, M., (1998). Spatial and temporal variations of herbicide (triazines and phenylureas) concentrations in the catchment basin of the Marne river (France). *Science of the Total Environment, 224*(1–3), 93–107.

Gerhardt, K. E., Gerwing, P. D., Huang, X. D., & Greenberg, B. M., (2015). Microbe-assisted phytoremediation of petroleum impacted soil: A scientifically proven green technology. *Handbook of Oil Spill Science Technology* (pp. 407–427). Wiley, Hoboken.

Ghosh, M., & Singh, S. P., (2005). A review on phytoremediation of heavy metals and utilization of it's by-products. *Asian J. Energy Environ, 6*(4), 18.

Goel, A., & Aggarwal, P., (2007). Pesticide poisoning. *National medical journal of India, 20*(4), p.182.

Gong, Y. P., Ni, Z. Y., Xiong, Z. Z., Cheng, L. H., & Xu, X. H., (2017). Phosphate and ammonium adsorption of the modified biochar based on *Phragmites australis* after phytoremediation. *Environmental Science and Pollution Research, 24*(9), 8326–8335.

Gosavi, K., Sammut, J., Gifford, S., & Jankowski, J., (2004). Macroalgal biomonitors of trace metal contamination in acid sulfate soil aquaculture ponds. *Science of the Total Environment, 324*(1–3), 25–39.

Hamzah, A., Hapsari, R. I., & Wisnubroto, E. I., (2016). Phytoremediation of cadmium-contaminated agricultural land using indigenous plants. *International Journal of Environmental & Agriculture Research, 2*(1), 8–14.

Hansen, D., Duda, P. J., Zayed, A., & Terry, N., (1998). Selenium removal by constructed wetlands: Role of biological volatilization. *Environmental Science & Technology, 32*(5), 591–597.

Hayes, T. B., Collins, A., Lee, M., Mendoza, M., Noriega, N., Stuart, A. A., & Vonk, A., (2002). Hermaphroditic, demasculinized frogs after exposure to the herbicide atrazine at low ecologically relevant doses. *Proceedings of the National Academy of Sciences, 99*(8), 5476–5480.

Hayes, T., Haston, K., Tsui, M., Hoang, A., Haeffele, C., & Vonk, A., (2003). Atrazine-induced hermaphroditism at 0.1 ppb in American leopard frogs (*Rana pipiens*): Laboratory and field evidence. *Environmental Health Perspectives, 111*(4), 568–575.

Herzig, R., Nehnevajova, E., Pfistner, C., Schwitzguebel, J. P., Ricci, A., & Keller, C., (2014). Feasibility of labile Zn phytoextraction using enhanced tobacco and sunflower: Results of five-and one-year field-scale experiments in Switzerland. *International Journal of Phytoremediation, 16*(7, 8), 735–754.

Hong, M. S., Farmayan, W. F., Dortch, I. J., Chiang, C. Y., McMillan, S. K., & Schnoor, J. L., (2001). Phytoremediation of uranium-contaminated soils: Role of organic acids in triggering uranium hyperaccumulation in plants. *Environ. Sci. Technol., 35*, 1231–1239.

Huang, J. W., & Cunningham, S. D., (1996). Lead phytoextraction: species variation in lead uptake and translocation. *New Phytologist, 134*(1), 75–84.

Jacobs, A., De Brabandere, L., Drouet, T., Sterckeman, T., & Noret, N., (2018). Phytoextraction of Cd and Zn with *Noccaea caerulescens* for urban soil remediation: Influence of nitrogen fertilization and planting density. *Ecological Engineering, 116*, 178–187.

Järup, L., (2003). Hazards of heavy metal contamination. *British Medical Bulletin, 68*(1), 167–182.

Kang, J. W., (2014). Removing environmental organic pollutants with bioremediation and phytoremediation. *Biotechnology Letters, 36*(6), 1129–1139.

Kapungwe, E. M., (2013). Heavy metal contaminated water, soils and crops in peri-urban wastewater irrigation farming in Mufulira and Kafue towns in Zambia. *Journal of Geography and Geology, 5*(2), 55.

Karavangeli, M., Labrou, N. E., Clonis, Y. D., & Tsaftaris, A., (2005). Development of transgenic tobacco plants overexpressing maize glutathione S-transferase I for chloroacetanilide herbicides phytoremediation. *Biomolecular Engineering, 22*(4), 121–128.

Katoh, M., Risky, E., & Sato, T., (2017). Immobilization of lead migrating from contaminated soil in rhizosphere soil of barley (*Hordeum vulgare* L.) and hairy vetch

(*Vicia villosa*) using hydroxyapatite. *International Journal of Environmental Research and Public Health, 14*(10), p.1273.

Kayser, A., Wenger, K., Keller, A., Attinger, W., Felix, H. R., Gupta, S. K., & Schulin, R., (2000). Enhancement of phytoextraction of Zn, Cd, and Cu from calcareous soil: The use of NTA and sulfur amendments. *Environmental Science & Technology, 34*(9), 1778–1783.

Kokyo, O., Tao, L., Hongyan, C., Xuefeng, H., Chiquan, H., Lijun, Y., & Yonemochi, S., (2013). Development of profitable phytoremediation of contaminated soils with biofuel crops. *Journal of Environmental Protection, 2013.*

Korte, F., Kvesitadze, G., Ugrekhelidze, D., Gordeziani, M., Khatisashvili, G., Buadze, O., Zaalishvili, G., & Coulston, F., (2000). Organic toxicants and plants. *Ecotoxicology and Environmental Safety, 47*(1), 1–26.

Kumar, R., Kumar, M., Shah, A., Bhat, S. A., Wani, M. A., & Ram, D., (2016). Modeling of soil loss using USLE through remote sensing and geographical information system in micro-watershed of. *Journal of Soil & Water Conservation, 15*(1), 40–45.

Larsen, M., Trapp, S., & Pirandello, A., (2004). Removal of cyanide by woody plants. *Chemosphere, 54*(3), 325–333.

Li, H., Li, X., Xiang, L., Zhao, H. M., Li, Y. W., Cai, Q. Y., Zhu, L., Mo, C. H., & Wong, M. H., (2018). Phytoremediation of soil co-contaminated with Cd and BDE-209 using hyperaccumulator enhanced by AM fungi and surfactant. *Science of the Total Environment, 613*, 447–455.

Lin, Y. T., Weng, C. H., & Lee, S. Y., (2012). Spatial distribution of heavy metals in contaminated agricultural soils exemplified by Cr, Cu, and Zn. *Journal of Environmental Engineering, 138*(3), 299–306.

Lokhande, V. H., & Suprasanna, P., (2012). Prospects of halophytes in understanding and managing abiotic stress tolerance. In: *Environmental Adaptations and Stress Tolerance of Plants in the Era of Climate Change* (pp. 29–56). Springer, New York, NY.

Lone, M. I., He, Z. L., Stoffella, P. J., & Yang, X. E., (2008). Phytoremediation of heavy metal polluted soils and water: Progresses and perspectives. *J. Zhejiang Univ. Sci. B, 9*, 210–220.

Macdonald, D. D., Carr, R. S., Calder, F. D., Long, E. R., & Ingersoll, C. G., (1996). Development and evaluation of sediment quality guidelines for Florida coastal waters. *Ecotoxicology, 5*(4), 253–278.

Mahar, A., Wang, P., Ali, A., Awasthi, M. K., Lahori, A. H., Wang, Q., Li, R., & Zhang, Z., (2016). Challenges and opportunities in the phytoremediation of heavy metals contaminated soils: A review. *Ecotoxicology and Environmental Safety, 126*, 111–121.

McKone, T. E., & Ryan, P. B., (1989). Human exposures to chemicals through food chains: An uncertainty analysis. *Environmental Science & Technology, 23*(9), 1154–1163.

Meagher, R. B., (2000). Phytoremediation of toxic elemental and organic pollutants. *Current Opinion in Plant Biology, 3*(2), 153–162.

Memon, A. R., Aktoprakligil, D., Özdemir, A., & Vertii, A., (2001). Heavy metal accumulation and detoxification mechanisms in plants. *Turkish Journal of Botany, 25*(3), 111–121.

Mench, M., Lepp, N., Bert, V., Schwitzguébel, J. P., Gawronski, S. W., Schröder, P., & Vangronsveld, J., (2010). Successes and limitations of phytotechnologies at field scale: Outcomes, assessment and outlook from COST Action 859. *Journal of Soils and Sediments, 10*(6), 1039–1070.

Micó, C., Recatalá, L., Peris, M., & Sánchez, J., (2006). Assessing heavy metal sources in agricultural soils of an European Mediterranean area by multivariate analysis. *Chemosphere, 65*(5), 863–872.

Mishra, A., Sharma, S. D., Pandey, R., & Mishra, L., (2004). Amelioration of a highly alkaline soil by trees in northern India. *Soil use and Management, 20*(3), 325–332.

Munns, R., (2005). Genes and salt tolerance: Bringing them together. *New Phytologist, 167*(3), 645–663.

Nagajyoti, P. C., Lee, K. D., & Sreekanth, T. V. M., (2010). Heavy metals, occurrence and toxicity for plants: A review. *Environmental Chemistry Letters, 8*(3), 199–216.

Nakajima, H., Fujimoto, N., Yamamoto, Y., Amemiya, T., & Itoh, K., (2019). Response of secondary metabolites to Cu in the Cu-hyperaccumulator lichen *Stereocaulon japonicum. Environmental Science and Pollution Research, 26*(1), 905–912.

National Research Council, (2003). Committee on bioavailability of contaminants in soils and sediments. *Bioavailability of Contaminants in Soils and Sediments: Processes, Tools and Applications.*

Nicholson, F. A., Smith, S. R., Alloway, B. J., Carlton-Smith, C., & Chambers, B. J., (2003). An inventory of heavy metals inputs to agricultural soils in England and Wales. *Science of the Total Environment, 311*(1–3), 205–219.

Ozawa, T., Miura, M., Fukuda, M., & Kakuta, S., (2009). Cadmium tolerance and accumulation in a halophyte *Salicornia europaea* as a new candidate for phytoremediation of saline soils (Vol. 60, pp. 1–8). *Scientific Report of the Graduate School of Life and Environmental Sciences,* Osaka Prefecture University.

Panda, A. K., (1996). Bio-accumulation of nickel and zinc by water hyacinth and water lettuce. *Indian Journal of Environmental Health, 38*(1), 51–53.

Pandey, V. C., & Bajpai, O., (2019). Phytoremediation: From theory toward practice. In: *Phytomanagement of Polluted Sites* (pp. 1–49). Elsevier.

Pandey, V. C., & Singh, V., (2019). Exploring the potential and opportunities of current tools for removal of hazardous materials from environments. In: *Phytomanagement of Polluted Sites* (pp. 501–516). Elsevier.

Pandey, V. C., Singh, J. S., Singh, R. P., Singh, N., & Yunus, M., (2011). Arsenic hazards in coal fly ash and its fate in Indian scenario. *Resources, Conservation and Recycling, 55*(9, 10), 819–835.

Parida, A. K., & Das, A. B., (2005). Salt tolerance and salinity effects on plants: A review. *Ecotoxicology and Environmental Safety, 60*(3), 324–349.

Pilon-Smits, E., (2005). Phytoremediation. *Annu. Rev. Plant Biol., 56,* 15–39.

Pivetz, B. E., (2001). *Phytoremediation of Contaminated Soil and Ground Water at Hazardous Waste Sites.* US Environmental Protection Agency, Office of Research and Development, Office of Solid Waste and Emergency Response.

Qadir, A., Hameed, M. A., Zafar, M. S. B., Farooqi, Z. U. R., Younas, F., Hussain, M. M., & Mohy-Ud-Din, W., (2021). *Plant and Environment.*

Rabhi, M., Ferchichi, S., Jouini, J., Hamrouni, M. H., Koyro, H. W., Ranieri, A., Abdelly, C., & Smaoui, A., (2010). Phytodesalination of a salt-affected soil with the halophyte *Sesuvium portulacastrum* L. to arrange in advance the requirements for the successful growth of a glycophytic crop. *Bioresource Technology, 101*(17), 6822–6828.

Rabhi, M., Hafsi, C., Lakhdar, A., Hajji, S., Barhoumi, Z., Hamrouni, M. H., Abdelly, C., & Smaoui, A., (2009). Evaluation of the capacity of three halophytes to desalinize their rhizosphere as grown on saline soils under nonleaching conditions. *African Journal of Ecology, 47*(4), 463–468.

Ramanjaneyulu, A. V., Neelima, T. L., Madhavi, A., & Ramprakash, T., (2017). Phytoremediation: An overview. In *Applied Botany* (pp. 42–84). American Academic Press.

Rascio, N., & Navari-Izzo, F., (2011). Heavy metal hyperaccumulating plants: How and why do they do it? And what makes them so interesting? *Plant Science, 180*(2), 169–181.

Raskin, I., & Ensley, B. D., (2000). *Phytoremediation of Toxic Metals*. John Wiley and Sons.

Rehman, M. Z., Rizwan, M., Ali, S., Ok, Y. S., Ishaque, W., Nawaz, M. F., Akmal, F., & Waqar, M., (2017). Remediation of heavy metal contaminated soils by using *Solanum nigrum*: A review. *Ecotoxicology and Environmental Safety, 143*, 236–248.

Sachs, J., (1865). *Handbuch der Experimental-Physiologie der Pflanzen: Investigations into the General Living Conditions of Plants and the Functions of Their Organs* (Vol. 4). W. Engelmann.

Salt, D. E., Smith, R. D., & Raskin, I., (1998). Phytoremediation. *Annual Review of Plant Biology, 49*(1), 643–668.

Sarwar, N., Malhi, S. S., Zia, M. H., Naeem, A., Bibi, S., & Farid, G., (2010). Role of mineral nutrition in minimizing cadmium accumulation by plants. *Journal of the Science of Food and Agriculture, 90*(6), 925–937.

Schnepp, R., (2006). Cyanide: Sources, perceptions, and risks. *Journal of Emergency Nursing, 32*(4), S3–S7.

Schulz, R., (2004). Field studies on exposure, effects, and risk mitigation of aquatic nonpoint-source insecticide pollution: A review. *Journal of Environmental Quality, 33*(2), 419–448.

Shah, A. I., Din, D. M. U., Bhat, R. A., Singh, J. P., Singh, K., & Bhat, S. A., (2020). Prospectives and challenges of wastewater treatment technologies to combat contaminants of emerging concerns. *Ecol. Eng., 152*, 105882 https://doi.org/10.1016/j.ecoleng.2020.105882.

Sharma, P., & Pandey, S., (2014). Status of phytoremediation in world scenario. *International Journal of Environmental Bioremediation & Biodegradation, 2*(4), 178–191.

Shekhawat, V. P., Kumar, A., & Neumann, K. H., (2006). Bio-reclamation of secondary salinized soils using halophytes. In: *Biosaline Agriculture and Salinity Tolerance in Plants* (pp. 147–154). Birkhäuser Basel.

Shiyab, S., Chen, J., Han, F. X., Monts, D. L., Matta, F. B., Gu, M., Su, Y., & Masad, M. A., (2009). Mercury-induced oxidative stress in Indian mustard (*Brassica juncea* L.). *Environmental Toxicology: An International Journal, 24*(5), 462–471.

Singh, B. R., Gupta, S. K., Azaizeh, H., Shilev, S., Sudre, D., Song, W. Y., Martinoia, E., & Mench, M., (2011). Safety of food crops on land contaminated with trace elements. *Journal of the Science of Food and Agriculture, 91*(8), 1349–1366.

Song, B., Xu, P., Chen, M., Tang, W., Zeng, G., Gong, J., Zhang, P., & Ye, S., (2019). Using nanomaterials to facilitate the phytoremediation of contaminated soil. *Critical reviews in Environmental Science and Technology, 49*(9), 791–824.

Stephenson, C., & Black, C. R., (2014). One step forward, two steps back: The evolution of phytoremediation into commercial technologies. *Bioscience Horizons: The International Journal of Student Research, 7.*

Suciu, I., Cosma, C., Todică, M., Bolboacă, S. D., & Jäntschi, L., (2008). Analysis of soil heavy metal pollution and pattern in central Transylvania. *International Journal of Molecular Sciences, 9*(4), 434–453.

Suman, J., Uhlik, O., Viktorova, J., & Macek, T., (2018). Phytoextraction of heavy metals: A promising tool for clean-up of polluted environment? *Frontiers in Plant Science, 9*, 1476.

Taebi, A., Jeirani, K., Mirlohi, A., & Bafghi, A. Z., (2008). Phytoremediation of cyanide-polluted soils by non-woody plants. *Journal of Science and Technology of Agriculture and Natural Resources.*

Tanji, K. K., (2002). Salinity in the soil environment. In: *Salinity: Environment-Plants-Molecules* (pp. 21–51). Springer, Dordrecht.

Trapp, S., Köhler, A., Larsen, L. C., Zambrano, K. C., & Karlson, U., (2001). Phytotoxicity of fresh and weathered diesel and gasoline to willow and poplar trees. *Journal of Soils and Sediments, 1*(2), 71–76.

Ul Zaman, M., Bhat, S., Sharma, S., & Bhat, O., (2018). Methods to control soil erosion: A review. *Int. J. Pure Appl. Biosci., 6*(2), 1114–1121. 10.18782/2320-7051. 6462.

Ullah, A., Heng, S., Munis, M. F. H., Fahad, S., & Yang, X., (2015). Phytoremediation of heavy metals assisted by plant growth promoting (PGP) bacteria: A review. *Environmental and Experimental Botany, 117*, 28–40.

USEPA (United States Environmental Protection Agency), (2000). National risk management authority. *Introduction to Phytoremediation.* Environmental Protection Agency.

Vangronsveld, J., Herzig, R., Weyens, N., Boulet, J., Adriaensen, K., Ruttens, A., Thewys, T., etal., (2009). Phytoremediation of contaminated soils and groundwater: Lessons from the field. *Environmental Science and Pollution Research, 16*(7), 765–794.

Várallyay, G., (1994). Climate change, soil salinity and alkalinity. In: *Soil Responses to Climate Change* (pp. 39–54). Springer, Berlin, Heidelberg.

Vejan, P., Abdullah, R., Khadiran, T., Ismail, S., & Nasrulhaq, B. A., (2016). Role of plant growth promoting rhizobacteria in agricultural sustainability—A review. *Molecules, 21*(5), 573.

Wang, W., Vinocur, B., & Altman, A., (2003). Plant responses to drought, salinity and extreme temperatures: Towards genetic engineering for stress tolerance. *Planta, 218*(1), 1–14.

Wei, S., Zhou, Q., & Wang, X., (2005). Identification of weed plants excluding the uptake of heavy metals. *Environment International, 31*(6), 829–834.

Wilson, P. C., Whitwell, T., & Klaine, S. J., (2000). Phytotoxicity, uptake, and distribution of 14C-simazine in *Acorus gramenius* and *Pontederia cordata. Weed Science, 48*(6), 701–709.

Yan, A., Wang, Y., Tan, S. N., Yusof, M. L. M., Ghosh, S., & Chen, Z., (2020). Phytoremediation: A promising approach for revegetation of heavy metal-polluted land. *Frontiers in Plant Science, 11.*

Ye, W. L., Khan, M. A., McGrath, S. P., & Zhao, F. J., (2011). Phytoremediation of arsenic contaminated paddy soils with *Pteris vittata* markedly reduces arsenic uptake by rice. *Environmental Pollution, 159*(12), 3739–3743.

Zhang, H., Dang, Z., Zheng, L. C., & Yi, X. Y., (2009). Remediation of soil co-contaminated with pyrene and cadmium by growing maize (*Zea mays* L.). *International Journal of Environmental Science & Technology, 6*(2), 249–258.

Zhu, J. K., (2001). Plant salt tolerance. *Trends in Plant Science, 6*(2), 66–71.

ZuccaRini, P., (2008). Ion uptake by halophytic plants to mitigate saline stress in *Solanum lycopersicon* L., and different effect of soil and water salinity. *Soil and Water Research, 3*(2), 62–73.

CHAPTER 11

Anaerobic Biovalorization for Paper Mill Wastewater Treatment

NETHI NAGA HARI SAIRAM,[1] JARIPITI TRIVIKRAMA RAJU,[2] RATNALA SUDHA RANI,[1] BODASINGI KRISHNA KANTH,[3] PAKKIRANNA SIVAMMA,[4] and SHAKEEL AHMAD BHAT[5]

[1]*Department of Soil and Water Engineering, Dr. NTR College of Agricultural Engineering, ANGRAU, Bapatla, Andhra Pradesh, India*

[2]*Department of Soil and Water Engineering, College of Technology and Engineering, MPUAT, Udaipur, Rajasthan, India*

[3]*Department of Farm Machinery and Power Engineering, Dr. NTR College of Agricultural Engineering, ANGRAU, Bapatla, Andhra Pradesh, India*

[4]*Department of Agricultural Processing and Food Engineering, Dr. NTR College of Agricultural Engineering, ANGRAU, Bapatla, Andhra Pradesh, India*

[5]*College of Agricultural Engineering and Technology, Sher-e-Kashmir University of Agricultural Sciences and Technology of Kashmir, Shalimar Campus, Srinagar – 190025, Jammu and Kashmir, India, E-mail: wakeelbhat@gmail.com*

ABSTRACT

Wastewater treatment is one of the major challenges in developing countries. The majority of the industries disposes the effluent in the natural

Advanced Technologies for Water Quality Treatment and Management. Mehraj U. Din Dar, PhD, Aamir Ishaq Shah, MTech, Shakeel Ahmad Bhat, PhD & Syed Rouhullah Ali, PhD (Eds.)

streams and adjoining water bodies. Among those industries, pulp, and paper mills cause the major degradation of water bodies through their effluents containing organic compounds like lignocellulosic material as well as chemicals like chlorinated compounds and other acids. This paper mill industry is mainly chosen because of the enormous amount of water required as well as it has certain benefits of producing a potential source of biogas using wastewater. To degrade the sludge decomposition from the waste, anaerobic biovalorization treatment is generally used because of the high loading rate and can produce energy using biomass. The complete illustration regarding anaerobic treatment, enhancement strategies to improve the anaerobic treatment, production of the value-added products through this treatment were briefly discussed. Therefore, the anaerobic treatment plays a major role in not only generating biogas but also transforming the sludge into useful material by adding it as a fertilizer, as well as producing chemicals, etc.

11.1 INTRODUCTION

The pulp and paper industry are the prime industries in the world, (Bajpai, 2017) which contributes essential products such as paperboard, paper, and insulation material to individuals across the world are China and United States being the largest paper producing countries. Each year, more than 400 million metric tons of paper and cardboard are produced globally. On average, an individual uses 60 kg paper/year approximately but the usage ranges between 265 kg and 7 kg in the United States and some African countries. India has a relatively low paper consumption of about 9 kg/year/person which is about 2% of the global output. The market is projected to raise extensively because of the fast-growing population.

Due to several reasons, the market of pulp and paper industry was down. The main cause is the falling product prices, rising production costs and risky environmental conditions. Also, the industry is subjected to a lot of criticism owing to detrimental environmental impacts. According to Chellipan et al. (2012), 60,000 L of water is essential to produce 1 ton of paper which generates bulky volumes of wastewater. However, a substantial amount of organic content in the wastewater is used to produce biogas. Hence, creativeness in the industrial process, as well as waste valorization, play a vital role for the development and nourishment of the paper industry (Gottumukkala et al., 2016). The large amount of waste that is produced from pulp and paper industry is a major challenge connected with this industry. The comprehensive analysis on pulp and paper mill waste biorefineries is less investigated, and progress is critically important in the process of pulp or paper mill waste recovery into chemical and fuel use and for the overall sustainability of this industry.

11.2 PULP AND PAPER MILL WASTE

The paper and pulp mill waste plays a major role in generating a massive quantity of wastewater. These wastes are obtained as wastewater and sludge. The wastewater is generated from different procedure involved in paper production. Mud is the major essential which is generated due to the technical process involved in the wastewater treatment. These wastes become a potential problem because of its organic nature as well as usable raw material for energy creation (Bajpai, 2017).

11.3 WASTEWATER

Besides chemical and steel industries, Pulp, and paper industries are the third-largest wastewater generating industry (Gottumukkala, 2016). The different techniques entailed in pulp and paper mills are wet debarking, mechanical/chemical pulping, bleaching, and paper-making. The step-wise production of the paper mill waste is shown in Figure 11.1.

The first stage is the removal of the hard-outer covering of wood, generally termed as debarking. There are two methods for debarking wood. They are:

1. **Wet Debarking:** The procedure of water removal includes thawing, removal of barks and the washing of wood logs. This process comprises of 5% of the chemical requirement for oxygen and 16% of the total wastewater for suspended solids (SS) (Thompson et al., 1991).
2. **Dry Debarking:** The technique of dry debarking involves washing and thawing. This method decreases 10% chemical oxygen demand (COD)/biochemical oxygen demand. The quantity of water required is very low in this process.

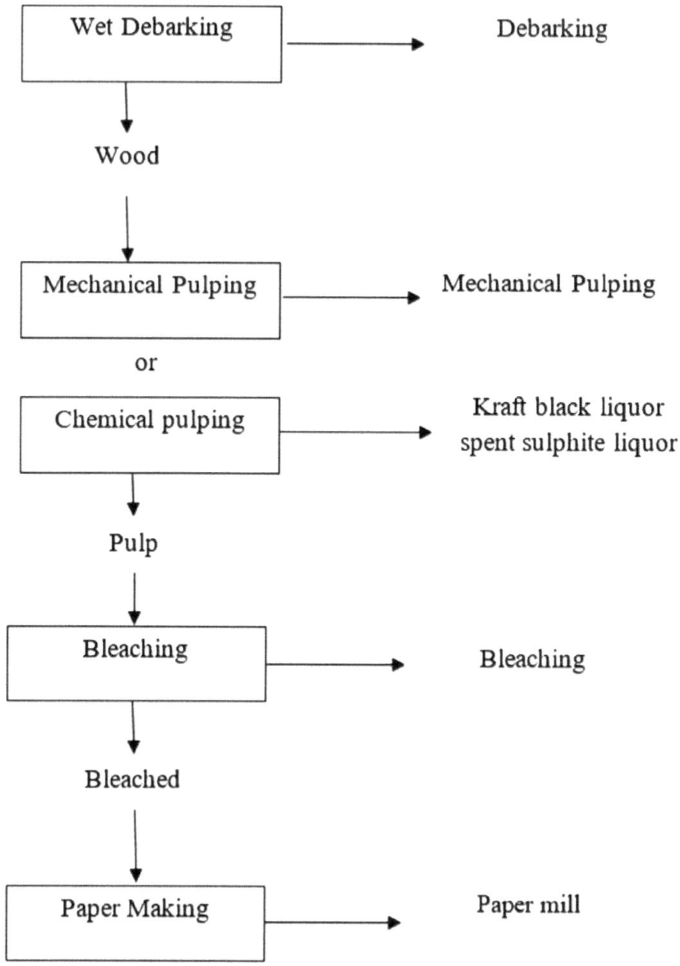

FIGURE 11.1 Foremost wastewater causes from pulp and paper mills.
Source: Kumar et al. (2020).

The wet debarking method is generally preferred to dry debarking due to quality reasons (Pokherel and Veeraraghavan, 2004).

The second stage is the pulping, which aims at separating fibers from lignin and conversion of fibers into the slurry which can be used as source material for the papermaking production. The most prevalent methods used for pulping are 1. Mechanical pulping and 2. Chemical pulping. Chemical pulping is widely used when compared to mechanical pulping.

The third stage is the bleaching step used for the whitening of papers. The color of the pulp is generally detached by means of chemicals such as chlorine, sodium hypochlorite, chlorine dioxide or ozone (Ashrafi et al., 2015). In pulping and bleaching processes of industrial plant, the maximal amount of wastewater is created by the overall analysis.

11.4 SLUDGE

Waste from this manufacturing contains of chemical substances which are toxic to the various ecosystems. The effluent treatment process produces a huge load of solid waste (primary and secondary sludge) which comprises inorganic and organic chemicals (Veluchamy and Kalamdhad, 2017; Shah et al., 2020). Hemicellulose, celluloses, lignin, their derivatives, and other fibers are the major organic compounds present in the wastewater. The primary sludge in this industry consists of nearly 40–95% of fibers and 5–60% of ash content (Alda, 2008). Various chemicals like chlorine and its derivatives were found which makes the sludge unfit for application on agricultural fields.

11.5 PROCESS INVOLVED IN THE PAPER PRODUCTION

Two sorts of paper, i.e., high, and low grades, are prepared based on the pulp quality. Around 43% of overall paper output is high, and 57% of paper pulp is reused (Santos and Lobo, 2012). Generally, two methods are used for pulp production.

11.5.1 PULP MAKING

The pulp process involves numerous steps from the debarking of wooden logs through the washing of logs with water and then chipping them into small parts. These chips are stored in containers for feed (Bajpai, 2015).

Several treatments have been evolved to split up biomass into the individual components of lignin, hemicellulose, and cellulose. Solubilization or separation of lignocellulosic elements leads to the separation. Pulping operations are divided into chemical pulping and mechanical pulping, as shown in Figure 11.2, based upon the approach and equipment utilized.

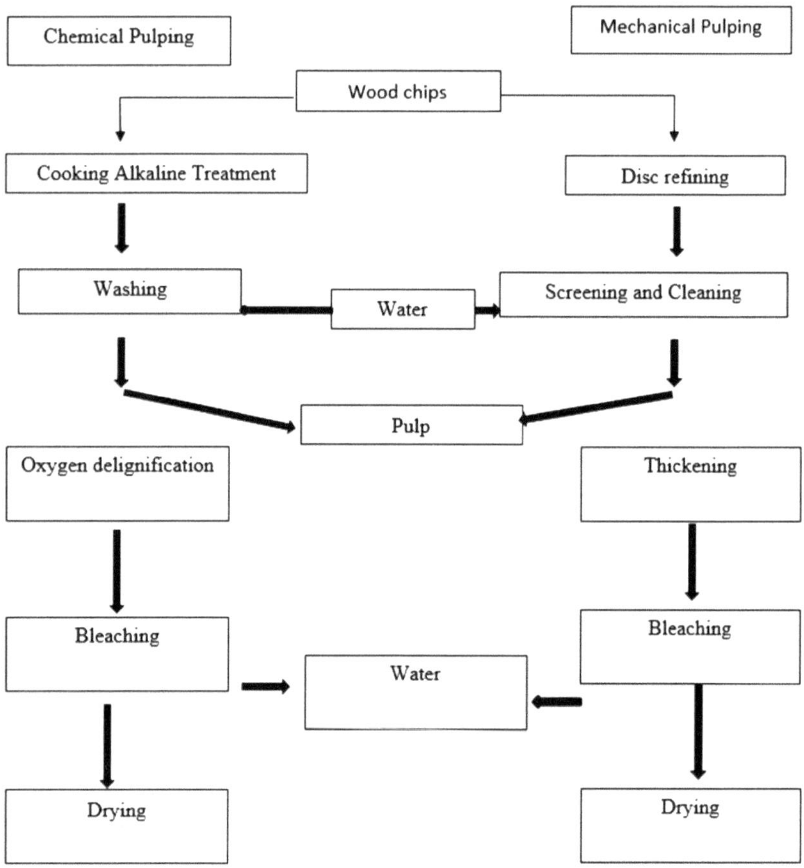

FIGURE 11.2 Flowchart of the mechanical and chemical pulping process.
Source: Kumar et al. (2020).

11.5.2 CHEMICAL PULPING

Initially, wood chips are heated with chemicals containing reactive species like bisulfate ions (sulfite process) and hydroxide ions-hydro sulfide ions (Kraft process) at higher temperatures (150–200°C) and an increased pressure of 100–135 psi. The very complicated structure of lignocelluloses breaks down and the fiber is separated from lignin. Due to the manufacturing of pulp, the kraft pulp production method in pulp industry is dominating and extensively practiced (Monica et al., 2009). The following are detailed several procedures used for chemical pulping:

1. Kraft Pulping: In this process, sodium hydroxide and sodium sulfide are used as cooking chemicals. The chips are cooked in the digester for the period of 2–3 in order to separate the lignin. The lignin and hemicellulose get dissolved in the water leaving behind the cellulose fiber in the solid form due to the chemical combination with high pressure and at high temperature. Cellulosic fiber is then thoroughly cleaned with water and further bleached for whitening of pulp (Mendes et al., 2009). The pulp produced is of high strength which yields about 50%.
2. Sulfide Pulping: In this process, calcium/magnesium bisulfite and sulfuric acid act as cooking liquor for biomass treatment. This Acid-Base combination initiates the process for the separation of lignin from the fibers, which results in leaching the lignin from the fibers (Evutuguin, 2016). Pulp yield is higher (approximately 55%), but the strength is very less when compared with kraft pulp.
3. Soda Pulping: In this process, sodium hydroxide is used because it increases the pH so that the lignin is easily dissolved and fibers can be separated (Holm, 2018). This biomass is treated with 14%–16% of alkaline solution at 140–170°C. This method is predominantly used for nonwoody biomass resources.

11.5.3 MECHANICAL PULPING

The mechanical pulping is generally done with the aid of machines that withdraws fibers from the lignocellulosic material. The fibers yield is high (approximately 97%) compared to chemical methods (40–45%), but the quality of fibers is inferior to the chemical one (Ewjik et al., 2017). Some of the mechanical pulping processes are listed below:

i. Stone groundwood;
ii. Refiner mechanical pulp;
iii. Thermomechanical pulping;
iv. Chemo thermal mechanical pulping.

After the pulping process, bleaching is done so that to enhance the color of the fibers which is termed whitening. The washed-out pulp is kept at low temperatures for the preservation of fiber before paper making (Ahvazi et al., 2007).

11.5.4 PAPERMAKING

Papermaking is a historical practice that passes through several phases through the slurry of fibers. These steps include refining/beating, wire mesh, pressing, drying, size pressing/metering, calendaring, reeling, and the final product (Biermann, 1996a). In the papermaking process, the desiccated pulp is mixed with water in which the slurry is prepared in the pulp. Then the pulp is beaten in the beater/refiner (machine equipped with the stationary metal bar and the rotating metal bar). This refiner is mainly used to shorten the fibers. Thereafter, to improve the quality of the paper additives like talc, clay, and resin are added. This process is known as sizing (Rittmann and McCarty, 2001; Bajpai, 2008).

The stock is finally transferred into the head of the Fourdrinier drying machine. The wire, Dryer, and press sections remove 99.5% of the water contained in the pulp (Biermann, 1996b). After that, pulp forms a web over the mesh which is conveyed for drying in the dryer. This dryer consists of Yankee cylinders comprised of well-polished metal cylinders of 2.4–4.5 m diameter used to eliminate the residual water from the pulp piece (Smook, 1992a).

Eventually, the dried sheet is passed through the various highly pressurized cylinders to make it flat, eliminating wire symbols and lumpy creations to produce the final sheet uniform. This whole process is called calendaring. The end product is rolling and cut in rolls (Smook, 1992b). The steps involved in the papermaking is shown in Figure 11.3.

11.6 MAJOR CONTAMINANTS PRESENT IN THE PULP AND THE PAPER MILL WASTEWATER

These mills cause the effluent load consists of solid, liquid, and gaseous pollutants (Ugurlu and Karaoglu, 2009). The main challenge involved in this industry is to deal with the huge amount of wastewater per ton of paper produced. The wastewater is dumped into the water bodies which leads to the degradation of the environment with its brown color and odor which causes a threat to aquatic life. In the papermaking process, approximately 55% of the wood weight is going as waste that contributes a large amount of organic load to the wastewater (Young and Chase, 1965). This organic content leads to high COD, BOD, suspended, and dissolved solids in the wastewater discharge. Also, there are certain chemical compounds like chlorine, stilbenes, resin acids, chlorinated lignins, lignin, phenols, dioxins, and furans (Singh, 2015). The major pollutants entailed in this wastewater are tannins, resin acids, and halogenated compounds.

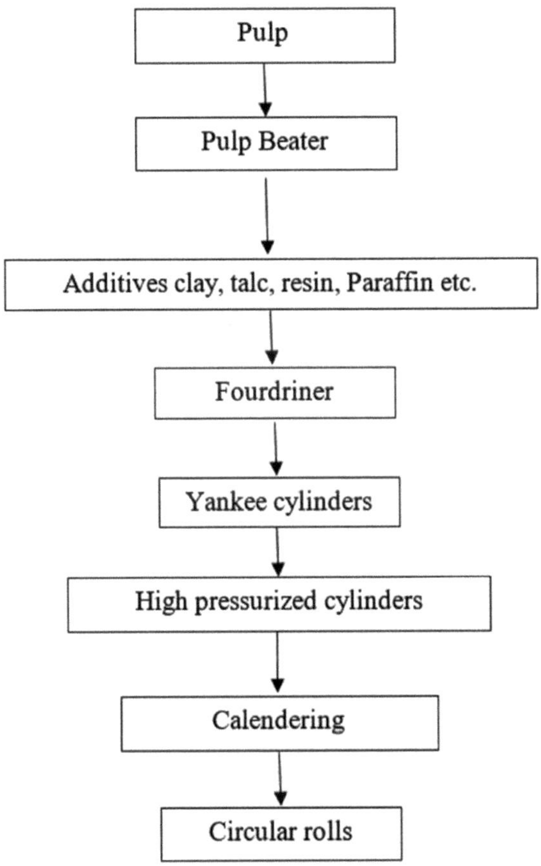

FIGURE 11.3 Flowchart of the papermaking process.
Source: Kumar et al. (2020).

11.6.1 TANNINS

These are the polymeric chemical compounds whose molecular masses range between 500 g/mol and 3,000 g/mol. It consists of several benzenes' rings with various –OH groups which are soluble in water (White, 1957). These compounds generally impart color to the water due to their polar nature. Water also retains less oxygen and decreases light penetration via these fluids. This process severely affects the aquatic life, leading to reduced growth, long term exposure causes death (Loomis and Battaile, 1966).

11.6.2 RESIN ACIDS

These resin acids generally originate in the pulp and paper mill wastewater in which the majority of these acids are tricyclic diterpenoids containing the carboxylic acid function group (Martin et al., 1999). The most commonly found resin acids are the habitats and the primaries. The resin acids concentrations are very less in the chemical pulping wastewater than in the mechanical pulping. Taylor et al. (1988) showed the noticeable effect of resin acids on the bluegill, salmon species, fathead minnows, rainbow trout and sunfish. The resin acids effected on vertebrates is two to threefold higher compared to invertebrates (Zanella et al., 1983; Servizi et al., 1986).

11.6.3 HALOGENATED COMPOUNDS

The pulp involved in paper mill process is processed for the brown color removal using chlorine as a bleaching agent (Solomon, 1996). This step plays a crucial role in the contamination of the wastewater using halogen compounds. This wastewater also contains the furans and dioxins which are produced while pulp washing after bleaching with the pentachlorophenol (PCP) (Adergani et al., 2016). It also contains chlorinated lignin compounds, chlorinated phenolics as well as non-phenolics. Even though these are present in a minute concentration, they can cause acute toxicity to the aquatic ecosystem (Leach, 1980).

11.7 ANAEROBIC TREATMENT OF PULP AND PAPER MILL WASTE

Anaerobic digestion is a process that occurs naturally such that the organic substances are decomposed by the micro-organisms under anaerobic conditions. During the anaerobic conditions, less energy is consumed compared to the aerobic process. The energy harnessed during the process is rewarded by the production of the 1.16 kWh/kg COD energy removal in the methane form. The anaerobic process is simple, economical, less energy and area, less sludge production, higher organic loading rate when associated with the aerobic procedure, that leads a vital role in wastewater treatment technology. The flow chart of anaerobic treatment of pulp and paper mill waste is shown in Figure 11.4.

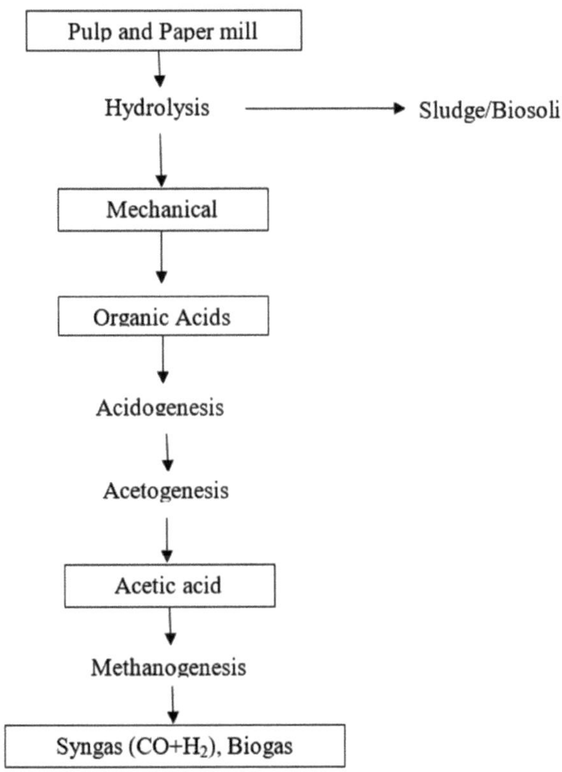

FIGURE 11.4 Steps involved in the anaerobic treatment process.
Source: Kumar et al. (2020).

11.7.1 ANAEROBIC TREATMENT OF WASTEWATER

For one ton of paper production, 16,000 gallons of water is required, so that a considerable quantity of organic wastewater is generated which can be harnessed as a potential energy source called biogas. To convert to biogas, Methanogens are generally used to decompose the waste and convert to water and gas as a by-product. There are different types of reactors used for wastewater treatment:

- Up-flow anaerobic sludge blank reactor (UASB) (Gomez, 2011);
- Anaerobic filter (Show and Tay, 1999);
- Expanded granular sludge bed reactor (ESGB) (Kato et al., 1994);
- Continuous stirred tank reactor (CSTR) and anaerobic contact process (ACP) (Zhang et al., 2015).

11.7.1.1 UP-FLOW ANAEROBIC SLUDGE BLANK REACTOR (UASB)

The UASB reactor is maximum chosen reactor for the anaerobic wastewater treatment. It has the capability that they can work on the large organic loading rates. The influent is bottom fed and the accumulation of sludge takes place at the bottom and agitation takes place due to the gas formation (Gomez, 2011). The gas is accumulated and received at the top and preserved in the gas storage tank which is used for energy generation. The treated wastewater discharged at the tank top, which is again processed further. The UASB reactor removes 80–93% of COD removal from the inflow (Chinnaraj and Rao, 2006). The schematic diagram of the UASB is shown in Figure 11.5.

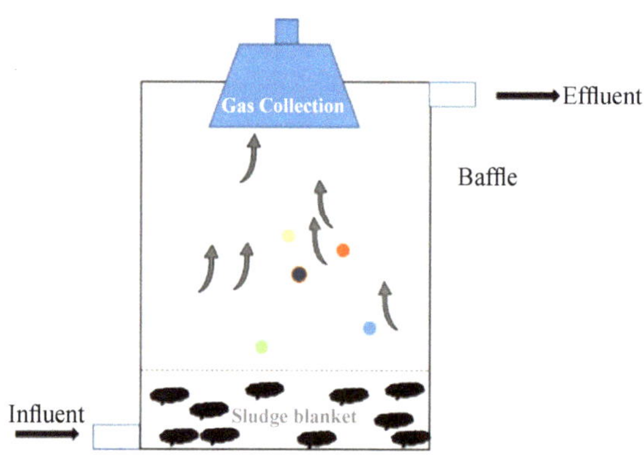

FIGURE 11.5 Schematic diagram of UASB process.
Source: Kumar et al. (2020).

11.7.1.2 ANAEROBIC FILTER

In this type of filter, a column is used which is surrounded with supportive media such as glass, pall ring, open-pored, baked clay, PVC, coral/mussel shells and polypropylene rings (Show and Tay, 1999). This media is useful for developing bacterial progress and exopolysaccharides production. In this secondary treatment process, anaerobic digestion is combined, and the produced biogas is recovered which is used for energy purposes. Depending on the influent flow, it is divided into two types: (i) up-flow filter; and (ii) downward filter. The Schematic diagram of the Anaerobic filter is shown in Figure 11.6.

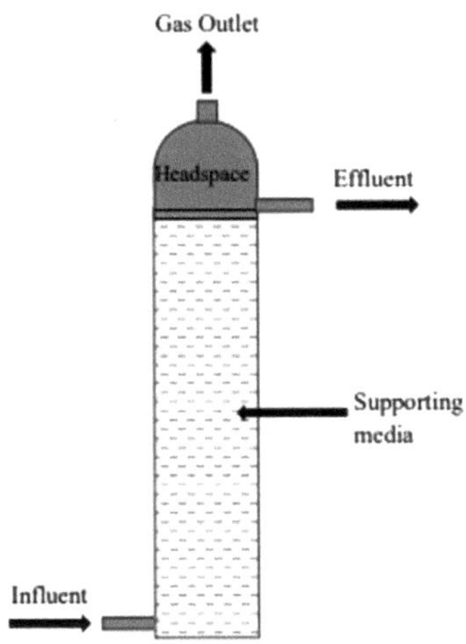

FIGURE 11.6 Schematic diagram of anaerobic filter.
Source: Kumar et al. (2020).

11.7.1.3 EXPANDED GRANULAR SLUDGE BED FILTER

The granules are fluidized by the action of influence or recirculation of gas, which is a perpendicularly extended improved UASB reactor (Jian-long and Jing, 2005). In the one-stage, three-stage separators which are present on reactor top, the wastewater, biogas, and biomass are separated. When compared with USAB, it has more potential to treat wastewater containing COD< 1 at a low temperature of 10°C and improved mass transfer characteristics (Kato et al., 1994). The Schematic diagram of the Anaerobic filter is shown in Figure 11.7.

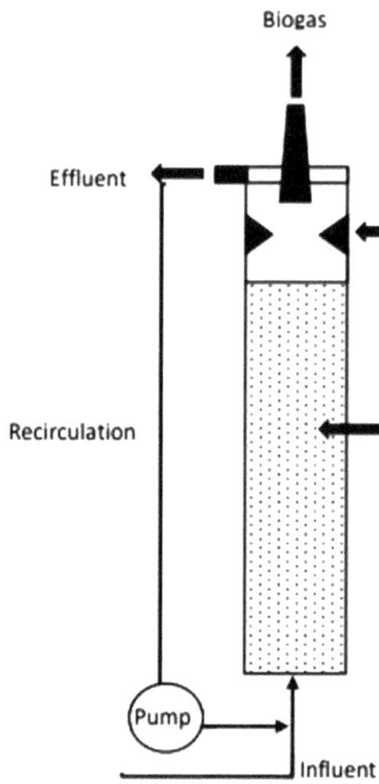

FIGURE 11.7 Schematic diagram of EGSB reactor.
Source: Singh et al. (2006).

11.7.1.4 CONTINUOUS STIRRED REACTOR AND ANAEROBIC CONTACT PROCESS (ACP)

It is a batch type reactor consists of a rotor used to agitate the constituents mix properly in the tank. Generally, ACP, and CSTR are two popular types of agitated reactors used. In both the reactors, the main principle involved is stirring the waste inside the anaerobic tank and gas production. When it comes to ACP, the effluent recirculation is done back into the tank itself (Zhang et al., 2015). Due to proper mixing, constituents in the tank are homogenous throughout *t* (McCarty and Smith, 1986). The Schematic diagram of the continuous stirred reactor and ACP is shown in Figure 11.8.

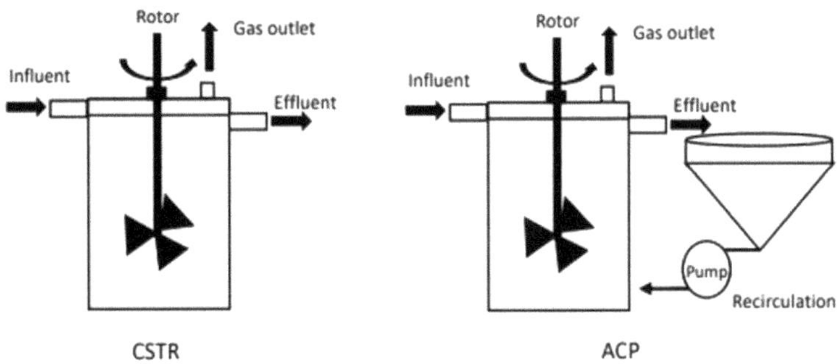

FIGURE 11.8 Schematic diagram of CSTR and ACP.
Source: Kumar et al. (2020).

11.7.2 ANAEROBIC DIGESTION OF SLUDGE

The amount of sludge produced from this industry is very large due to half of the lignocellulosic material is converted into pulp for the papermaking process and the remaining is produced as waste (Balwaik and Raut, 2011). The dry sludge of 40–50 kg is produced for 1 ton of paper manufacture (Mahmood and Elliott, 2006). After the wastewater treatment, the solid part left is called the biosolid (sludge). This sludge has much water and is not very dry, i.e., 0.5–2% of overall weight. The large organic load and lower sludge production usually make anaerobic treatment preferable in pulp mill storage.

The digestion of anaerobic waste in the world is intriguing and is utilized to convert sludge into value-added goods. The amount of methane generated in the reactor is measured by sludge degradation. Lin et al. (2009) described the process of sludge bioconversion and focused on the importance of pretreatment in broadening sludge biogas production potential. There is an increase of 183% in in pre-treatment procedure in methane generation compared to untreated sludge.

11.8 STRATEGIES TO ENHANCE THE ANAEROBIC PROCESS OF WASTE TREATMENT

11.8.1 CODIGESTION OF WASTE

This is an extraordinary method of integrating the pulp and paper mill waste with the simply recyclable substrate. It is usually done to preserve the optimal balance of biodegradation nutrients. This sludge has a necessary amount of carbon but is low in nitrogen. A suitable C/N ratio must be maintained by providing enough nitrogen for the efficient decomposition of nitrogen limiting waste (Thanarasu et al., 2018). Several studies have shown that anaerobic digestion and biogas generation from paper mill waste has a positive code show. Teghammar et al. (2013) demonstrated that the decrease of hydraulic retention time (HRT) from 20–25 days leads to the 15–34% rise in the yield of methane when nitrogen loaded industrial waste was combined with sludge. There are various studies investigated by different researchers that the quantity of gas attained with and without codigestion was observed to be 246±10 L CH_4/Kg VS (20-day HRT) and 201±18 L CH_4/Kg VS (60-day HRT), respectively (Chatterjee et al., 2018). The codigestion of this waste and monosodium glutamate waste liquor was reported that there was an optimum feed of mesophilic bacteria and generated 200 mL/g VS methane (Lin et al., 2011). The same group of researchers examined the highest output of methane: the effective mixing of food waste with pulp and paper mill sludge in 1:1 reached 256 ml/g VS with a COD deduction of 94%. This combination demonstrated that the buffer and the pH were retained in the 5.8–8.4 range (Lin et al., 2012). Municipal wastewater and paper mill sludge output increased in 19 days by 50% (Hagelqvist, 2013). In addition, the pig manure and the 1:3 paper sludge in combination and 250 mL/g VS in 14 days obtained methane from the pulp and paper mill (Parameswaran and Rittmann, 2012). Waste coding has speeded up the degradation process, decreased COD, and increased methane generation.

11.8.2 PRETREATMENT

Before anaerobic digestion sludge, pre-treatments were generally carried out for lignin removal. Lignin, hemicellulose, and cellulose are the three major constituents of lignocellsic waste. The cellulose in the lignin is a fragrant, composite heteropolymer. Lignin has a complicated three-dimensional assembly which works as an inhibitor to the enzyme activity of the sludge of a paper mill. Elimination of inhibitors for methanogenic microorganisms and simple digestion of LINCALLUSIC materials. In the case of digestion of sludge different strategies such as lignin depolymerization, breakdown of covalent ties and cellulose reduction can also be boosted (Taherzadeh and Karimi, 2008; Hendriks and Zeeman, 2009). In subsections, many techniques for biomass pre-seeding are explored.

11.8.2.1 MILLING AND THE IRRADIATION

In this process, milling, and irradiation are used to reduce the particle size and crystallinity for enhancing the enzyme hydrolysis and biomass biodegradability (Harmsen et al., 2010; Menon and Rao, 2012). The milled generated lignin represents lignin that is natural but the extraction of lignin is low that cause structural changes (Harmsen et al., 2010; Duval and Lawoko, 2014). Microwave along with alkali, acid, and H_2O_2 was observed to raise the efficacy of pre-treatment by accelerating reaction (Menon and Rao, 2012; Moretti et al., 2014).

11.8.2.2 STEAM EXPLOSION

The biomass is processed for a period under high-pressure saturated stream and immediately depressurized and cooled down to the biomass, resulting in an explosion of the water. This results in the breakdown of hemicellulose and lignin conversion for different biomass residues (Harmsen et al., 2010; Menon and Rao, 2012; Moretti et al., 2014).

11.8.2.3 AMMONIA FIBER EXPLOSION AND AMMONIA RECYCLE PERCOLATION

The biomass is handled at a high pressure, then abrupt reduction in the pressure and the biomass with ammonia fiber-explosion (AFX) ammonia solution in ammonia-recycle percolation (ARP) (Harmsen et al., 2010; Menon and Rao, 2012). It is extremely effective, leading to a disruption in the decrystallization, dealignment, depolymerization, and deacetylation of cellulose lignin-carbon linkage (Menon and Rao, 2012; Moretti et al., 2014). For enzyme activity, the surface area in this treatment has been enhanced, which does not generate biological process inhibitors and the recycling and utilization of ammonia. This method is very effective for Herbaceous crops and grasses (Harmsen et al., 2010; Menon and Rao, 2012).

11.8.2.4 ORGANOSLOV LIGNIN

Biomass is treated with organic solvents and acids and further acid or alkaline is increased to enhance the Solubilization of pulping and lignin (Harmsen et al., 2010; Lange et al., 2018). Lignin derived from organosolve is highly pure and relies on the parameters of processing. Lignin is sulfur-free; however, the resulting pulp is low-quality in comparison to others like pulping or sulfite pulping (Lange et al., 2018; Moretti et al., 2014).

11.8.2.5 OTHER PROCESSES

In the pretreatment process, conversion of the lignocellulosic biomass into value-added products played a major role. Ionic liquids have been applied for the pretreatment of biomass. Ionic liquids have been applied for the pretreatment of the biomass components. At room temperature, they are liquids and composed of only ionic species. The cellulose is amorphous because of this treatment and enhances enzymatic hydrolysis. This technique is straightforward to run, energy-efficient, and cleaner than other widely applied technologies (Harmsen et al., 2010; Menon and Rao, 2012). Dilute and strong acid procedures, CO_2 explosion, liquid hot water pre-treated, deep eutectic solvents, etc., are many additional pre-treatment methods (Harmsen et al., 2010; Moretti et al., 2014).

11.8.2.6 BIOLOGIC PRETREATMENT

For pre-treatment of lignocellulosic biomass, microorganisms or their enzymes are generally used. In this biological treatment, microorganisms including white, brown, and soft red fungus and bacteria are produced by wood-feeding. The main degraders of lignocellulosic biomass are fungi and bacteria which can cause alteration in the chemical configuration and the biomass structure (Menon and Rao, 2012). Solubilization of lignin and fewer inhibitors generation takes place in this process. The benefits of this treatment are low energy consumption, mild operation conditions, no chemical treatment and environmental friendliness. The major disadvantages are the very slow processing, large space requirement, requirement of proper growth conditions, contamination issues and slow hydrolysis rate (Harmsen et al., 2010; Menon and Rao, 2012; Moretti et al., 2014).

11.9 OPTIMUM CONDITION FOR THE ANAEROBIC DIGESTION

In addition to pretreatment and codigestion, various operating conditions should be considered for sludge digestion. The temperatures, pH, alkalinity, C/N ratio and heavy and total ammonia levels of such anaerobe species are extremely sensitive to sulfur. These factors should thus be checked correctly throughout digestion (Kamali et al., 2016).

11.10 BIOAUGMENTATION

The prospective degraders of lignocellulosic biomass may efficiently convert waste into methane by different microorganisms (Kamali et al., 2016). The use in anaerobic reactor of efficient microbial strains degrading biomass and methane-producing systems can boost digestion and the generation of biogas.

11.11 ANAEROBIC DIGESTION AND PRODUCTION OF VALUE-ADDED PRODUCTS

During wastewater treatment the produced sludge has a high-water content. Sludge is usually burned or composted, thus. This precious resource is usually not used as the incineration technique. Researchers will address waste recovery for the creation of value-added goods with technological advancements. The following are the most important value-added goods.

11.11.1 BIOGAS

The main reason for using the anaerobic digestion process is the biogas production. The production of electricity or heat that can be generated using biogas. Biogas contains a mixture of methane (60%–70%), carbon dioxide (30%–40%), and nitrogen (less than 1%), water vapor (1%–2% v/v) and trace compounds like mercaptans (RSH), siloxanes, and hydrogen sulfide

(H_2S) (Rasi et al., 2007). Biogas can be used for operating furnaces, boilers, heaters, etc. The usage of gas turbines and internal combustion engines, Biogas can be transformed into electrical energy. The harmful trace gases such as H_2S are need to be removed to avoid corrosion for turbine metal parts (Schweigkofler and Niessner, 1999). The biogas upgrading the methane with a purity of >95% then biomethane is extracted from biogas. The effective substitute of natural gas for smooth running of vehicles is clean methane is because of its wide acceptability (Zak et al., 2018). Saha et al. (2011) analyzed that the 90% methane production increase in using various pretreatment methods such as ultrasonication, chemo-mechanical, and microwave methods in pulp and paper sludge utilization.

11.11.2 BIOHYDROGEN

Hydrogen is one of the cleanest burning fuels which have a high calorific value fuel (142 kJ/kg) (Zilouei and Taherdanak, 2015). The Hydrogen production than can be done using various methods such as chemical, electrochemical, and biologic methods. In order to transform the biomass into valuable chemicals and fuels such as nitrogen, methane, and oxygen anaerobic digestion is adopted. The production of Biologic hydrogen can be divided into two categories: light-independent (dark fermentation) and light-dependent (Bakonyi et al., 2014; Wang et al., 2014). Dark fermentation process is widely used for the production of hydrogen because of its flexible operation and no need for external energy (Zilouei and Taherdanak, 2015). The high BOD in the pulp and paper mill waste is beneficial for the production of hydrogen. Vaez et al. (2017) reported that using dark fermentation 62.2 mL/g COD of hydrogen generation is generated after the alkaline hydrolysis of paper mill effluent. Lin et al. (2013) showed that, usage of food waste and sludge in an equal ratio in anaerobic codigestion, the generation of the hydrogen production is about 64.48 mL/g volatile solid. Lakshmidevi and Muthukumar (2010) showed the utmost amount of production of hydrogen is about 2.03 mol H_2/mol sugar.

11.11.3 BIOFERTILIZER

The huge quantity of sludge is generated as a by-product during pulp making and treatment of this wastewater. This sludge act as a fertilizer which contains a considerable quantity of macronutrients (NP) needed for plant growth. Ribeiro et al. (2010) described that the paper sludge act as a potential fertilizer containing phosphorus (167–370 mg P/kg), organic matter (11%–47%), and nitrogen (38–2,560 mg N/kg). Fahim et al. (2019) concluded that the pulp and paper mill primary sludge is more effective than secondary sludge on the growth of *Microtus sachalinensis* and Abelmoschus esculentus plants due to the large availability of macronutrients. According to US EPA Factsheet for Biosolid Application (2000), 120 pounds of nitrogen/acre/year is needed for corn crop. The major challenge that is faced while applying low nutrient sludge is nutrient immobilization. In order to reduce the impact of nutrient immobilization composting, delayed crop planting, use of legumes and supplemental fertilization (Camberato et al., 2006). Vasconcelos and Cabral (1993) proved that Pulp Sludge and Paper Sludge may be employed as an efficient fertilizer because of high bio-nutrient content.

11.11.4 VOLATILE FATTY ACIDS (VFAS)

During the anaerobic degradation of organic compounds as well as acidogenic fermentation period, there is a generation of volatile fatty acids (VFAs). During the fermentation process, the most common VFAs produced are butyric acid, propanoic acid and acetic acid. The VFAs play a major role for the production of polyhydroxy alkenoates (PHAs) and useful chemicals. Bengtsson et al. (2008) reported the VFA generation was based on the pH and retention time using paper and pulp mill wastewater. The important fermentation products obtained are butyrate, propionate, and acetate. Sharma et al. (2013) concluded that the electrochemical synthesis method is used for the decrease of VFAs to alcohol.

11.12 CHALLENGES ASSOCIATED WITH ANAEROBIC TREATMENT AND MITIGATION STRATEGIES

Pulp and paper mill trash consists of valuable components of organic waste. Anaerobic treatment is usually employed to degrade these chemicals. UASB and CSTR anaerobic reactors as well as their modifiers are used mostly for anaerobic digestion of waste. The anaerobic therapy is utilized and more successful, however this technique presents numerous problems. Below are some major problems.

11.12.1 C/N RATIO

Pulp and paper waste, since organic entities like lignin, cellulose, and phenol-like substances contain a significant quantity of carbon. The presence of extra carbon changes the C/N ratio needed by the anaerobic digester to properly operate and metabolize microbials. This causes extra ungated loop to accumulate, which stops the entire treatment process due to sluggish anaerobic processes and reduces waste decay. Moreover, the C/N ratio impacts the microorganisms responsible for this anaerobically therapy for the hydrolysis and methanogenesis process severely. During the early stage of growth, a rise in VFAs effects hydrolytic bacterial development causes imbalanced C-N ratio (Mao et al., 2017). For the microbial population growth in the tank, the C/N ratio has an important function. The report also investigated the influence of ammonia on the anaerobic process by Yenigun and Demirel (2013), which found that an extensive ammonia concentration has a detrimental impact on methanogenesis, which leads to a poor biogas production. The C/N ratio may be controlled and balanced with the use of coding technology. It is a procedure by adding some additional digestive waste to increase the processing rate of digester. Waste is selected to keep the C/N ratio optimal for the correct microbial activity.

11.12.2 ENERGY RECOVERY FROM BIOGAS

In anaerobic chamber, the biogas produced is around 40% that is both dissolved and inaccessible. Dissolved methane is washed away with bulk liquid effluent. One of the biggest problems is to exploit the entire biogas generated. Also important are ongoing biogas generation in the anaerobic digester. Due to the higher costs of biogas technology as a fuel and the current engine changes, biogas is a problem for commercial usage. Other obstacles to this technology include the government's interest in biogas in the generation of energy, cheap investment, economic practicability, dependency on local conditions, biogas stockpiling and supply (Salomon and Lora, 2009). Thus, the research and development focus on equipment such as biogas burners that make biogas inexpensive and viable. The usage of biogas in enhanced biogas will be diversified in order to improve the methane (>95.5%), using appropriate and cost-efficient biogas upgrading or enrichment technology. Methane is highly heated, having 40–80% of biogas (25.1–28.7 MJ Nm^3). In pulp and paper industries, biogas may be used for the fueling of boilers, kilns, and firms. Power equipment in pulp and paper mills can be supplied using energy generated by turbines. The reduction in the anaerobically chamber of dissolved methane (Hartley and Lant, 2006) can be achieved by employing degasification membranes using biogas micro aeration method. Luo et al. (2012) have utilized over 86% methane; however, this technology has not been applied on a wide scale because of the high membranes' costs.

11.12.3 HYDROGEN SULFIDE (H_2S) FROM PULP AND PAPER MILLS

Kraft processes are frequently employed in pulp and paper mills to pretreat lignocellulosic biomass. In addition to NaOH a cooking liquor, i.e., Na_2S is used, however a significant amount of Na_2S is left in the liquid that causes a foul stench. H_2S, extremely malodorous and metal corrosive, are the most important gas produced. In addition, sulfide ion interacts with different cellulose and lignin side-chain radicals, forming tiny quantities of methyl mercaptans and dimethyl sulfide. According to calculations, roughly 6.2 pounds of sulfur are emitted for one ton of pulp (Hansen, 1962). Pulp mill employees exposed to the dangerous H_2S gas suffer from headaches and take longer leave than unexposed people (Kangas et al., 1984). Besides this, a major concern related to hydraulic sulfide is the corrosion of the metallic facilities in the digestors. To manage the odor of H_2S, it is the most often utilized way to add some other gas or chemical. But this approach does not eliminate the gas source, i.e., the bacteria that reduce sulfur (SRB). The reactor's H_2S control can only be carried out with the injection of certain chemicals to prevent SERB bacterial reactions. The iron that interacts with H_1S and generates iron sulfide and removes H_2S from the gas is one of the best additions. Cherosky and Li (2013) highlighted reduced anaerobic digestion inhibition owing to H_2S. $FeCl_3$ was utilized by Romero-Guiza et al. (2014) and the H_2S decrease was approximately 97% .

11.13 CONCLUSIONS

The amount of organic waste that comes from pulp and paper mills is complicated and comprises lignin model compounds, lignin, chlorolignin, cellulose fibers and hemicelluloses that enhance the waste stream's COD, BODs, and total suspended solids (TSS). Biological treatment is used to reduce the BOD and COD to allowable levels in order to handle this waste. One of the most used biological techniques is the activated sludge process because of the excellent treatment effectiveness. The biggest problem with this approach is the large volume of sludge produced. Anaerobic treatment of these wastes is one of the most famous technical methods for the wastewater treatment in order to utilize a cost-efficient and environmentally beneficial technology. Wastewater may efficiently be treated and converted to value added products such as biogas, biohydrogen, biofertilizers, and VFAs with the help of anaerobic treatment. Bio valuation of waste from these mills will offset the total cost and environmental effect of pulp and paper mills. Biogas is anaerobic treatment's potential product, of which the methane is between 40% and 70%. The most widely utilized anaerobic digesters are UASB, CSTR, and their many variations. Microbial development and its functioning within these reactors are the driving factors of this technology. Methanogens are very susceptible to temperature, C/N, pH, and inhibitors. The optimal circumstances for the functional microorganisms, which can be maintained with the initial pH monitors and alters, the correct C/N ratio maintained and the hazardous gas monitoring. More targeted study to identify effective microbes/consortiums should be done to increase the anaerobic process development. Improving biogas production will help you to grasp the functional details at the micro level which can be applied at the digester level by better knowing the structure of the microbe population using metagenomics, met proteomics and metabolic methods. The biovalorization of effluent waste also entails the use of Genetically Engineered Anaerobic Microbes. A crucial consideration of essential factors, procedure plan and anaerobic digestion process community structure would allow waste and bioconversion from the pulp and paper mill into different bioproducts and increase biomethane quality and quantity.

KEYWORDS

- **anaerobic contact process**
- **anaerobic treatment**
- **biovalorization**
- **continuous stirred tank reactor**
- **digestion**
- **expanded granular sludge bed reactor**
- **paper mill**
- **wastewater**

REFERENCES

Adergani, B. A., Eskandaril, S., Hashemi, A., & Shekarchi, M., (2016). Evaluation of pentachlorophenol residues in some hygienic papers prepared from virgin and secondary pulp by electron capture gas chromatographic method. *Journal of Chemical Health Risks, 6*, 23–31. https://doi. org/10.22034/jchr.2016.544125.

Ahvazi, B., Radiotis, T., Bouchard, J., & Goel, K., (2007). Chemical pulping of steam-exploded mixed hardwood chips. *Journal of Wood Chemistry and Technology, 27*, 49–63. https://doi.org/10.1080/02773810701486618.

Alda De, J. A. G. O., (2008). Feasibility of recycling pulp and paper mill sludge in the paper and board industries. *Resources, Conservation and Recycling, 52*, 965–972. https://doi. org/10.1016/j.resconrec.2008.02.005.

Ashrafi, O., Yerushalmi, L., & Haghighat, F., (2015). Wastewater treatment in the pulp and paper industry: A review of treatment processes and the associated greenhouse gas emission. *Journal of Environmental Management, 158*, 146–157. https://doi. org/10.1016/j.jenvman.2015.05.010.

Bajpai, P., (2008). *Chemical Recovery in Pulp and Paper Making* (p. 166). PIRA International, UK. https://doi.org/10.1007/978-3-319-00545-4_2.

Bajpai, P., (2015). *Pulp and Paper Industry* (1st edn.). https://doi.org/10.1016/ C2014-0-02795-5.

Bajpai, P., (2017). *Anaerobic Technology in Pulp and Paper Industry*. https://doi. org/10.1007/978-981-10-4130-3.

Bakonyi, P., Nemestothy, N., Simon, V., & Belafi-Bako, K., (2014). Review on the start-up experiences of continuous fermentative hydrogen producing bioreactors. *Renewable and Sustainable Energy Reviews, 40*(0), 806e813. https://doi.org/10.1016/j.rser.2014.08.014.

Balwaik, S. A., & Raut, S. P., (2011). Utilization of waste paper pulp by partial replacement of cement in concrete. *International Journal of Engineering Research and Applications, 2*, 300–309.

Bengtsson, S., Hallquist, J., Werker, A., & Welander, T., (2008). Acidogenic fermentation of industrial wastewaters: Effects of chemostat retention time and pH on volatile fatty acids production. *Biochemical Engineering Journal, 40*, 492–499. https://doi.org/10.1016/j.bej.2008.02.004.

Biermann, C. J., (1996a). Paper manufacture. *Handbook of Pulping and Papermaking* (p. 209). San Diego. https://doi.org/10.1016/b978-0-12-097362- 0.x5000-6.

Biermann, C. J., (1996b). Refining and pulp characterization. In: *Handbook of Pulping and Papermaking* (pp. 137–157). https://doi.org/10.1016/ b978-012097362-0/50010-8.

Camberato, J. J., Gagnon, B., Angers, D. A., Chantigny, M. H., & Pan, W. L., (2006). Pulp and paper mill by-products as soil amendments and plant nutrient sources. *Canadian Journal of Soil Science, 86*, 641–653. https://doi.org/10.4141/s05-120.

Chatterjee, P., Lahtinen, L., Kokko, M., & Rintala, J., (2018). Remediation of sedimented fiber originating from pulp and paper industry: Laboratory scale anaerobic reactor studies and ideas of scaling up. *Water Research, 143*, 209e217. https://doi.org/10.1016/j.watres.2018.06.054.

Chelliapan, S., Mahat, S. B., Din, F., Yuzir, A., & Othman, N., (2012). Anaerobic digestion of paper mill wastewater. *Iranica Journal of Energy and Environment, 3*, 85–90. (Special Issue on Environmental Technology). https://doi.org/10.5829/idosi.ijee.2012.03.05.14.

Cherosky, P., & Li, Y., (2013). Hydrogen sulphide removal from biogas by bio-based iron sponge. *Biosystems Engineering, 114*, 55–59. https://doi.org/ 10.1016/j.biosystemseng.2012.10.010.

Chinnaraj, S., & Rao, G. V., (2006). Implementation of an UASB anaerobic digester at bagasse-based pulp and paper industry. *Biomass and Bioenergy, 30*, 273–277. https://doi.org/10.1016/j.biombioe.2005.10.007.

Duval, A., & Lawoko, M., (2014). A review on lignin-based polymeric, micro-and nano-structured materials. *Reactive and Functional Polymers*, 78–96. https://doi.org/10.1016/j.reactfunctpolym.2014.09.017.

Evtuguin, D. V., (2016). Sulphite pulping. In: *Lignocellulosic Fibers and Wood Handbook* (pp. 225–244). https://doi.org/10.1002/9781118773727.ch8.

Ewijk, S. V., Stegemann, J. A., & Ekins, P., (2017). Global life cycle paper flows, recycling metrics, and material efficiency. *Journal of Industrial Ecology, 22*(4), 686–693. https://doi.org/10.1111/jiec.12613.

Fahim, S., Nisar, N., Ahmad, Z., & Asghar, Z., (2019). Managing paper and pulp industry by-product waste utilizing sludge as a bio-fertilizer. *Polish Journal of Environmental Studies, 28*, 83e90. https://doi.org/10.15244/pjoes/83614.

Gomez, R. R., (2011). *Licentiate Thesis on "Upflow Anaerobic Sludge Blanket Reactor: Modeling*". Royal Institute of Technology, Stockholm, Sweden.

Gottumukkala, L. D., Haigh, K., Collard, F. X., Rensburg, E. V., & Gorgens, J., (2016). Opportunities and prospects of biorefinery-based valorization of pulp and paper sludge. *Bioresource Technology, 215*, 37–49. https://doi.org/10.1016/j.biortech.2016.04.015.

Hagelqvist, A., (2013). Batchwise mesophilic anaerobic co-digestion of secondary sludge from pulp and paper industry and municipal sewage sludge. *Waste Management, 33*, 820–824. https://doi.org/10.1016/j.wasman.2012.11.002.

Hansen, G. A., (1962). Odor and fallout control in a kraft pulp mill. *Journal of the Air Pollution Control Association, 12*(9), 409–436. https://doi.org/ 10.1080/00022470.1962.10468107.

Harmsen, P. F. H., Huijgen, W., Bermudez, L., & Bakker, R., (2010). *Literature Review of Physical and Chemical Pretreatment Processes for Lignocellulosic Biomass*. Report /,

1184. Food & Biobased Research, Wageningen. http://edepot.wur.nl/150289 (accessed on 7 July 2022).

Hartley, K., & Lant, P., (2006). Eliminating non-renewable CO_2 emissions from sewage treatment: An anaerobic migrating bed reactor pilot plant study. *Biotechnology and Bioengineering, 95*, 384–398. https://doi.org/10.1002/bit.2092.

Hendriks, A. T. W. M., & Zeeman, G., (2009). Pretreatments to enhance the digestibility of lignocellulosic biomass. *Bioresource Technology, 100*(1), 10–18. https://doi.org/10.1016/j.biortech.2008.05.027.

Holm, A., (2018). *The Effect on Wood Components During Soda Pulping (Master's Thesis).* Department of Chemistry and Chemical Engineering, Chalmers University of Technology, Gothenburg, Sweden.

Jianlong, W., & Jiang, K., (2005). The characteristics of anaerobic ammonium oxidation (ANAMMOX) by granular sludge from an EGSB reactor. *Process Biochemistry, 40*(5), 1973–1978. https://doi.org/10.1016/j.procbio.2004.08.001.

Kamali, M., Gameiro, T., Costa, M. E. V., & Capela, I., (2016). Anaerobic digestion of pulp and paper mill wastes e an overview of the developments and improvement opportunities. *Chemical Engineering Journal, 298*, 162–182. https://doi.org/10.1016/j.cej.2016.03.119.

Kangas, J., Jappinen, P., & Savolainen, H., (1984). Exposure to hydrogen sulphide, mercaptans and sulfur dioxide in the pulp industry. *American Industrial Hygiene Association Journal, 45*, 87–90. https://doi.org/10.1080/15298668491400647.

Kato, M. T., Field, J. A., & Lettinga, G., (1994). The anaerobic treatment of low strength wastewaters in UASB and EGSB reactors. *Water Science and Technology, 36*(6, 7), 375–382. https://doi.org/10.2166/wst.1997.0613.

Kumar, M., Morya, R., Gupta, A., & Thakur, I. S., (2020). Anaerobic biovalorization of pulp and paper mill waste. *Bipolarization of Wastes to Renewable Chemicals and Biofuel*, 41–61. https://doi.org/10.1016/B978-0-12-817951-2.00003-1.

Lakshmidevi, R., & Muthukumar, K., (2010). Enzymatic saccharification and fermentation of paper and pulp industry effluent for biohydrogen production. *International Journal of Hydrogen Energy, 5*(8), 3389–3400. https://doi.org/10.1016/j.ijhydene.2009.12.165.

Lange, J., Muller, F., Takors, R., & Blombach, B., (2018). Harnessing novel chromosomal integration loci to utilize an organosolv-derived hemicellulose fraction for isobutanol production with engineered *Corynebacterium glutamicum. Microbial Biotechnology, 11*(1), 257–263. https://doi.org/ 10.1111/1751-7915.12879.

Leach, J. M., (1980). Loadings and effects of chlorinated organics from a bleached pulp mill. In: *Canadian Environmental Protection Act, Water Chlorination: Environmental Impact and Health Effects* (Vol. 3, pp. 325–334).

Lin, Y., Wang, D., Liang, J., & Li, G., (2012). Mesophilic anaerobic co-digestion of pulp and paper sludge and food waste for methane production on a fed-batch basis. *Environmental Technology, 33*(23), 2627–2633. https://doi.org/10.1080/09593330.2012.673012.

Lin, Y., Wang, D., Qing, L. Q., & Xiao, M., (2011). Mesophilic batch anaerobic co-digestion of pulp and paper sludge and monosodium glutamate waste liquor for methane production in a bench-scale digester. *Bioresource Technology, 102*(4), 3673–3678. https://doi.org/10.1016/ j.biortech.2010.10.114.

Lin, Y., Wang, D., Wu, S., & Wang, C., (2009). Alkali pretreatment enhances biogas production in the anaerobic digestion of pulp and paper sludge. *Journal of Hazardous Materials, 170*(1), 366–373. https://doi.org/10.1016/j.jhazmat.2009.04.086.

Lin, Y., Wu, S., & Wang, D., (2013). Hydrogen-methane production from pulp & paper sludge and food waste by mesophilic thermophilic anaerobic co-digestion. *International Journal of Hydrogen Energy, 38*(35), 15055–15062. https://doi.org/10.1016/j.ijhydene.2012.01.051.

Loomis, W. D., & Battaile, J., (1966). Plant phenolic compounds and the isolation of plant enzymes. *Phytochemistry, 5*, 423–438. https://doi.org/ 10.1016/s0031-9422(00)82157-3.

Luo, G., Johansson, S., Boe, K., Xie, L., Zhou, Q., & Angelidaki, I., (2012). Simultaneous hydrogen utilization and in situ biogas upgrading in an anaerobic reactor. *Biotechnology and Bioengineering, 109*, 1088e1094. https://doi.org/10.1002/bit.24360.

Mahmood, T., & Elliott, A., (2006). A review of secondary sludge reduction technologies for the pulp and paper industry. *Water Research, 40*, 2093–2112. https://doi.org/10.1016/j.watres.2006.04.001.

Mao, C., Tong, Z., Xiaojiao, W., Yongzhong, F., Guangxin, R., & Gaihe, Y., (2017). Process performance and methane production optimizing of anaerobic co-digestion of swine manure and corn straw. *Scientific Reports, 7*, 9379. https://doi.org/10.1038/s41598-017-09977-6.

Martin, V. J., Yu, Z., & Mohn, W. W., (1999). Recent advances in understanding resin acid biodegradation: Microbial diversity and metabolism. *Archives of Microbiology, 172*, 131–138. https://doi.org/10.1007/s002030050752.

McCarty, P. L., & Smith, D. P., (1986). Anaerobic wastewater treatment. *Environmental Science & Technology, 20*(12), 1200–1206. https://doi.org/ 10.1021/es00154a002.

Mendes, C. V. T., Carvalho, M. G. V. S., Baptista, C. M. S. G., Rocha, J. M. S., Soares, B. I. G., & Sousa, G. D. A., (2009). Valorization of hardwood hemicelluloses in the kraft pulping process by using an integrated biorefinery concept. *Food and Bioproducts Processing, 87*(3), 197–207. https://doi.org/ 10.1016/j.fbp.2009.06.004.

Menon, V., & Rao, M., (2012). Trends in bioconversion of lignocellulose: Biofuels, platform chemicals & biorefinery concept. *Progress in Energy and Combustion Science, 38*, 522–550.

Monica, E., Gellerstedt, G., & Henriksson, G., (2009). *Pulp and Paper Chemistry and Technology*. De Gruyter, Berlin, Boston. Retrieved from: https://www.degruyter.com/view/serial/41891 (accessed on 7 July 2022).

Moretti, M. M. S., Bocchini-Martins, D. A., Nunes, C., Villena, M. A., Perrone, O. A., Silva, R., Boscolo, M., & Gomes, E., (2014). Pretreatment of sugarcane bagasse with microwaves irradiation and its effects on the structure and enzymatic hydrolysis. *Applied Energy*, 189–195. https://doi.org/ 10.1016/j.apenergy.2014.02.020.

Parameswaran, P., & Rittmann, B. E., (2012). Feasibility of anaerobic co-digestion of pig waste and paper sludge. *Bioresource Technology, 124*, 163–168. https://doi.org/10.1016/j.biortech.2012.07.116.

Pokhrel, D., & Viraraghavan, T., (2004). Treatment of pulp and paper mill wastewater a review. *The Science of the Total Environment, 333*, 37–58. https://doi.org/10.1016/j.scitotenv.2004.05.017.

Rasi, S., Veijanen, A., & Rintala, J., (2007). Trace compounds of biogas from different biogas production plants. *Energy, 32*(8), 1375–1380. https:// doi.org/10.1016/j.energy.2006.10.018.

Ribeiro, P., Albuquerque, A., Quinta-Nova, L., & Cavaleiro, V., (2010). Recycling pulp mill sludge to improve soil fertility using GIS tools. *Resources, Conservation and Recycling, 54*(12), 1303–1311. https://doi.org/10.1016/j.resconrec.2010.05.009.

Rittmann, B. E., & McCarty, P. L., (2001). *Environmental Biotechnology: Principles and Applications* (p. 326). McGraw-Hill Education.

Romero-Guiza, M. S., Peces, M., Astals, S., Benavent, J., Valls, J., & Mata-Alvarez, J., (2014). Implementation of a prototypal optical sorter as the core of the new pretreatment configuration of a mechanical biological treatment plant treating OFMSW through anaerobic digestion. *Applied Energy, 135*, 63–70. https://doi.org/10.1016/j.apenergy.2014.08.077.

Saha, M., Eskicioglu, C., & Marin, J., (2011). Microwave, ultrasonic and chemo-mechanical pretreatments for enhancing methane potential of pulp mill wastewater treatment sludge. *Bioresource Technology, 102*, 7815–7826. https://doi.org/10.1016/j.biortech.2011.06.053.

Salomon, K. R., & Lora, E. E. S., (2009). Estimate of the electric energy generating potential for different sources of biogas in Brazil. *Biomass and Bioenergy, 33*(9), 1101–1107. https://doi.org/10.1016/j.biombioe.2009.03.001.

Santos, M. O., & Lobo, B. A., (2012). Integrated pulp and paper mill planning and scheduling. *Computers & Industrial Engineering, 63*, 1–12. https:// doi.org/10.1016/j.cie.2012.01.008.

Schweigkofler, M., & Niessner, R., (1999). Determination of siloxanes and VOC in landfill gas and sewage gas by canister sampling and GC-MS/AES analysis. *Environmental Science and Technology, 33*, 3680–3685. https://doi.org/10.1021/es9902569.

Servizi, J. A., Martens, D. W., Gordon, R. W., Kutney, J. P., Singh, M., Dimitriadis, E., Hewitt, G. M., et al., (1986). Microbiological detoxification of resin acids. *Water Pollution Research Journal of Canada, 21*, 119–129.

Shah, A. I., Dar, M. U. D., Bhat, R. A., Singh, J. P., Singh, K., & Bhat, S. A., (2020). Prospectives and challenges of wastewater treatment technologies to combat contaminants of emerging concerns. *Ecological Engineering, 152*, 105882.

Sharma, M., Aryal, N., Sarma, P., & Vanbroekhoven, K., (2013). Bioelectrocatalyzed reduction of acetic and butyric acids via direct electron transfer using a mixed culture of sulfate-reducers drives electrosynthesis of alcohols and acetone. *Chemical Communications, 49*(58), 6495–6497. https://doi.org/10.1039/c3cc42570c.

Show, K. S., & Tay, J. H., (1999). Influence of support media on biomass growth and retention in anaerobic filters. *Water Research, 33*(6), 1471–1481. https://doi.org/10.1016/s0043-1354(98)00352-2.

Singh, S., (2015). Study of wastewater effluent characteristics generated from paper industries. *Journal of Basic and Applied Engineering Research, 2*, 1505–1509.

Smook, G. A., (1992a). Overview of pulping methodology. In: *Handbook for Pulp & Paper Technologists* (2nd edn., p. 36). Angus Wilde Publications, Vancouver.

Smook, G. A., (1992b). Paper manufacture e wet end operations. In: *Handbook for Pulp & Paper Technologists* (2nd edn., p. 228). Angus Wilde Publications, Vancouver.

Solomon, K. R., (1996). Chlorine in the bleaching of pulp and paper. *Pure and Applied Chemistry, 68*(9), 1721–1730. https://doi.org/10.1351/ pac199668091721.

Taherzadeh, M. J., & Karimi, K., (2008). Pretreatment of lignocellulosic wastes to improve ethanol and biogas production: A review. *International Journal of Molecular Sciences, 9*(9), 1621–1651. https://doi.org/10.3390/ijms9091621.

Taylor, B. R., Yeager, K. L., Abernathy, S. G., & Westlake, G. F., (1988). *Scientific Criteria Document for Development of Provincial Water Quality Objectives and Guidelines: Resin Acids*. Ontario Ministry of the Environment. Water Resources Branch, Toronto, Ontario.

Teghammar, A., Castillo, M. P., Ascue, J., Niklasson, C., & Horvath, I. S., (2013). Improved anaerobic digestion by the addition of paper tube residuals: Pretreatment, stabilizing, and synergetic effects. *Energy &Fuels, 27*, 277–284. https://doi.org/10.1021/ef301633x.

Thanarasu, A., Periyasamy, K., Devaraj, K., Periyaraman, P., Palaniyandi, S., & Subramanian, S., (2018). Tea powder waste as a potential co-substrate for enhancing the methane production in anaerobic digestion of carbon-rich organic waste. *Journal of Cleaner Production, 199*, 651–658. https://doi.org/10.1016/j.jclepro.2018.07.225.

Thompson, G., Swain, J., Kay, M., & Forster, C. F., (1991). The treatment of pulp and paper mill effluent: A review. *Bioresource Technology, 77*, 275–286. https://doi.org/10.1016/ S0960-8524(00)00060-2.

Ugurlu, M., & Karaoglu, M. H., (2009). Removal of AOX, total nitrogen and chlorinated lignin from bleached Kraft mill effluents by UV oxidation in the presence of hydrogen peroxide utilizing $TiO_{(2)}$ as photocatalyst. *Environmental Science and Pollution Research, 16*, 265–273. https://doi.org/ 10.1007/s11356-008-0044-x.

USEPA Biosolids Technology Fact Sheet, (2000). *Land Application of Sewage Sludge and Biosolids.* https://doi.org/10.1201/9781420032116.ch1.

Vaez, E., Taherdanak, M., & Zilouei, H., (2017). Dark hydrogen fermentation from paper mill effluent (PME): The influence of substrate concentration and hydrolysis. *Environmental Energy and Economic Research, 1*(2), 163–170. https://doi.org/10.22097/ eeer.2017.47243.

Vasconcelos, E., & Cabral, F., (1993). Use and environmental implications of pulp-mill sludge as an organic fertilizer. *Environmental Pollution, 80*, 159e162. https://doi. org/10.1016/0269-7491(93)90142-b.

Veluchamy, C., & Kalamdhad, A. S., (2017). Influence of pretreatment techniques on anaerobic digestion of pulp and paper mill sludge: A review. *Bioresource Technology, 245*, 1206–1219. https://doi.org/10.1016/j.biortech.2017.08.179.

White, T., (1957). Tannins their occurrence and significance. *Journal of Food Science and Agriculture, 8*, 377–384. https://doi.org/10.1002/ jsfa.2740080702.

Yeniguin, O., & Demirel, B., (2013). Ammonia inhibition in anaerobic digestion: A review. *Process Biochemistry, 48*, 901–911. https://doi.org/10.1016/ j.procbio.2013.04.012.

Young, H. E., & Chase, A. J., (1965). Fiber weight and pulping characteristics of the logging residue of seven tree species in Maine. In: *Maine Agricultural Experiment Technical Bulletin* (Vol. 17). Retrieved from: https://digitalcommons.library.umaine.edu/ aes_techbulletin/176/ (accessed on 7 July 2022).

Zak, M., Bendova, H., Friess, K., Bara, J. E., & Izak, P., (2018). Single-step purification of raw biogas to biomethane quality by hollow fiber membranes without any pretreatment-an innovation in biogas upgrading. *Separation and Purification Technology, 203*, 36–40.

Zanella, E., (1983). Effect of pH on acute toxicity of dehydroabietic acid and chlorinated dehydroabietic acid to fish and daphnia. *Bulletin of Environmental Contamination and Toxicology, 30*, 133–140.

Zhang, Y., & Angelidaki, I., (2015). Counteracting ammonia inhibition during anaerobic digestion by recovery using submersible microbial desalination cells. *Biotechnology and Bioengineering, 112*, 1478–1482. https://doi.org/10.1002/bit.25549.

Zilouei, H., & Taherdanak, M., (2015). Biohydrogen from lignocellulosic wastes. *Biofuel and Biorefinery Technologies*, 253–288. https://doi.org/ 10.1007/978-3-319-14033-9_7.

Index

B

D

F

G

H

X

Z